Remote Sensing of Atmospheric Conditions for Wind Energy Applications

Remote Sensing of Atmospheric Conditions for Wind Energy Applications

Special Issue Editors

Charlotte Bay Hasager
Mikael Sjöholm

MDPI • Basel • Beijing • Wuhan • Barcelona • Belgrade

MDPI

Special Issue Editors

Charlotte Bay Hasager   Mikael Sjöholm
Technical University of Denmark   Technical University of Denmark
Denmark   Denmark

Editorial Office
MDPI
St. Alban-Anlage 66
4052 Basel, Switzerland

This is a reprint of articles from the Special Issue published online in the open access journal *Remote Sensing* (ISSN 2072-4292) from 2018 to 2019 (available at: https://www.mdpi.com/journal/remotesensing/special_issues/Wind_Energy_RS)

For citation purposes, cite each article independently as indicated on the article page online and as indicated below:

LastName, A.A.; LastName, B.B.; LastName, C.C. Article Title. *Journal Name* **Year**, *Article Number, Page Range.*

ISBN 978-3-03897-942-5 (Pbk)
ISBN 978-3-03897-943-2 (PDF)

Cover image courtesy of Mikael Sjöholm.

Contents

About the Special Issue Editors

Charlotte Bay Hasager is working at the Department of Wind Energy, Technical University of Denmark (DTU). She completed her MSc in 1992 and PhD in 1996 at the University of Copenhagen at the Institute of Geography. Employed at Risø National Laboratory from 1993 to 2012, she has since been employed at DTU. Starting out as a PhD student in 1993, she went on to become a Post Doc and Scientist. In 2001, she became a Senior Scientist, with special competences in 2008, and became a team leader in 2012. She was a visiting scholar at Penn State University, Department of Meteorology for six months in 1995. Some of the assignments she has completed include as the President for Atmospheric Sciences Division, European Geosciences Union, 2007–2011, Co-chair for Wind Energy Community of Practice, GEOSS 2005 to present, Member of the steering committee for the Danish Space Consortium, 2005 to 2010, External examiner at the University of Copenhagen, Roskilde and Aalborg University, 1997 to present. She has been a member of IEEE since 2006. Her key research areas include: Offshore winds using satellite and ground-based remote sensing for wind profiles, wind power meteorology, wind resources, and turbulence; experimental field work and theoretical work on atmospheric boundary layer structure and dynamics over heterogeneous land, coast and sea; and micro-meteorology, data analysis, statistics, and satellite remote sensing image processing. Since 1999, Dr. Hasager has been coordinating more than 30 science and innovation projects, both international and Danish, within her research area, and she has published 55 peer-reviewed journal articles, 11 book chapters, and numerous conference articles, with 74 publications in Web of Science and 25 in h-index. Since 2008, Dr. Hasager has been coordinating the PhD Summer School: Remote Sensing for Wind Energy annual event and coordinated the educational website VirtuelGalathea3.dk on cross-disciplinary STEM education for 7th to 12th grade, with more than 1,000,000 visitors.

Mikael Sjöholm is working at the Department of Wind Energy, Technical University of Denmark (DTU). Having completed his MSc in Engineering Physics in 2001 and PhD in 2007 at Lund University on the thesis "Laser Spectroscopic Analysis of Atmospheric Gases in Scattering Media", supervised by Professor Sune Svanberg, he then went on to work at Risø National Laboratory from 2007 to 2012 as a Post Doc and Scientist, before moving on to DTU as a Scientist until 2014 and a Senior Scientist thereafter. His key research areas include laser-based measurement techniques for remote sensing of wind, turbulence, precipitation, gases, and other atmospheric quantities. Dr. Sjöholm has contributed in the development, testing and application of the DTU-initiated WindScanner concept for remote measurements of wind and turbulence in three dimensions around wind turbines by multiple synchronized scanning lidars. He is working on projects about ground-based, and wind turbine-mounted lidars as well as lidars for wind tunnels, and he has almost two decades of experience in experimental field work using various lidar technologies for gas measurements, as well as for wind and turbulence measurements. Dr. Sjöholm is the co-inventor of two patents about laser-based gas measurements in scattering materials that are explored by Astra Zeneca plc and GasPorOx AB, respectively. In addition, he is the co-inventor of several patent applications. He currently has 52 peer-reviewed documents in his Scopus profile, with a corresponding h-index of 18. He has co-supervised two PhD projects and several Master's thesis projects and is a lecturer at the PhD Summer School: Remote Sensing for Wind Energy annual event coordinated by Dr. Hasager.

remote sensing

MDPI

Editorial

Editorial for the Special Issue "Remote Sensing of Atmospheric Conditions for Wind Energy Applications"

Charlotte Bay Hasager * and Mikael Sjöholm

Department of Wind Energy, Technical University of Denmark, Frederiksborgvej 399, 4000 Roskilde, Denmark; misj@dtu.dk
* Correspondence: cbha@dtu.dk

Received: 27 March 2019; Accepted: 28 March 2019; Published: 1 April 2019

check for updates

Abstract: This Special Issue hosts papers on aspects of remote sensing for atmospheric conditions for wind energy applications. The wind lidar technology is presented from a theoretical view on the coherent focused Doppler lidar principles. Furthermore, wind lidar for applied use for wind turbine control, wind farm wake, and gust characterizations are presented, as well as methods to reduce uncertainty when using lidar in complex terrain. Wind lidar observations are used to validate numerical model results. Wind Doppler lidar mounted on aircraft used for observing winds in hurricane conditions and Doppler radar on the ground used for very short-term wind forecasting are presented. For the offshore environment, floating lidar data processing is presented as well as an experiment with wind-profiling lidar on a ferry for model validation. Assessments of wind resources in the coastal zone using wind-profiling lidar and global wind maps using satellite data are presented.

Keywords: Doppler wind lidar; wind energy; aerosol; wind turbine; wind farm; wake; control; complex terrain; offshore

1. Introduction

Wind power is an important ingredient in the energy mix for achieving the objectives of the Paris Climate Change agreement and the Sustainable Development Goals. Next to hydropower, wind energy is the renewable energy source that contributes the most to the electricity generation worldwide with around 6%. By the end of 2018, the installed wind power capacity surpassed 600 GW with a growth of nearly 54 GW during 2018 [1].

The levelized cost of energy from wind power is competitive with that of the conventional energy sources at wind-favorable land sites. However, there is still a need to further understand and efficiently use available wind resources. This is the key motivation for research in wind energy where efforts are ongoing to lower the cost of wind energy at offshore sites, in complex terrain, and in forested areas. The wake effect within and between wind farms and wind-power forecasting are areas with increasing importance because of the need to accurately predict wind power. There is, therefore, a need for reliable, robust, and accurate measurements and datasets to further improve our understanding of the physical conditions in which wind turbines and wind farms operate and for flow model evaluation.

Nowadays, remote sensing observations are used widely in wind energy applications. During recent years, remote sensing technologies for wind have been improved in terms of accuracy and costs. The Research Infrastructure WindScanner.eu [2,3] and other new lidar advancements for the measurement of atmospheric wind and turbulence have evolved and lidars are used for research in many application fields. Site assessment based on lidar is being progressively achieved. Commercial acceptance of lidars, including floating lidars, for wind resource assessment and power performance testing is taking place.

It has been an amazingly fast development of the modern wind lidars spurred by the electro-optical developments in the telecom industry. The wind energy lidar application era started in 2003 with the testing of a continuous-wave lidar near a tall meteorological mast onshore [4]. Still, to the editors' knowledge, no entire book has yet been written on remote sensing for wind energy applications, but a compendium for education at PhD level [5] is available as well as a chapter in a book about Energy Forecasting [6].

This Special Issue, "Remote Sensing of Atmospheric Conditions for Wind Energy", from 2019 is successor to the Special Issue, "Remote Sensing for Wind Energy" [7], from 2016. There are 15 articles in the present special issue, of which 13 are on wind lidar, while in the first special issue, 9 out of 11 articles were on wind lidar. In conclusion, the wind lidar technology is dominant within remote sensing research for wind energy application at present.

In the first special issue [7], 11 articles on remote sensing for wind energy were presented. Three articles on wind lidar for wind farm wake application were given. Kumer et al. [8] focused on characterization of turbulence in the wake using lidar. Doubrowa et al. [9] focused on the uncertainty on mean winds within wind farm wake using lidar, while Dooren et al. [10] presented a methodology to reconstruct the 2D horizontal wind fields in wind farm wake based on lidars.

Along the lines of obtaining the best possible data from wind lidar, a methodology for field calibration of nacelle-based lidar was presented by Borraccino et al. [11] and a comparison of lidar data observed using three scanning lidars directed to one point versus a profiling lidar installed in complex terrain was presented by Pauscher et al. [12,13]. The latter result is important for site assessment in complex terrain.

Site assessment using lidar was also presented for other types of terrain. Floors et al. [14] presented the scanning lidar for wind resource assessment in the coastal offshore zone and Kim et al. [15] presented the use of lidar for wind assessment for high-rise building planning in urban areas.

The wind gust detection using lidar was presented by Bos et al. [16] and Vasiljević et al. [17] presented the so-called long-range WindScanner incorporating three synchronized lidar beam scanners as well as a summary of lidar field experiments.

Two studies were on satellite remote sensing. Chang et al. [18] presented on land surface temperature increase in very stable atmospheric conditions during nighttime at a large-scale wind farm in China and Hasager et al. [19] presented on ocean surface wind speeds climatology during 25 years in the South China Sea and the Atlantic North Sea.

The following Section delivers a summary of all the 15 articles published in the current special issue. Thirteen of the articles present on wind lidar. The listing of contributed articles starts with the overview article on the way towards acceptance of wind lidar within the wind energy community followed by contributions on the Doppler wind lidar technology, lidar for wind turbine control, wind turbine wake measuring using lidar, gust characterization based on lidar, and lidar for use in complex terrain. Airborne wind lidar used for measuring in hurricane conditions and ground-based Doppler radar for very short-term periods follow. Next come contributions on offshore applications including floating lidar, lidar on board a ferry, and lidars installed at the coast for characterizing coastal offshore winds. Finally, a global perspective using satellite wind data is given.

2. Overview of Contributions

Within the wind energy community, the use of wind lidar is an important new focus area with a wide spectrum of applications including site assessment, power performance testing, controls and loads, and complex flows as presented in the overview given by Clifton et al. [20]. That contribution is a status update for the International Energy Agency (IEA) Wind Task 32 called "Wind Lidar Systems for Wind Energy Deployment" that since 2012 at international level aims to identify and mitigate barriers to the adoption of lidar for wind energy applications. Already achieved are several recommended practices and expert reports that have contributed to the adoption of ground-based, nacelle-based, and floating lidar by the wind industry. It is concluded that despite progress in identifying barriers to the adoption of wind lidar for wind energy applications—and mitigating some of them—there remains a significant amount of research to be done in this field.

The coherent focused continuous-wave Doppler wind lidar fundamental equations and principles are presented in Hill [21]. The comprehensive overview in general aims at bringing forward the classical radar/lidar lessons to the broader community presently applying coherent Doppler lidars. In particular, the behavior that may be observed from a modern coherent lidar used at short ranges (e.g., in a wind tunnel) and/or with weak aerosol seeding where only very few scatterers are present in the probe volume is explained. Results on simulation of few-scatterer and multiscatterer lidar experiments are revisited and in addition, a discussion of some problems (and solutions) for Doppler-sign-insensitive lidars is presented.

Simley et al. [22] present the topic of optimizing lidars for wind turbine control. A large body of work on the optimization of lidar beam configuration via the resulting controller performance, time domain assessments of measurement accuracy, and direct frequency domain calculations of measurement coherence and measurement error is presented. Simley et al [22] present results considering beam configuration optimization for rotor effective wind speed. Various lidar types including coherent continuous wave lidar and pulsed four-beam lidar are included. Also important is the lidar data availability for the feedforward pitch controller for rotor speed regulation.

Wind turbine wake characterization based on observations using two nacelle-mounted pulsed scanning Doppler lidars at 2.5 MW wind turbines is presented by Carbajo Fuertes et al. [23]. One lidar measured the inflow while the other measured the downwind wake region. The observations were dedicated to quantify the growth rate of the wake width, the near- and far-wake extent, and the velocity deficit. The observations were compared to an analytical wake model, with good results for the velocity deficit and wake expansion. It was observed that higher turbulence intensity in the inflow resulted in shorter near-wake length and in faster recovery of the velocity deficit.

Carbajo Fuertes and Porté-Agel [24] used a virtual lidar approach to assess how accurate it appears possible to observe with lidar in the full-scale wind turbine wake. The study was based on Large-Eddy Simulation (LES) model results for a wake. The performance of a virtual lidar performing stacked step-and-stare plan position indicator (PPI) scans within the volume was calculated and the volumetric reconstruction of the winds and the accuracy were assessed for the average velocity. As an outcome, optimization of the angle resolution that minimizes the total error was provided.

Zhou [25] contributed a study using a coherent Doppler lidar data set of three hours to study wind gusts on a scale from 100 m to 1000 m. The method proposed to extract gusts from a wind field and track their movement utilizes the "peak over threshold method", Moore–Neighbor tracing algorithm, and Taylor's frozen turbulence hypothesis. The prediction model was used to estimate the impact of gusts with respect to arrival time, the probability of arrival locations, the span-wise deviation of the gusts, and the gust size. Finally, the method was used to estimate the impact of gust on the production on a hypothetical wind farm.

Mayor and Dérian [26] refuted statements in Zhou [25] that argued on impracticality of motion estimation methods to derive two-component vector wind fields from single scanning aerosol lidar data. Two image-based motion estimation methods, namely the cross-correlation and wavelet-based optical flow methods, were demonstrated to be also practical to use for wind gust detection and impact prediction on wind turbines. The characteristics and performances of the cross-correlation and wavelet-based methods were compared to a two-dimensional variational method applied to radial velocity fields from a single scanning Doppler lidar. In conclusion, the wavelet-based method and two-dimensional variational method have much in common and both are practical to use.

In complex terrain, ground-based wind-profiling lidars are used for observing the vertical wind profile but with additional uncertainty as compared to use in flat terrain. Hofsäß et al. [27] analyzed a wind lidar data set and compared to wind speeds observed at a 100 m tall meteorological mast near an escarpment of 150 m height in Germany. Three different methods to optimize the lidar data reconstruction were evaluated. It was found that a linear approach performed the best. Furthermore, the influences of the opening angle of the scanning cone and the scanning duration on data quality were analyzed in the data set. The opening angle had importance, while the scanning duration did not.

Risan et al. [28] also investigated flow in complex terrain using lidar. This study was based on a pulsed lidar measuring towards a ridge in Norway. The lidar data were used for comparison to Computational Fluid Dynamic (CFD) model results using two methods: one was a hybrid Reynolds-Averaged Navier Stokes (RANS)/Large-Eddy Simulation (LES) model, the other was a more traditional RANS model. The lidar data 10-minute mean wind speeds were very accurate compared to sonic wind data, while the 1-second turbulence had lower accuracy compared to sonic data. However, both the mean wind speed along the line of sight and the turbulence data were useful for comparing the performance of the two flow models. The first model performed well but overestimated the turbulence at the ridge, while the other model failed to estimate the turbulence over the ridge.

Zhang et al. [29] presented analysis of airborne Doppler wind lidar observation from a tropical storm called Erika (2015) in the US. The lidar was installed on a P3 Hurricane Hunter aircraft. The observations were compared with two other types of vertical profiling data sets. One was the dropsonde measurements, the other Doppler radar. Lidar and dropsonde wind speeds correlation was high and root-mean-square error and bias were low. The wind lidar enlarged the sampling size and spatial coverage of boundary layer winds, and was found valuable for real-time intensity forecasts and for understanding boundary layer structure and dynamics, and can be used for offshore wind energy applications. The lidar operates best in rain-free and low-rain conditions while the radar performs better in rain. Thus the lidar and radar are complementary in observing winds.

Valldecabres et al. [30] investigated Dual-Doppler radar data from two radars located on the coastline and observing several kilometers offshore covering an offshore wind farm in the North Sea. The radar-based wind data are analyzed to produce very short-term wind power forecasts, around five minutes ahead. The wind variations observed upstream are used in an advection Lagrangian persistence technique to forecast the density of wind speeds at the target turbines. The radar-based forecasts outperformed the persistence and climatology benchmarks when predicting the power generated. The radar-based forecasts were corrected for induction effects. It is important that a sufficiently large spatial coverage of the inflow for a turbine is observed to produce a reliable density forecast.

The winds offshore in the coastal zone were investigated using two vertical profiling pulsed lidars by Shimada et al. [31]. The instruments were located at the coast and 400 m out, mounted on a long pier in Japan. Six months of observations were available for the analysis that focused on the effect of fetch for winds blowing from land to sea. Also, the winds from the ocean towards land were analyzed. The effect of fetch at several levels up to 200 m were quantified and a strong gradient was noted at the lower heights but less pronounced at higher levels. The data set would be valuable for comparison to model results of the coastal wind climate.

Gottschall et al. [32] also investigated offshore coastal winds using lidar observations. The study was based on a pulsed vertical profiling lidar installed on board a ferry that daily passes a distance of several hundreds of kilometers in the Baltic Sea. The lidar data were motion-corrected using relevant motion measurements. The data availability was good and observations reached as high as 250 m. The lidar data were subsequently used for comparison to a mesoscale numerical weather prediction model, and the correlation of wind speeds was good. The ferry-based lidar data are from a trajectory so the collocation of lidar data and model results prior to comparison had to be done.

Gutiérrez-Antuñano et al. [33] focused on motion correction of data from a floating vertical profiling coherent continuous wave lidar observing in the Netherlands. The comparison of the observations from the floating lidar uncorrected and corrected were done to sonic anemometer measurements and data from a fixed vertical wind-profiling lidar nearby during 60 days of observations. Both 10-minute mean wind speed and turbulence were analyzed. The proposed motion correction method combines a software-based velocity-azimuth display and motion simulator and a statistical recursive procedure. It assumes simple-harmonic motional conditions such that only one motional amplitude and period is needed and wind direction is neglected. The comparison was good for both wind speed and turbulence in particular comparing lidar to lidar.

Guo et al. [34] analyzed the archive of satellite ocean wind speed products based on two scatterometers, ASCAT and QuikSCAT, and one passive polarimetric microwave, WindSat, satellite. Firstly, the wind products were compared to ocean buoy data during several years and differences in the biases between the products and buoy winds were noted. Next, the collections of data sets were combined and mean wind speed was calculated in all grid cells. A simple extrapolation to 100 m was done and wind power density was estimated.

3. Conclusions

This Special Issue highlights the use of remote sensing for atmospheric conditions for wind energy research. Wind lidar is the dominant remote sensing method giving new opportunities for observing winds and turbulence and the wind lidar technology is applied broadly. The lidar observations prove useful for validating models as well as for characterizing the boundary layer structure and dynamics. This is particularly valuable in complex terrain and offshore. The flow near wind turbines and wind farms influenced by the operating wind turbines (i.e., wake and inflow) is observed well by remote sensing and remote sensing data are useful for forecasting the power at very short time scales.

Author Contributions: C.B.H. and M.S. prepared the editorial together.

Funding: This research received no external funding.

Acknowledgments: The Guest Editors of this Special Issue would like to thank all authors for their valuable contributions. We also would like to thank the distinguished reviewers for their insightful inputs and the Remote Sensing editorial staff for excellent support throughout the completion of this volume. The Guest Editors dedicate this Special Issue to the too early deceased in 2012 Colorado-based lidar researcher Rod Frehlich who paved the path for coherent lidar signal processing before the onset of the widespread use of Doppler lidars in the rapidly expanding wind energy community.

Conflicts of Interest: The authors declare no conflict of interest.

References

1. World Wind Energy Association. Available online: https://wwindea.org/information-2/information/ (accessed on 22 March 2019).
2. WindScanner.eu. Available online: http://www.windscanner.eu/ (accessed on 22 March 2019).
3. Mikkelsen, T.; Siggaard Knudsen, S.; Sjöholm, M.; Angelou, N.; Pedersen, A.T. WindScanner.eu—A new Remote Sensing Research Infrastructure for On- and Offshore Wind Energy. In Proceedings of the International Conference on Wind Energy: Materials, Engineering and Policies (WEMEP-2012), Hyderabad, India, 22 December 2012.
4. Jørgensen, H.E.; Mikkelsen, T.; Mann, J.; Bryce, D.; Coffey, A.; Harris, M.; Smith, D. Site wind field determination using a CW Doppler lidar - comparison with cup anemometers at Risø. In Proceedings of the Special Topic Conference: The Science of Making Torque from Wind, Delft, The Netherlands, 19 April 2004; Delft University of Technology: Delft, The Netherlands, 2004; pp. 261–266.
5. Peña, A.; Hasager, C.B.; Badger, M.; Barthelmie, R.J.; Bingöl, F.; Cariou, J.-P.; Emeis, S.; Frandsen, S.T.; Harris, M.; Karagali, I. Remote Sensing for Wind Energy. DTU Wind Energy. Available online: http://orbit.dtu.dk/files/111814239/DTU_Wind_Energy_Report_E_0084.pdf (accessed on 22 March 2019).
6. Gryning, S.-E.; Mikkelsen, T.K.; Baehr, C.; Dabas, A.; Gómez Arranz, P.; O'Connor, E.; Rottner, L.; Sjöholm, M.; Suomi, I.; Vasiljević, N. Measurement methodologies for wind energy based on ground-level remote sensing. In *Renewable Energy Forecasting: From Models to Applications*; Kariniotakis, G., Ed.; Woodhead Publishing: Cambridge, UK, 2017. [CrossRef]
7. Special Issue "Remote Sensing for Wind Energy" 2016 A special issue of Remote Sensing (ISSN 2072-4292). Available online: https://www.mdpi.com/journal/remotesensing/special_issues/wind_energy_sensing (accessed on 22 March 2019).

8.   Kumer, V.-M.; Reuder, J.; Oftedal Eikill, R. Characterization of Turbulence in Wind Turbine Wakes under Different Stability Conditions from Static Doppler LiDAR Measurements. *Remote Sens.* **2017**, *9*, 242. [CrossRef]

9.   Doubrawa, P.; Barthelmie, R.J.; Wang, H.; Pryor, S.C.; Churchfield, M.J. Wind Turbine Wake Characterization from Temporally Disjunct 3-D Measurements. *Remote Sens.* **2016**, *8*, 939. [CrossRef]

10. Van Dooren, M.F.; Trabucchi, D.; Kühn, M. A Methodology for the Reconstruction of 2D Horizontal Wind Fields of Wind Turbine Wakes Based on Dual-Doppler Lidar Measurements. *Remote Sens.* **2016**, *8*, 809. [CrossRef]

11. Borraccino, A.; Courtney, M.; Wagner, R. Generic Methodology for Field Calibration of Nacelle-Based Wind Lidars. *Remote Sens.* **2016**, *8*, 907. [CrossRef]

12. Pauscher, L.; Vasiljević, N.; Callies, D.; Lea, G.; Mann, J.; Klaas, T.; Hieronimus, J.; Gottschall, J.; Schwesig, A.; Kühn, M.; et al. An Inter-Comparison Study of Multi- and DBS Lidar Measurements in Complex Terrain. *Remote Sens.* **2016**, *8*, 782. [CrossRef]

13. Pauscher, L.; Vasiljevic, N.; Callies, D.; Lea, G.; Mann, J.; Klaas, T.; Hieronimus, J.; Gottschall, J.; Schwesig, A.; Kühn, M.; et al. Erratum: Pauscher, L., et al. An Inter-Comparison Study of Multi- and DBS Lidar Measurements in Complex Terrain. *Remote Sens.* **2016**, *8*, 782 . *Remote Sens.* **2017**, *9*, 667. [CrossRef]

14. Floors, R.; Peña, A.; Lea, G.; Vasiljević, N.; Simon, E.; Courtney, M. The RUNE Experiment—A Database of Remote-Sensing Observations of Near-Shore Winds. *Remote Sens.* **2016**, *8*, 884. [CrossRef]

15. Kim, H.-G.; Jeon, W.-H.; Kim, D.-H. Wind Resource Assessment for High-Rise BIWT Using RS-NWP-CFD. *Remote Sens.* **2016**, *8*, 1019. [CrossRef]

16. Bos, R.; Giyanani, A.; Bierbooms, W. Assessing the Severity of Wind Gusts with Lidar. *Remote Sens.* **2016**, *8*, 758. [CrossRef]

17. Vasiljević, N.; Lea, G.; Courtney, M.; Cariou, J.-P.; Mann, J.; Mikkelsen, T. Long-Range WindScanner System. *Remote Sens.* **2016**, *8*, 896. [CrossRef]

18. Chang, R.; Zhu, R.; Guo, P. A Case Study of Land-Surface-Temperature Impact from Large-Scale Deployment of Wind Farms in China from Guazhou. *Remote Sens.* **2016**, *8*, 790. [CrossRef]

19. Hasager, C.B.; Astrup, P.; Zhu, R.; Chang, R.; Badger, M.; Hahmann, A.N. Quarter-Century Offshore Winds from SSM/I and WRF in the North Sea and South China Sea. *Remote Sens.* **2016**, *8*, 769. [CrossRef]

20. Clifton, A.; Clive, P.; Gottschall, J.; Schlipf, D.; Simley, E.; Simmons, L.; Stein, D.; Trabucchi, D.; Vasiljević, N.; Würth, I. IEA Wind Task 32: Wind Lidar Identifying and Mitigating Barriers to the Adoption of Wind Lidar. *Remote Sens.* **2018**, *10*, 406. [CrossRef]

21. Hill, C. Coherent Focused Lidars for Doppler Sensing of Aerosols and Wind. *Remote Sens.* **2018**, *10*, 466. [CrossRef]

22. Simley, E.; Fürst, H.; Haizmann, F.; Schlipf, D. Optimizing Lidars for Wind Turbine Control Applications—Results from the IEA Wind Task 32 Workshop. *Remote Sens.* **2018**, *10*, 863. [CrossRef]

23. Carbajo Fuertes, F.; Markfort, C.D.; Porté-Agel, F. Wind Turbine Wake Characterization with Nacelle-Mounted Wind Lidars for Analytical Wake Model Validation. *Remote Sens.* **2018**, *10*, 668. [CrossRef]

24. Carbajo Fuertes, F.; Porté-Agel, F. Using a Virtual Lidar Approach to Assess the Accuracy of the Volumetric Reconstruction of a Wind Turbine Wake. *Remote Sens.* **2018**, *10*, 721. [CrossRef]

25. Zhou, K.; Cherukuru, N.; Sun, X.; Calhoun, R. Wind Gust Detection and Impact Prediction for Wind Turbines. *Remote Sens.* **2018**, *10*, 514. [CrossRef]

26. Mayor, S.D.; Dérian, P. Comments on "Wind Gust Detection and Impact Prediction for Wind Turbines". *Remote Sens.* **2018**, *10*, 1625. [CrossRef]

27. Hofsäß, M.; Clifton, A.; Cheng, P.W. Reducing the Uncertainty of Lidar Measurements in Complex Terrain Using a Linear Model Approach. *Remote Sens.* **2018**, *10*, 1465. [CrossRef]

28. Risan, A.; Lund, J.A.; Chang, C.-Y.; Sætran, L. Wind in Complex Terrain—Lidar Measurements for Evaluation of CFD Simulations. *Remote Sens.* **2018**, *10*, 59. [CrossRef]

29. Zhang, J.A.; Atlas, R.; Emmitt, G.D.; Bucci, L.; Ryan, K. Airborne Doppler Wind Lidar Observations of the Tropical Cyclone Boundary Layer. *Remote Sens.* **2018**, *10*, 825. [CrossRef]

30. Valldecabres, L.; Nygaard, N.G.; Vera-Tudela, L.; Von Bremen, L.; Kühn, M. On the Use of Dual-Doppler Radar Measurements for Very Short-Term Wind Power Forecasts. *Remote Sens.* **2018**, *10*, 1701. [CrossRef]

31. Shimada, S.; Takeyama, Y.; Kogaki, T.; Ohsawa, T.; Nakamura, S. Investigation of the Fetch Effect Using Onshore and Offshore Vertical LiDAR Devices. *Remote Sens.* **2018**, *10*, 1408. [CrossRef]

32. Gottschall, J.; Catalano, E.; Dörenkämper, M.; Witha, B. The NEWA Ferry Lidar Experiment: Measuring Mesoscale Winds in the Southern Baltic Sea. *Remote Sens.* **2018**, *10*, 1620. [CrossRef]
33. Gutiérrez-Antuñano, M.A.; Tiana-Alsina, J.; Salcedo, A.; Rocadenbosch, F. Estimation of the Motion-Induced Horizontal-Wind-Speed Standard Deviation in an Offshore Doppler Lidar. *Remote Sens.* **2018**, *10*, 2037. [CrossRef]
34. Guo, Q.; Xu, X.; Zhang, K.; Li, Z.; Huang, W.; Mansaray, L.R.; Liu, W.; Wang, X.; Gao, J.; Huang, J. Assessing Global Ocean Wind Energy Resources Using Multiple Satellite Data. *Remote Sens.* **2018**, *10*, 100. [CrossRef]

remote sensing

MDPI

Project Report

IEA Wind Task 32: Wind Lidar Identifying and Mitigating Barriers to the Adoption of Wind Lidar

Andrew Clifton [1,*], Peter Clive [2], Julia Gottschall [3], David Schlipf [4], Eric Simley [5], Luke Simmons [6], Detlef Stein [7], Davide Trabucchi [8], Nikola Vasiljevic [9] and Ines Würth [10]

[1]   WindForS, University of Stuttgart, Allmandring 5b, 70569 Stuttgart, Germany
[2]   Wood-Clean Energy, 2nd Floor, St. Vincent Plaza, 319 St. Vincent Street, Glasgow G2 5LP, UK;
      peter.clive@woodplc.com
[3]   Fraunhofer Institute for Wind Energy Systems IWES, Am Seedeich 45, 27572 Bremerhaven, Germany;
      julia.gottschall@iwes.fraunhofer.de
[4]   Stuttgart Wind Energy, University of Stuttgart, Allmandring 5b, 70569 Stuttgart, Germany;
      schlipf@ifb.uni-stuttgart.de
[5]   Envision Energy USA Ltd., 1201 Louisiana St. Suite 500, Houston, TX 77002, USA;
      eric.simley@envision-energy.com
[6]   DNV GL—Measurements, 1501 9th Avenue, Suite 900, Seattle, WA 98001, USA; Luke.Simmons@dnvgl.com
[7]   Multiversum GmbH, Shanghaiallee 9, 20457 Hamburg, Germany; d.stein@multiversum.consulting
[8]   ForWind, University of Oldenburg, Küpkersweg 70, 26129 Oldenburg, Germany;
      davide.trabucchi@uni-oldenburg.de
[9]   Department for Wind Energy, Technical University of Denmark, Frederiksborgvej 399, 4000 Roskilde,
      Denmark; niva@dtu.dk
[10] Stuttgart Wind Energy, University of Stuttgart, Allmandring 5b, 70569 Stuttgart, Germany;
      wuerth@ifb.uni-stuttgart.de
*   Correspondence: clifton@windfors.de; Tel.: +49-711-6856-8325

Received: 24 January 2018; Accepted: 23 February 2018; Published: 6 March 2018

Abstract: IEA Wind Task 32 exists to identify and mitigate barriers to the adoption of lidar for wind energy applications. It leverages ongoing international research and development activities in academia and industry to investigate site assessment, power performance testing, controls and loads, and complex flows. Since its initiation in 2011, Task 32 has been responsible for several recommended practices and expert reports that have contributed to the adoption of ground-based, nacelle-based, and floating lidar by the wind industry. Future challenges include the development of lidar uncertainty models, best practices for data management, and developing community-based tools for data analysis, planning of lidar measurements and lidar configuration. This paper describes the barriers that Task 32 identified to the deployment of wind lidar in each of these application areas, and the steps that have been taken to confirm or mitigate the barriers. Task 32 will continue to be a meeting point for the international wind lidar community until at least 2020 and welcomes old and new participants.

Keywords: wind energy; resource assessment; power performance testing; wind turbine controls; complex flow; Doppler lidar

---

1. Introduction

Wind lidar can measure the line of sight (LOS) wind speed at distances from a few centimeters to several kilometers. Depending on their deployment, the LOS speed is obtained by means of lidar systems firmly sitting on the ground, floating in the water or orbiting around the Earth. The first commercial wind lidar systems targeted at wind energy applications appeared in the early 2000s [1]. Because of their costs and ease of installation, lidar have become accepted as an alternative to the traditional mast-based wind sensors for site assessment and power performance testing, as evidenced by their inclusion in international standards (e.g., [2]). They are popular offshore because they reduce the need for a fixed platform (e.g., [3]). Additionally, because they can measure upwind of operating turbines, wind lidars are being used for feed-forward control of wind turbines [4]. However, despite their advantages, wind lidars have not replaced traditional anemometry in everyday use.

IEA Wind Task 32 "Wind Lidar"—herein referred to as "Task 32"—was set up in 2011 to support the deployment of lidar for wind energy applications. Its members identify the barriers to the deployment of lidar for wind energy and develop roadmaps to removing those barriers that can then be implemented by the wind energy lidar community. Task 32 is one of several international collaborative research tasks that are enabled by the International Energy Agency (IEA) Technology Collaboration Programme (IEA Wind TCP). IEA Wind is a vehicle for member countries to exchange information on the planning and execution of large-scale wind system projects and to undertake co-operative research and development projects called Tasks or Annexes.

In 2015, participants in Task 32 identified four main application areas where an international collaboration between researchers, device manufacturers, end users and other stakeholders could mitigate barriers. These application areas include site assessment, wind turbine power performance, wind turbine controls, and the use of lidar to measure complex flow. The application areas are currently at different technology readiness levels: lidar systems are already used by industry for site assessment and power performance, and so the barriers are more related to implementation. For loads and control as well as for use in complex flow applications, there are still open research questions to solve (Figure 1). This paper describes the barriers that Task 32 identified to the deployment of wind lidar in each of these application areas, and the steps that have been taken to confirm or mitigate the barriers.

The barriers identified in this paper are generally described in terms of specific "use cases", for example for the use of nacelle-mounted, forward-looking lidar for wind turbine control. A use case has three elements:

1. **Data requirements.** A fundamental aspect of the "use case" approach is the articulation of the needs that are being fulfilled by the measurement campaign in advance, without reference to any assumptions about instrument capabilities, to ensure an outcome-driven approach to measurement and analysis campaign design is adopted. This contrasts with the constraint-driven approach previously adopted in response to the limitations of instruments that preceded lidar, such as met masts.
2. **Measurement method.** Given the diversity of methods that can be implemented using lidar, it is important that a technique is selected that both fulfills the data requirements and is amenable to complete documentation, such as risk assessments and method statements, to support repeatability and reproducibility of measurements.
3. **Measurement situation.** A given measurement method will perform with respect to accuracy in different ways under different sets of circumstances. It is important that the situation in which the lidar is operated is documented and its influence on lidar performance is understood. Assumptions entailed by wind flow reconstruction algorithms may be invalid under certain sets of circumstances. The result of a calibration of a lidar operated in accordance with a given method in a given situation is only transferable to situations (i.e., can be used later or elsewhere) that are similar in all important respects (those that influence lidar performance) where the same method is used.

The adoption of a use case approach to describing lidar operations ensures all the relevant information is captured to ensure lidar is operated in a consistent manner that is fit for purpose and supports calibrations and uncertainty evaluations that are transferable.

Readers are asked to note that, because of commercial considerations, it is not always possible to provide quantitative information on business practices within the wind energy industry. Therefore, some material in this text is by necessity qualitative. Unless otherwise stated, such material is based on anecdotal evidence or from presentations and discussions at IEA Wind Task 32 and other meetings.

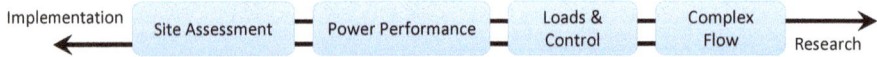

Figure 1. Barriers to the use of lidar technology in wind energy applications are mostly related to implementation and research.

2. Lidar for Site Assessment

Site assessment is the process of quantifying the wind and weather conditions on a potential wind farm site on land or offshore. The objective of a site assessment is to measure parameters such as wind speed, turbulence intensity, gusts, and wind direction at multiple heights, and also the temperature, pressure, and precipitation at the site. This information is used to derive the information needed to layout the plant (for example, wind shear or veer), help select suitable wind turbines, and estimate power production before a wind farm is built [5,6]. The basic value of such studies is in providing the data required for a wind energy development; further value can be added by reducing the uncertainty of the data, which reduces financing costs [7,8].

2.1. On Land

Wind lidar have many applications in wind energy project development. For example, profiling lidar can often measure wind speed and direction up to 300 m above ground and so can be used to confirm hub-height wind speeds based on vertical extrapolation from meteorological (met) tower measurements at lower heights above ground [9]. Profiling wind lidar can also directly measure the wind speed at multiple heights across the potential turbine rotor disk [10]. Scanning devices may measure up to 10 km away, enabling measurements across a site from one location (see, e.g., [11–13]).

These applications require measuring LOS wind speed with the lidar and converting them in to a wind speed and direction. This process—known as wind field reconstruction—requires a model of the flow. In simple terrain, a very basic homogeneous flow model is assumed, thus a wind vector for the whole flow can be estimated by probing the wind field from different directions to give multiple different line of sight speed, azimuth and elevation data. The wind vector is derived by fitting the measured LOS wind speeds to a wind vector, assuming horizontal homogeneity [14].

An initial barrier to the adoption of wind lidar was a lack of guidelines or standards for specific applications and situations. In order to support the adoption of wind lidar, Task 32 worked with IEA Wind Task 11 to develop recommended practices for the use of remote sensing for resource assessment in simple terrain [15]. This internationally recognized, community-led recommended practice and other industry guidelines helped users gain confidence in lidar.

Another early barrier to the use of lidar was its purchase cost. Initially, manufacturers argued that the total cost of ownership of a lidar was comparable to two or more tall met masts [8], and that value could be gained by reducing uncertainty. Since then, the purchase price of lidar hardware has come down. This may be related to the increasing availability of high-quality photonics and electronics at lower prices, better manufacturing processes, and competition between manufacturers. The cost of lidar is therefore no longer seen as a hard barrier to deployment, but even so a reduction in the lifetime cost of lidar would make it even more compelling for wind energy applications. Although Task 32 does not address cost directly, the Task helps users exchange ideas and experiences with wind lidars and thereby helps reduce the learning curve and increase the potential value that can be extracted from the lidar data.

Because of the way they obtain and process data, the turbulence information obtained from a lidar differ from those obtained by point measurements from a cup, or smaller volumes such as sonic anemometers [16,17]. Task 32 members produced a comprehensive report that summarized these issues and potential solutions in 2015 [18]. The potential for differences between turbulence metrics obtained by lidar and other devices does not appear to have become a barrier to adoption.

As lidar has become more widely used and wind energy developments have moved into more complex terrain, questions have been raised around the use of lidar in complex terrain. This is because flow reconstruction in complex terrain is challenging because the flow is no longer homogeneous (e.g., [19]). Furthermore, heterogeneity makes measurements from traditional anemometers less representative of an area as well. Early thoughts on the barriers to the use of lidar in complex terrain, and potential ways to mitigate its effect, were summarized in a 2015 Task 32 report [11]. The most common approach currently to measuring in complex terrain conditions is to use lidar and mast measurements together with appropriate flow models (e.g., [20]). Other mitigation methods that can be used with other types of complex flow are described in Section 6.

A Task 32 workshop in November 2017 explored the current barriers to the use of lidar for site assessment in complex terrain. The workshop used a series of desktop planning studies around the use of wind lidar in complex terrain. Many of the barriers that were identified also hold for simpler terrain:

- **The role of flow models**: One way to improve wind field reconstruction is to use flow models to fit measurements to a modelled wind field, especially in complex terrain. If this approach becomes common it would mean that the adoption of lidars is directly related to the validation and acceptance of flow models.
- **Unknown uncertainty**: Current approaches quantify wind speed uncertainty as the root mean square difference between the wind speed measured by lidars, and that measured by cup anemometers. This requires either that a cup anemometer have no other sensitivity than to the wind speed, or that all external factors are known and can be accounted for especially if they change during a measurement. This uncertainty model may hold in simple, flat terrain with low turbulence and no precipitation, icing, or vertical flow component. However, in complex terrain or where there is significant variation in external conditions, the uncertainty of a lidar could be considered undefined as both the cup and lidar have unknown sensitivities outside of a narrow ideal range.
- **Lacking or misleading guidelines**: Existing recommended practices and standards do not cover the whole range of potential applications and fail to explain what should be done to achieve satisfactory measurements in complex terrain.
- **Lack of experts**: Deploying and operating wind lidar requires training and experience. There are simply not enough experienced users available to support the many possible uses of wind lidar. This delays its deployment and increases costs.

The November 2017 workshop and other events noted that complex terrain can make site access difficult and that complex terrain is often associated with a lack of access to reliable power. Because these issues are equally a problem for towers, they have not been a focus of Task 32 work. Importantly, the reliability and availability of the lidar itself was not identified as an issue, which reflects progress made in the last decade in developing lidar as a commercial product.

2.2. Offshore

Offshore lidar applications for site assessments can be subdivided into installations on fixed platforms (Section 2.2.1) and on (or in) floating structures such as buoys (Section 2.2.2).

2.2.1. Fixed Lidar Offshore

Wind lidars mounted on existing offshore platforms near to project sites can be a viable alternative to conventional offshore met towers (see, e.g., [21]). Potential deployment locations include offshore meteorological stations as well as oil or natural gas rigs, whether disused or still in operation. Some modifications may be required to make the lidar system reliable in the offshore environment.

Based on the review of an extensive body of onshore and offshore evidence, as reported from various industry stakeholders, the use of industry-proven ground-based lidars operated on an offshore stationary platform is considered a benign scenario, provided that no significant flow distortion from the platform or its components might affect the lidar measurements. Industry-proven lidars are lidar types that are commercially available and have an accepted track record onshore. An example of the steps required for a lidar type to be considered "industry-proven" is given in Section 2.2.2 of [22]. This benign scenario further increases the potential competitiveness of fixed lidar to conventional offshore met masts.

As with any use of lidars, it is best practice to carry out an appropriate pre- and post-deployment verification program onshore. Furthermore, a sufficient length of data set and data coverage rates are key objectives in wind resource measurement campaigns. Hence, lidar deployments should span a similar period as those undertaken with conventional anemometry. In addition, as with onshore deployments, it is crucial to deploy a lidar system offshore with a sufficient power supply and an appropriate operations and maintenance (O&M) program. If these conditions are met, it is anticipated that similar amounts and quality of wind data would be obtained from industry-proven lidars mounted on a stationary platform, as those from a classical offshore mast (see, e.g., [21]).

Traditional offshore met masts have relatively complex structure and low porosity, which results in increased drag and flow distortion around the mast compared to guyed, land-based towers. Such effects are also poorly predicted using current standards [23]. Although the platform itself may cause localized effects, these flow distortions rarely extend to the measurement height [21], and thus lidar wind measurements may even be more accurate than a mast. This further adds to the advantages of a wind lidar in the offshore environment. The main barrier to adoption is to collect evidence for these advantages.

Lidar on fixed offshore platforms can also be used to calibrate floating lidar systems. This method was included in a IEA recommended practices document [24].

2.2.2. Floating Lidar

Floating Lidar Systems (FLSs) are essentially lidar units mounted on buoys. The first FLS were developed and tested in the late 2000s to meet the wind industry's needs for data for offshore wind resource assessment. They offered the potential for reduced costs compared to fixed met masts, similar data, and the ability to measure at the same or even greater heights above water. The first barrier to their adoption was a lack of experience with such systems and a question of whether or not such devices would survive and deliver the required data. Initial results were promising, which together with the potential cost savings helped drive their adoption [25].

One early barrier to the adoption of floating lidar technology was the lack of objective measures of the performance and maturity of an FLS. To address this, the Carbon Trust Offshore Wind Accelerator programme (OWA) developed a "Roadmap for the commercial acceptance of floating lidar technology" [22], which included key performance indicators (KPI). The Roadmap defined several objective maturity stages based on the FLS performance measured using the KPIs and respective acceptance criteria.

Another early challenge was how to effectively deploy and use an FLS for different use cases. Therefore, Task 32 started an initiative in 2012 to collect recommended practices (RP) for the application of FLSs. A first collection was published as a state-of-the-art report in early 2016, and then further developed with the support of the Carbon Trust as part of a project within the OWA programme. Finally, a new IEA Wind Recommended Practices document (RP 18, "Floating Lidar Systems") that combined aspects of all of these documents was published in autumn 2017 [24].

Today, there are about 10 different FLS providers, offering quite different designs. A schematic diagram of one design is shown in Figure 2. Most current FLSs use industry-proven lidars that were originally designed for use on land. Experience has shown that these lidars can be deployed on floating platforms with minor adjustments such as an offshore-qualified casing or bird deterrents.

At least six systems have so far reached the Carbon Trust's "pre-commercial" maturity stage. KPIs for commercial maturity are not yet fully defined, but will be developed in 2018 as part of an ongoing Carbon Trust initiative.

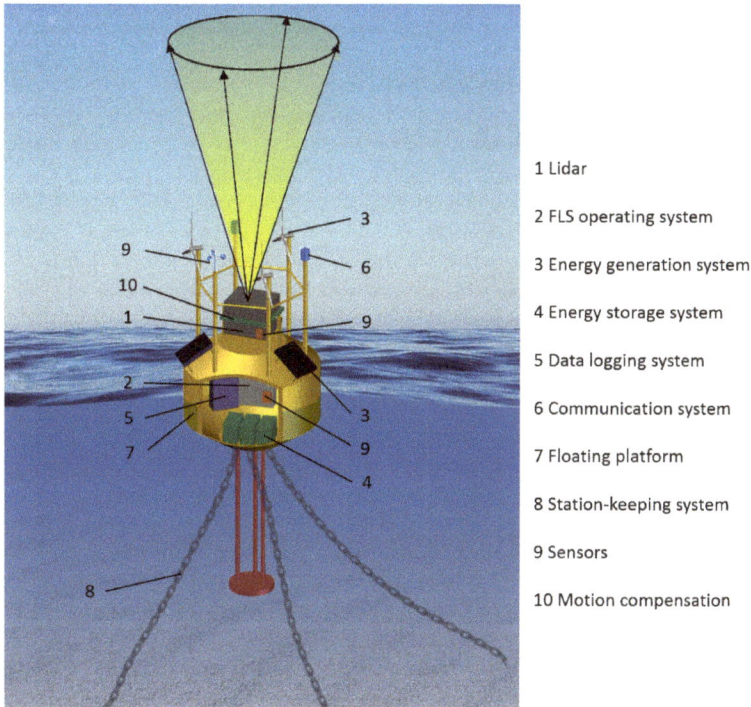

Figure 2. Schematic drawing of an FLS and its components [3].

1 Lidar

2 FLS operating system

3 Energy generation system

4 Energy storage system

5 Data logging system

6 Communication system

7 Floating platform

8 Station-keeping system

9 Sensors

10 Motion compensation

Despite progress, some barriers remain to the adoption of FLSs. These include:

- **Motion:** In an FLS, the movement of the sea imparts motion on the platform and the lidar, which makes it challenging to maintain the accuracy of the wind speed and direction measured by the FLS [26,27].
- **Reliability:** An FLS is often deployed in remote locations in extremely challenging environments, which necessitates robust, autonomous, and reliable measurement, power supply, data logging, and communication systems [25].
- **Acceptance:** During a Task 32 Workshop in February 2016 on "Floating Lidar Systems: Current Technology Status and Requirements for Improved Maturity", the acceptance of the (mature) technology by the industry was identified as a remaining barrier to adoption.

A common understanding is that future activities for the further promotion of floating lidar technology need to focus even more on the interests of the end-users. This is needed to reach full acceptance of the technology by demonstrating a validated performance with respect to the final wind resource estimates and at the most attractive costs. When these objectives are met, FLSs have the potential to serve a significant part of the market for offshore wind resource assessments.

2.3. Lidar Verifications

Given the importance of accurate wind measurements, it is industry best practice to verify a specific lidar device at an appropriate onshore flat terrain (or offshore) test site before and—should inconsistent behavior be observed during the measurement campaign—after the measurement campaign. The purpose of lidar wind data comparisons against co-located cup data from a met mast is to assess the capability of the lidar device to measure wind speed and direction to the same level of accuracy as what is obtained with conventional anemometry. This verification process ensures traceability back to classical anemometry, and provides a lidar unit specific standard uncertainty of wind measurements.

Current best practice is therefore to verify the lidar against anemometers on a tall conventional met mast. This mast should be documented as satisfying the guidelines of the International Electrotechnical Commission (IEC) standard 61400-12-1, Ed 2 [2] for low uncertainty anemometry, which allows traceability and confidence in the results. Such a mast is often known as "IEC compliant". While measurements from high quality cup anemometers are still seen as the norm against which any new measurement device should be judged, alternative reference data are seriously discussed and developed. For example, verification against a so-called "Golden Lidar" (a lidar unit of the same type that has repeatedly been calibrated against a more accurate reference) may also prove to be sufficient. However, the mechanical cup anemometer has been, and continues to be, the industry standard for measuring the wind speed at potential wind farm development sites.

The usually applied direct comparison of the readily reconstructed wind speed and direction (reconstructed from the line-of-sight measurements along the laser beams) with the corresponding reference quantity is called a Black Box verification. It is "black" as no insight into the wind data reconstruction process (i.e., neither of lidar geometry nor of algorithms) is needed. Only the final lidar wind data output is used for comparison to the reference measurements. In contrast to that, a White Box verification treats components of instrument function and reconstruction algorithm, individually. The White Box approach can be applied if assumptions entailed by the reconstruction algorithm (e.g., flow homogeneity within the lidar's measurement volume) are valid for the test site (e.g., benign flow conditions over non-complex terrain), and by extension, for sites for which the test results are considered to be valid. Weaknesses of this approach are that these assumptions may not be valid, and that multiple reference instruments are used to verify the wind speed and direction (cups and vanes). Strengths of the white box approach include greater flexibility in the test setup: all the details of the final deployment are not replicated during the test of the system's individual components and subsystems. However, if a feature of the final deployment is not adequately represented either in the test setup or the assumptions on which wind field reconstruction is based, an unanticipated bias or uncertainty can arise which cannot be observed without Black Box testing. There is a strong need in the wind industry to foster the acceptance of this technology by reducing the measurement uncertainty, as still too high uncertainties represent a barrier using lidar technology. It is often the reference uncertainty from the cup calibrations which dominates the overall uncertainty of the lidar verifications, compared to the uncertainty from the actual comparison. Thus, to mitigate this in the future, it is important to obtain a less conservative handle on reference uncertainties than that currently suggested by the IEC standard.

3. Lidar for Power Performance Testing of Wind Turbines

Wind lidars are attractive for power performance testing because they can be deployed temporarily on the ground or turbine nacelle to provide the required wind data, or they can be integrated into the turbine design and used to provide continuous performance monitoring. In addition, lidars have the ability to provide wind speed and direction data at multiple heights across the rotor to better characterize turbine response with respect to wind shear and wind veer. For nacelle or scanning lidars, there is potential for a single device to measure data over a range of horizontal positions which can allow modelling of turbine performance at any site and as subject to features (e.g., forest edges, buildings, etc.) in flat terrain. Task 32 has been active in the promotion and validation of new methods for defining wind speed for power performance using ground- and nacelle-based lidar.

3.1. Ground-Based Measurements

Ground-based lidar operation includes onshore applications where the lidar is situated on the ground or on a stable structure such as a building, and offshore operations where the lidar is situated on a fixed structure such as a met mast platform, a substation or an offshore wind turbine transition piece walkway. Task 32 has also supported validation of the uncertainty guidance published in Edition 2 (2017) of the IEC 61400-12-1 Standard [2]. Several barriers have been identified for the use of ground based lidar in power performance measurements, including:

- Calculation of a rotor equivalent wind speed (REWS) from lidar measurements
- Reduction of lidar uncertainty compared to a cup anemometer
- Application of lidar in complex terrain.

Task 32 has addressed these barriers in partnership with other groups. In 2014, Task 32 conducted a comparative exercise using common data to calculate REWS [10]. The exercise helped show where discrepancies might exist in the interpretation of the draft standard and gave industry participants an opportunity to apply and refine the method. Feedback from the exercise was incorporated into Ed. 2 of IEC 61400-12-1 to facilitate use of the REWS.

In 2016, Task 32 organized a second comparative exercise and follow-up workshop with support from the Power Curve Working Group. The purpose of the exercise was to estimate the uncertainty in power performance measurements when using a lidar as a standalone device or to normalize the hub height wind speed for wind shear and wind veer across the rotor using a REWS. This exercise started by using a time series dataset of 10-minute average wind speeds and required participants to create the binned power curves and uncertainty values for power and annual energy production. The main takeaway of the exercise was that based on current methods for calibration and classification of lidar, there would be additional uncertainty when using only a lidar as the primary wind measurement equipment. This is a potential barrier for the adoption of lidar which can be most easily be mitigated by refining the calibration and classification methods. Having a common understanding of the potential magnitude of additional uncertainty related to using lidar when following the uncertainty guidance in the IEC Standard should help focus efforts towards the main uncertainty contributors. The workshop discussed the results of the exercise but also provided a forum for different sections of the industry to present their experiences in the application of lidar for power performance. The results of the exercise have been submitted for publication at the conference Torque 2018.

3.2. Nacelle-Based Measurements

Nacelle-based lidar is one possible solution for power performance verification both onshore and offshore. The IEC 61400-50-3 standard is currently being developed to provide guidance relevant for this application. This will describe a method for nacelle-based lidar measurements suitable for power performance measurements both offshore and onshore. This new IEC standard is expected to be published by 2020.

The application of nacelle-based lidar shares similar barriers and needs to ground-based lidar. These are:

- Development of a common framework for wind field reconstruction.
- Determination of the optimal methods to calibrate and classify nacelle lidars.
- Quantification of the uncertainty.
- Application in complex terrain.

In 2017, Task 32 organized a workshop around the application of nacelle lidar for power performance. The workshop covered wind field reconstruction methods and calibration methods. There were four main takeaways from the workshop:

- The requirement to measure at 2.5D upstream for the coming wind turbines with very large rotor diameters (especially offshore) challenges the limitations of the measurement geometry of currently available nacelle lidars.

- The use of measurements at shorter range (inside the turbine's induction zone) that are then analyzed to find the freestream wind speed has been tested with good results [28].
- Nacelle lidars would also be a good solution for power performance testing in complex terrain, however several challenges need to be overcome.
- The installation of the lidar on the nacelle should be facilitated by collaboration between the lidar manufacturer (to make smaller and lighter lidars) and turbine OEMs (to include a dedicated place/bracket for the nacelle lidars in their turbine design).

In 2018, Task 32 is planning an additional workshop to support development of a common framework around wind field reconstruction and the application of nacelle lidar in complex terrain. It is expected that the content of these workshops will directly support developments in IEC PT 50-3.

4. Lidar for Turbine Control

Although wind serves as the "fuel" for wind turbines, changes in the wind inflow act as disturbances to the wind turbine which must be compensated for by the turbine's control system. Traditional feedback wind turbine control systems rely on measurements of generator speed to maximize or regulate power capture and reduce structural loads using pitch and torque control. Therefore, the controller can only react to wind disturbances after they impact the turbine. By using preview measurements of the approaching wind field from a nacelle-based lidar, pitch or torque commands from a "feedforward" controller can mitigate the impact of the wind disturbance on the turbine, improving power regulation and reducing loads.

Nacelle lidar systems were demonstrated in the field as early as 2003 [29], while simulation-based studies of lidar-assisted feedforward control began in 2005 [30]. Since then, lidar-assisted control (LAC) has been investigated for several applications including:

1. **Collective pitch control**, primarily targeting improvements in rotor speed regulation and reductions in tower base and blade root loads [31,32]
2. **Individual pitch control**, further improving blade and drive train component load reduction using lidar measurements of shear [31,33]
3. **Torque control** during below-rated operation to improve power capture [31,34]
4. **Combined torque and pitch control** to improve loads, especially during the transition between below-rated and above-rated operation [35,36]
5. **Yaw control** to improve power capture by improving rotor alignment with the wind direction [37,38].

Due to its simplicity and effectiveness at improving rotor speed regulation along with reducing tower and blade loads, lidar-assisted collective pitch control during above-rated operation has become one of the most popular categories of LAC investigated. Starting in 2012, successful field tests of lidar-assisted collective pitch controllers were performed, using the Controls Advanced Research Turbines (CART 2 and CART 3 turbines) at the National Renewable Energy Laboratory's National Wind Technology Center in Colorado, USA [39–41]. These field tests qualitatively verified many of the benefits observed in simulation by demonstrating improvements in rotor speed regulation [39–41], tower base load reduction [40,41], and a reduction in pitch actuation [41].

One important difference between LAC and the traditional applications (i.e., site assessment, power performance) is that the traditional applications focus on reproducing the point quantities measured by a met mast. For example, standard wind field reconstruction methods calculate the horizontal wind speed and direction from line-of-sight wind speeds. For LAC, new wind field reconstruction approaches have been developed to estimate rotor effective quantities, for example the rotor effective wind speed v_0 in Figure 3. Another difference is that, in traditional applications, the lidar data are usually post-processed offline and averaged over 10 min. For LAC, the signals are provided to the control system online with a high temporal resolution ($\geq$1 Hz) . Therefore, although similar to the REWS discussed in Section 3.1, the rotor effective wind speed used for LAC describes the instantaneous wind speed at the rotor, whereas the REWS is typically a 10-min average value. Further, uncorrelated frequencies are filtered out to avoid harmful control action (v_{0Lf} in Figure 3). Due to several issues such as the limitation to line-of-sight wind speeds or the evolution of the wind field from the measurement point to the rotor, the lidar system is only capable of estimating the rotor-effective quantities up to a certain frequency [42,43].

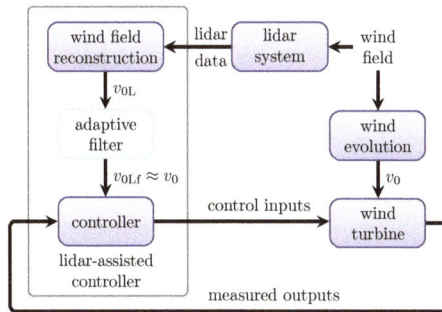

Figure 3. Basic control loop of lidar-assisted wind turbine control, based on [42].

4.1. Barriers

Research using both simulation and field testing shows meaningful structural load reduction with LAC, yet the technology has not been widely adopted in the wind industry. Several barriers preventing the widespread use of LAC have been identified by Task 32. The first barrier identified is the multidisciplinary nature of lidar-assisted control. Lidar manufacturers are responsible for supplying the lidar equipment while wind turbine manufacturers design the control algorithms that use the lidar measurements. However, there is often insufficient collaboration between the two parties; lidar systems may not necessarily be optimized for control applications (see above for some of the main differences), and assumptions about lidar systems made by control engineers might not match reality. For example, system cost, scan pattern, and availability requirements for control are different than for other applications. More communication about these needs to the lidar community would allow the most suitable products to be made available. On the other hand, myths about availability, system performance, and cost may be preventing wind turbine manufacturers from moving forward with LAC. More collaboration between wind turbine OEMs and lidar manufacturers could help address these concerns.

Second, the value creation of LAC is very difficult to assess. Structural load reduction can lead to cost reduction or allow an increase in annual energy production via larger rotors or taller towers while maintaining original design loads. However, a sensor such as a lidar introduces capital and O&M expenses. It can be challenging to determine whether the benefits outweigh the additional costs, leading to a reduction in the levelized cost of energy (LCOE).

Finally, guidelines for certification of wind turbines with LAC are lacking. As an example, the IEC 61400-1 design standard [44] defines the design load cases (DLCs) that must be simulated to assess structural loads, a necessary step in wind turbine type certification. However, LAC presents additional simulation requirements and several DLCs require further clarification when LAC is used. Without clear design standards for wind turbines using LAC, it is difficult for wind turbine manufacturers to understand how to include the technology in the design stage and assess the value creation.

4.2. Mitigation

To address the challenges caused by the multidisciplinary nature of LAC, the Task 32 workshop "Optimizing Lidar Design for Wind Turbine Control Applications" was held in Boston, MA in July 2016. The aim of the workshop was to bring together wind turbine OEMs, lidar suppliers, and researchers to identify the requirements of lidars for control applications and suggestions for optimizing lidar scan patterns. Although a shared opinion among workshop attendees was that more collaboration between lidar suppliers and wind turbine control engineers is needed, several useful suggestions and strategies for optimizing lidars for control applications were developed. For example, availability is very important for LAC applications; the benefits of LAC cannot be realized when the lidar is unavailable, potentially requiring the turbine to operate in a reduced-power "safe mode" to satisfy load requirements. Lidars for control applications should be designed for high availability, possibly adapting the scan pattern or data processing stage to improve availability when atmospheric conditions cause the signal-to-noise ratio of the backscattered light to become too low for reliable velocity estimates. Unique definitions of availability for control applications were discussed to address the short time scales (seconds) relevant to control. One useful metric for availability presented is the number of valid measurements obtained from the lidar scan during the previous few seconds. The lidar can be considered available when this value exceeds a threshold determined by a control engineer. Availability requirements unique to LAC can then be addressed by lidar manufacturers. Another topic discussed was the need to better anticipate and reduce the lifetime cost of the lidar, which is important for determining the value creation of LAC. One strategy for mitigating this barrier is to make lidar maintenance simple enough that it can be incorporated into standard wind turbine maintenance schedules without requiring special attention by lidar technicians.

Scan pattern optimization was addressed through an exercise in which computational tools were used to quickly assess the coherence and error between measurements from different scan patterns and the rotor effective wind speed of interest at the turbine. Tools such as these can be used to assess the performance vs. complexity of different scan patterns, allowing wind turbine manufacturers and lidar suppliers to find an appropriate tradeoff for a specific control application. A report for the Task 32 workshop "Optimizing Lidar Design for Wind Turbine Control Applications" is being prepared.

Task 32 has not yet focused on the problem of quantifying the reduction in LCOE through the use of LAC, but believes that this requires a systems engineering approach due to the complexity of the problem. Therefore, a collaboration with IEA Wind Task 37: Systems Engineering has been proposed, in which LAC can be applied to a reference wind turbine with a state-of-the-art baseline control system, allowing a number of parameters to be optimized with the objective of reducing LCOE.

To address the need for clear guidelines on certification to foster the widespread adoption of LAC, Task 32 recently held a workshop titled "Certification of Lidar-Assisted Control Applications" in January 2018 in Hamburg, Germany. The workshop, hosted by DNV GL, brought together wind turbine manufacturers, lidar suppliers, researchers, and certification bodies to develop ideas for modifications to existing design standards that should be made to address the use of LAC. Four categories relevant to certification were addressed: the lidar system, simulation models and load simulations, the control and protection system, and prototype measurements. The ideas generated during the workshop are to be incorporated into DNV GL's guidelines on wind turbine certification with LAC, planned for later in 2018.

5. Lidar for Load Verification

In addition to using lidar measurements for control, lidars have the potential to be used as part of the load verification procedure. Load verification consists of the comparison of simulated and measured structural loads on a wind turbine to verify the accuracy of the simulation environment, and is a step in obtaining certification of the prototype turbine by a certifying body.

The steps involved in a typical load verification campaign are as follows. Wind turbines are typically designed according to the IEC 61400-1 standards [44], where a range of wind conditions are prescribed. A load case matrix is created for different wind conditions and wind turbine components, and several load simulations are carried out. Eventually, extreme and fatigue loads are calculated, which provide a basis for dimensioning of the different wind turbine components. Subsequently, a prototype turbine is manufactured and installed at a test site for verification of the actual loads experienced by the turbine as compared to the simulated loads. The load measurements are usually carried out using strain gauges mounted on different wind turbine components, whereas the wind conditions are obtained using met mast anemometry. The instrumentation on the mast usually consists of cup anemometers and wind vanes installed on booms at three heights at some distance in front of the wind turbine. From the time series data of these instruments, first- and second-order statistics are estimated; amongst others mean wind speed, wind direction, wind speed profile, and turbulence intensity. Correspondingly statistics of the loads from the measured load time series are estimated. Eventually, the relationship between the wind and load statistics is determined. Simultaneously, new load simulations are carried out using wind conditions with the measured wind statistics (instead of the IEC standard wind conditions). The newly simulated loads are then compared with the measured loads, and subsequently verified to check whether the measured loads are smaller than the simulated loads. It is usually observed that the simulated loads do not compare very well with the measured loads, both in terms of accuracy and precision.

From the wind measurement side, two reasons have been attributed to the discrepancy between simulated and measured loads during the load verification stage; the first is the lack of availability of measurements with high spatial resolution that adequately represent the wind conditions across the entire rotor, and the second is the distance between the mast and the turbine, potentially introducing a mismatch between the measured wind conditions and those actually experienced by the turbine. Lidars have the potential to counter both these problems by allowing measurements at many different points relatively close to the turbine. The use of lidar to improve the estimation of the true wind statistics could yield better agreement between measured and simulated loads, and the improved load verification could ultimately lead to reducing the amount of material used for constructing certain components, thereby reducing the cost of wind turbines. Nevertheless, there are some barriers to the use of lidars for load verification.

5.1. Barriers

Lidars are proven to be quite accurate and precise for estimating first-order statistics such as the mean wind speed, wind direction, and the wind profile [45]. However, estimating second-order statistics required for load verification, such as turbulence intensity, from lidar measurements is not yet acceptable. In [18], two barriers to using a ground-based lidar in a VAD scanning mode have been identified; the first is the probe volume averaging along the lidar beam, and the second is the contamination due to the cross-correlation of different components of the wind vector. These two phenomena typically lead to underestimation and overestimation of the true turbulence intensity, respectively. The second barrier can be countered to a considerable extent by either using a different scanning configuration and data processing technique (e.g., six-beam method [18]), or by using three lidars intersecting at a point. The challenge of probe-volume filtering still remains to be tackled. In addition, due to the physical limitations of lidars, spatially interspersed measurements cannot be obtained at very high sampling rates. This presents a further challenge in capturing all the relevant turbulence scales that influence the loads on the wind turbine. Thus, the barriers to the use of ground-based lidars for load verification can be summarized as:

1. Difficulty in overcoming the cross-correlation of different wind components when estimating second-order statistics using a single lidar;
2. lidar probe-volume filtering affecting the second-order statistics of the measured wind velocities; and
3. obtaining measurements at a large number of points across the rotor disk with sufficient temporal resolution.

An alternative to using ground-based lidars is to use nacelle-based lidars. However, besides tackling the aforementioned challenges, a barrier to the use of nacelle lidars for load verification is that, to the authors' knowledge, a robust algorithm is yet to be developed for combining measurements in a vertical plane from nacelle lidars for estimating second-order statistics.

Lastly, the different configurations suggested above present economic challenges. Particularly the use of three lidar systems to overcome the cross-correlation of different wind components would increase the cost of measurements by a factor of three. It is therefore necessary to couple the benefits of using any configuration with the potential reduction in LCOE. A barrier to such coupling is that there is no robust model that provides a link between lidar-assisted load verification and LCOE.

5.2. Mitigation

Substantial efforts toward overcoming the barriers to the use of lidars for load verification remain as future work within Task 32. Within the research community, however, there has been some work related to the use of nacelle lidar measurements for load verification. In [46], the authors present a method for using measurements from nacelle lidars to recreate as closely as possible the full wind field that interacted with the turbine, as opposed to merely deriving second-order statistics from the measurements. The full reconstructed wind field can then be used to perform simulations for load comparison, with better agreement with field data expected. The authors consider how different nacelle lidar scan patterns affect the accuracy of the recreated wind fields, but have not yet included realistic sources of lidar measurement error such as probe-volume filtering and line-of-sight limitations.

6. Lidar in Complex Flow

Wind lidar's ability to measure wind profiles to greater heights than is possible with conventional met towers, to repeatedly sample large swathes using scanning lidars, and to retrieve multiple wind vectors from a single point using coordinated scanning lidars [47] have made them a popular choice for measuring flow over complex terrain [13], in turbine wakes, in the inflow to a turbine, and in urban areas [11,28,48]. These complex flows exhibit spatial heterogeneity and transient features introduced by terrain, patchy land cover, turbine or structural wakes, local meteorology, and other effects. This heterogeneity and the transient features can lead to difficulties in interpreting LOS wind speed data.

As with site assessment in complex terrain (Section 2.1), one way to mitigate the effect of flow heterogeneity is to analyze the measured data in conjunction with flow models that can account for the orography of the experiment site [49,50]. A similar methodology can be applied to scanning lidar measurements of different types of complex flows such as the induction zone upstream a wind turbine rotor, wind turbine wakes and low level jets and or various conditions of atmospheric stratification [28,48,51–54]. Physical models can also be used to reconstruct the wind field from lidar measurements. These models—such as flow complexity recognition (FCR) [55], MuLiWEA [56], and the LINCOM [57] algorithms—aim to determine the three-dimensional flow field from the lidar measurements, not just a few parameters such as wind speed and direction. These first attempts to combine lidar measurements and flow models have given promising results and suggest that this approach could help provide the most realistic wind field corresponding to the measurement environmental conditions.

The need for a flow model can be avoided by using three lidar to simultaneously sample the same point [58]. This approach, known as "multi lidar", removes the unknowns in the wind field reconstruction but only provides information at the intersecting measurement points. Two lidar—dual lidar—can be used if an assumption is made regarding one of the velocity components.

IEA Wind Task 32 started to address the issues associated with complex flow by collecting experts' know-how and methods applied within the research community into a summary of the state of the art [11]. Early on in that work, the authors realized that there was no clear definition of "complex-flow" and therefore recommended that practitioners should anticipate complex flow conditions when any of the following indicators are observed:

- Complex terrain
- Heterogeneity in the upwind surface roughness or presence of trees
- Presence of natural or artificial obstacles whose wake could reach the measurement volume

- Local meteorology or terrain condition that could make the flow within the measurement volume non-homogeneous
- Bias or uncertainty from the comparison between lidar and cup or sonic anemometer measurements that are unexpected according to standards (e.g., [2])

Because the impact of these indicators would vary by use case, it was decided not to create thresholds for the quantitative metrics. Instead, the goal of these indicators was to prompt users to be aware of the possible impact of complex flow on their measurements and make appropriate plans.

Several workshops about lidar measurements in complex flows followed. One was held in 2016 in conjunction with IEA Wind Task 31—Wakebench—to investigate the use of wind lidars to measure wind turbine wakes. Another was held in June 2017 to define use cases for lidar in complex flow situations. Lidar offers a variety of measurement methods exploiting capabilities that extend well beyond what is possible with met masts. One outcome of this workshop was the understanding that insufficient time is typically invested in understanding the relationship between the data requirements and the capabilities of the lidar. This occurs because the approach to measurement campaigns typically adopted in the wind energy industry is heavily conditioned by restrictions that were previously imposed by more limited instruments. Often, data requirements for lidar have been assumed on the basis of these more limited capabilities. A key challenge presented by lidar is the need to review basic objectives given the measurement capabilities now available. Furthermore, as described in Section 2, another workshop took place in November 2017 to explore the barriers to using lidar for measurements in complex terrain.

The following barriers and potential solutions have been identified at these workshops and by task participants:

- **Forecasting complex flows**: Potentially complex flow conditions need to be recognized before a measurement takes place, so that an appropriate measurement technique can be applied. Currently, there is no clear, objective definition of what counts as complex flow for lidar measurements in recommended practices or standards, which means that either all conditions should be considered potentially complex (which increases the cost of a campaign), or that marginal cases might be treated as simple to avoid increasing costs.
- **Detecting complex flows**: Complex flow conditions need to be recognized during a measurement or from the results, so that appropriate analysis methods can be used. Again, such conditions are not defined in standards, but might only be detected from lidar data after a measurement—or not even recognized.
- **Difficulty of multi-lidar measurements:** Using two or more lidar is challenging because of equipment cost, the need for very detailed campaign planning, and the difficulty of operating multiple lidar simultaneously.

  Task 32 tried to show that these obstacles are not insurmountable. Successful examples of multi-lidar measurements were presented during one of the workshops. In particular, it was demonstrated that concurrent lidar measurement could be applied for the investigation of the wind field spatial variability offshore but near coasts. Furthermore, the accuracy of scanning lidar was addressed to point out the importance of the scanner pointing precision and the uncertainty linked to dual Doppler wind field reconstruction methods.

  In a joint meeting of Task 31 and Task 32, a discussion about a possible benchmark of wake models based on multi-lidar measurements was initiated. With this activity, Task 32 intends to provide support to communities that could benefit from lidar data, but lack the required background knowledge.

- **Unclear methods for uncertainty estimation:** An estimate of the uncertainty associated with lidar wind field reconstruction in complex terrain is not straightforward. Two main solutions were proposed and discussed at the Task 32 workshops:

- Simulation of lidar measurements within high-fidelity wind field simulations. Simulations in a realistic (and therefore known) wind field can provide indications about the uncertainty to be expected during a field experiment [56,59–61].
- Application of the uncertainty propagation [62] to the wind field reconstruction method. This approach was introduced to deal with dual-Doppler data analysis [63,64] and further investigated for nacelle-based power-performance testing applications.

- **Lack of guidance:** Participants also noted that new users might find it difficult to use the examples, because of the slightly different approaches (e.g., scan patterns, flow models, data interpolation algorithm) applied to pursue similar objectives. To solve this issue, it was concluded that detailed definition of use cases could provide the needed guidance. From the data requirements, the conditions situation and the methodology that generally describe a use case, an unambiguous measurement and data analysis strategy could be outlined. If necessary, the field of application should be restricted. In this sense, Task 32 aims to extend the definition of the most common uses cases in the future.
- **Using lidar measurements and flow simulations together needs further development:** In general, the lidar community is cautious with regards to the combination of lidar measurements and flow simulations because of the complexity and wide scope of the topic and its very early level of development. Task 32 identified the need of a state of art review to support and speed up the development of this approach. Such document should describe the different method implemented to incorporate lidar measurement into flow simulations and, for each case, point out the assumptions used to develop the flow model and their field of applicability.

7. Future Challenges

The application areas discussed in Sections 2 to 6 were identified when Task 32 began its second phase in 2014. Since then, as wind lidar devices have become more advanced and accepted, other challenges have also become important. These challenges are discussed in this section and might be investigated by Task 32 in the future.

7.1. Uncertainty COmpared to Conventional Anemometers

Several sources of uncertainty impact how well we can determine the wind speed and wind direction based on a lidar's observations of the flow field. These are uncertainties in the LOS, pointing, ranging, and wind field reconstruction. Challenges related to wind field reconstruction are well known and have been described elsewhere in this report.

LOS uncertainty is related to the process of extracting the LOS wind speed from the acquired backscattered signal. Fundamentally, this uncertainty is tied to the characteristics of the lidar components dedicated to the generation of the laser beam and detection of the corresponding backscattered signal, atmospheric characteristics and a choice of the Doppler peak estimator. LOS uncertainty has been extensively studied since the first lidars were introduced in the late 1970s (e.g., [65]). However, as lidar technology becomes more widely adopted, IEA Wind Task 32 identified a need to revisit the LOS uncertainty and communicate it in a more approachable manner.

Several studies indicate that wind lidars are capable of acquiring wind information with lower uncertainty than the conventional cup or sonic anemometers mainly because the non-contact, remote measurement avoids flow distortion. Despite this, LOS uncertainty is currently determined by comparing the LOS measurements of wind lidar to reference measurements acquired by sensors mounted on met masts. This approach downplays the performance of wind lidars because of the relatively large uncertainties of mast based sensors and also results in a "black box" uncertainty, in that the physics behind the uncertainty is hidden.

Participants in Task 32 identified a physics-based way to determine the LOS uncertainty—also known as a "white box" method—for which the detailed understanding of contributions to the LOS uncertainty is essential. In addition, because lidars are a remote measurement, understanding where the LOS speed is acquired in the atmosphere is tied to quantifying the pointing and ranging uncertainty. A dedicated study of these two uncertainty sources has been presented in [63], while practical aspects of determining these uncertainties in field are given in [13,47].

Uncertainties arising from assumptions associated with wind field reconstruction cannot be neglected if the quantity that is ultimately the subject of the measurement is a reconstructed wind parameter such as wind speed. Comparisons of concurrent white and black box calibration results suggest these can often be the dominant source of uncertainty, with excellent white box results failing to indicate poorer black box performance. An uncertainty evaluation that neglects wind conditions that influence the validity of the wind field reconstruction is incomplete. Indeed, given the impossibility of anticipating all possible influences, evaluations are necessarily incomplete from this point of view. Completeness can only be achieved with respect to our prior knowledge of possible influences. In that case, uncertainty should not be interpreted as the absolute likelihood of a given outcome, but as a measure of information we obtain should a particular outcome occur, relative to our prior understanding.

7.2. Developing the Wind Lidar Ecosystem

As technologies become more accepted, popular, and prevalent, an ecosystem of related technologies, software, service providers and applications grows around them and adds value to the end user. A recent example is the growth of new businesses in the last 10 years around smartphones, where so-called app stores connect users with providers, data can be shared easily and securely, and third parties provide a wide range of services. Such rapid growth in use and services is enabled by clear use cases, data sharing, and the ability to rapidly try new applications. For wind lidar, only the first—the use cases—have stabilized.

7.2.1. Data Tools

A wind lidar creates wind data that are transformed into information to support decision making. Thus, choices made about how to set up and use lidar data directly impact the decision. Therefore, it is important to consider many factors when designing a measurement campaign such as information about the lidars that will be used (their capabilities and power requirements), local atmospheric conditions (wind, aerosols, etc.), infrastructure (electricity, telecommunications, access roads) and site restrictions. This information is used to derive an optimal lidar setup and lidar configuration, and there may be complex interdependencies (for examples see, [13]). It is therefore possible to envisage a tool or a suite of tools that can facilitate the deployment planning process.

As has been noted before, another issue is managing the data generated by a lidar measurement campaign. The general complexity of lidar data analysis, and the fact that different lidar systems store data in different formats with no or limited meta information, makes it harder to develop common tools for data processing and restricts implementation of novel methods such as machine learning.

In 2017, the e-WindLidar initiative [66] started with a focus on development of community-based tools for the facilitation of lidar data analysis, planning of lidar-based experiments, and lidar configuration. The first result of this initiative is a proposal for the universal lidar data format [67] which is in accordance with the FAIR principles [68]. In 2018, the e-WindLidar initiative will be disseminated through a range of activities within Task 32, including a workshop about the universal lidar data format and community-based tools for data analysis.

7.2.2. Modular Lidar

Commercial lidar have been designed around specific use cases, such as wind energy resource assessment from the ground or a floating platform, for power performance testing from a nacelle, or for broad area coverage. They are optimized for these use cases and are robust and reliable. However, they are difficult to use for other applications. It is rare to encounter extremely flexible lidar systems such as the long- or short- range WindScanner [13] and Stuttgart scanner [69] which allow the adaptation of the lidar for different applications. Although flexible systems might not be needed by industry, a high level of flexibility is essential for research groups to explore new use case for lidars.

During the IEA Task 32 workshop in 2014, an OpenLidar initiative was proposed. Several members of Task 32 then developed a concept for a platform for the open-source design, construction, and operation of wind lidar devices. A central aspect of the OpenLidar initiative is the development of modular lidar. A standardized modular lidar system architecture would support interoperability and enable the development and testing of new technology as modules could be replaced with new designs without having to redesign the entire lidar system. This approach would allow the same lidar system to be employed for different use cases. Figure 4 depicts a modular lidar concept developed by DTU Wind Energy.

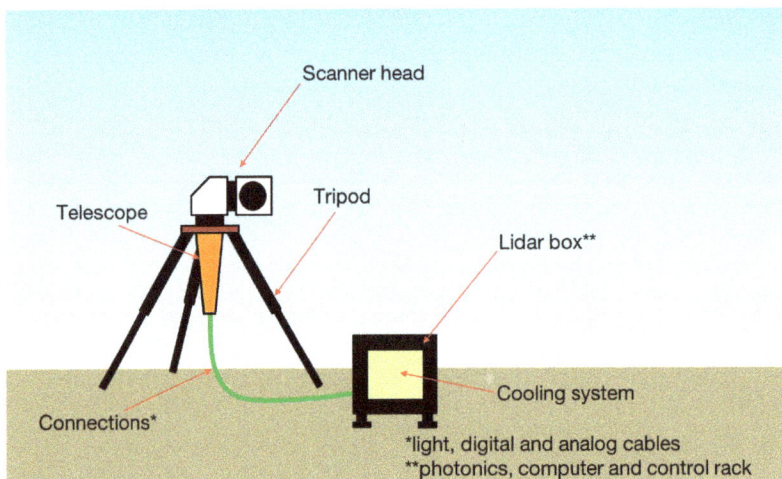

Figure 4. A concept of a modular scanning lidar (courtesy of DTU Wind Energy).

7.2.3. Turbine Integrated Lidars

Currently, lidar systems are installed individually on the nacelles of wind turbines. In the future, particularly for LAC, it will be important to integrate the lidar systems into the mechanical and controller design of wind turbines. This will lead to lidar becoming a standard sensor, much in the way that nacelle-mounted wind vanes are today. Challenges in this area will include the design of the lidar and its reliable integration into the turbine control system, which (like other lidar applications) will require support from standards.

8. Conclusions

Wind lidar is a maturing technology that helps to reduce the cost of wind energy through cheaper siting, and provides the possibility for increased energy capture and reduced loads. However, wind lidar is not uniformly accepted or used within the wind energy industry. IEA Wind Task 32 is an international collaboration that aims to identify the barriers to the adoption of lidar for site assessment, power performance testing, loads verification, lidar-assisted controls, and complex flows, and suggest ways to mitigate those barriers.

IEA Wind Task 32 has identified barriers that are common to all areas of wind lidar use. These barriers are relatively low and are all actively being addressed by the wind lidar community and Task 32. It should be noted that none of them prevent the use of lidar at this time. Instead, these are mostly barriers to extracting more valuable data from wind lidar:

- **Need for standards.** End users have more confidence in data when the collection and use of the data are supported by recommended practices and standards. Community-driven recommended practices are available for some applications of wind lidar, while internationally-recognized standards are only available for power performance testing. Standards for the use of lidar remote sensing are in development and will mitigate this barrier in the next few years.
- **Need for experts.** As with any emerging technology, there are a limited number of expert lidar users. This forms a barrier to the effective deployment of lidar for wind energy applications. Efforts are underway within the wind lidar community to embed more knowledge in planning and analysis tools, and to develop flexible and resilient processes, which will help mitigate this barrier in the near future.
- **Need for data tools.** Although commercial systems have condensed the amount of data delivered by a lidar to something similar to a conventional met tower, other applications can drown the user in data. The need to manage this flood forms a barrier to its rapid and effective use. Data management and processing tools are being developed to extract value from the data that should help mitigate this. The development of standards will also support this as standardized processes can be captured in such tools.
- **Need for better physics models.** Wind field reconstruction for wind lidars in complex flow and other applications requires flow models. Such models would also enable dynamic uncertainty estimates that include the effect of the lidar configuration, motion, external conditions and other factors on the lidar measurement and wind field reconstruction. While there have been some efforts in this area, more work is required. Importantly, work on on the other barriers will support the development of solutions to this challenge.

Despite progress in identifying barriers to the adoption of wind lidar for wind energy applications—and mitigating some of them—there remains a significant amount of research to be done in this field. There are also opportunities for product development, to explore new applications, and the potential for industry to provide value-adding commercial services. IEA Wind Task 32 has supported the wind lidar community since its inception in 2011 and will continue to do so for the foreseeable future. Interested parties are welcome to attend meetings and workshops and are invited to contact the operating agent for more information.

Supplementary Materials: IEA wind Task 32 is operated by the Chair of Wind Energy at the Institute of Aircraft design at the faculty of Aerospace Engineering at the University of Stuttgart. More information about IEA Wind can be found at www.ieawind.org. More details about IEA Wind Task 32, including minutes from the workshops and other documents, can be found at www.ieawindtask32.org.

Acknowledgments: IEA Wind Task 32 acknowledges the support and contributions of the entire wind lidar community. This publication was supported by the Open Access Publishing Fund of the University of Stuttgart.

Disclaimer: IEA Wind TCP functions within a framework created by the International Energy Agency. Views, findings, and publications of the IEA Wind TCP do not necessarily represent the views or policies of the IEA Secretariat or of all its individual member countries. IEA Wind TCP is part of IEA's Technology Collaboration Programme (TCP).

Author Contributions: Andrew Clifton led this paper and wrote the Abstract, Introduction, Section 2.1, Conclusions, and some section introductions, and edited all sections. Detlef Stein led Sections 2.2.1 and 2.3. Julia Gottschall and Ines Würth wrote Section 2.2.2. Luke Simmons led Section 3, which partly leveraged Workshop minutes written by Rozenn Wagner (DTU). Eric Simley wrote Sections 4 and 5 and used some unpublished text from Ameya Sathe (then DTU, now Ørsted) . David Schlipf contributed to Section 4 and wrote Section 7.2.3. Davide Trabucchi led Section 6. Peter Clive contributed to Sections 2.3, 3 and 6. Nikola Vasiljevic led Section 7. All authors contributed equally and are listed in the frontmatter in alphabetic order.

Conflicts of Interest: The authors declare no conflict of interest.

References

1. Emeis, S.; Harris, M.; Banta, R.M. Boundary-layer anemometry by optical remote sensing for wind energy applications. *Meteorologische Zeitschrift* **2007**, *16*, 337–347.

2. *Wind Energy Generation Systems—Part 12-1: Power Performance Measurements of Electricity Producing Wind Turbines*; Standard, International Electrotechnical Commission: Geneva, Switzerland, 2017.

3. Carbon trust. *Offshore Wind Accelerator Recommended Practices for Floating Lidar Systems*; Technical Report; Carbon Trust: London, UK, 2016.

4. Simley, E.; Pao, L.Y.; Frehlich, R.; Jonkman, B.; Kelley, N. Analysis of light detection and ranging wind speed measurements for wind turbine control. *Wind Energy* **2014**, *17*, 413–433.

5. Brower, M.C. *Wind Resource Assessment*; John Wiley & Sons, Inc.: Hoboken, NJ, USA, 2012.

6. Clifton, A.; Smith, A.; Fields, M.J. *Wind Plant Preconstruction Energy Estimates: Current Practice and Opportunities*; Technical Report TP-5000-64735; National Renewable Energy Laboratory: Golden, CO, USA, 2016.

7. Schwabe, P.; Feldman, D.; Fields, J.; Settle, E. *Wind Energy Finance in the United States: Current Practice and Opportunities*; Technical Report TP-5000-68227; National Renewable Energy Laboratory: Golden, CO, USA, 2017.

8. Boquet, M.; Callard, P.; Deve, N.; Osler, E. Return on Investment of a Lidar Remote Sensing Device. *DEWI Mag.* **2010**, *37*, 56–61.

9. Poveda, J.M.; Wouters, D. *Wind Measurements at Meteorological Mast IJmuiden*; Technical Report ECN-E–14-058; Energy Research Centre of the Netherlands (ECN): Petten, The Netherlands, 2015.

10. Wagner, R.; Cañadillas, B.; Clifton, A.; Feeney, S.; Nygaard, N.; Poodt, M.; Martin, C.S.; Tüxen, E.; Wagenaar, J. Rotor equivalent wind speed for power curve measurement: Comparative exercise for IEA Wind Annex 32. *J. Phys. Conf. Ser.* **2014**, *524*, 012108.

11. Clifton, A.; Boquet, M.; Burin Des Roziers, E.; Westerhellweg, A.; Hofsass, M.; Klaas, T.; Vogstad, K.; Clive, P.; Harris, M.; Wylie, S.; et al. *Remote Sensing of Complex Flows by Doppler Wind Lidar: Issues and Preliminary Recommendations*; Technical Report TP-5000-64634; National Renewable Energy Laboratory: Golden, CO, USA, 2015.

12. Risan, A.; Lund, J.; Chang, C.Y.; Sætran, L. Wind in Complex Terrain—Lidar Measurements for Evaluation of CFD Simulations. *Remote Sens.* **2018**, *10*, 59.

13. Vasiljević, N.; Palma, J.M.; Angelou, N.; Matos, J.C.; Menke, R.; Lea, G.; Mann, J.; Courtney, M.; Ribeiro, L.F.; Gomes, V.M. Perdigão 2015: Methodology for atmospheric multi-Doppler lidar experiments. *Atmos. Meas. Tech.* **2017**, *10*, 3463–3483.

14. Werner, C. *Doppler Wind Lidar—Range-Resolved Optical Remote Sensing of the Atmosphere*; Springer: Berlin/Heidelberg, Germany, 2005; Chapter 12, pp. 325–354.

15. Clifton, A.; Elliott, D.; Courtney, M. *IEA Wind RP 15. Ground-Based Vertically-Profiling Remote Sensing for Wind Resource Assessment*; Technical Report RP 15; IEA Wind: Paris, France, 2013.

16. Sathe, A.; Mann, J.; Gottschall, J.; Courtney, M.S. Can Wind Lidars Measure Turbulence? *J. Atmos. Ocean. Technol.* **2011**, *28*, 853–868.

17. Sathe, A.; Mann, J. A review of turbulence measurements using ground-based wind lidars. *Atmos. Meas. Tech.* **2013**, *6*, 3147–3167.

18. Sathe, A.; Mann, J.; Vasiljevic, N.; Lea, G. A six-beam method to measure turbulence statistics using ground-based wind lidars. *Atmos. Meas. Tech.* **2015**, *8*, 729–740.

19. Bradley, S.; Strehz, A.; Emeis, S. Remote sensing winds in complex terrain? a review. *Meteorol. Z.* **2015**, *24*, 547–555.

20. Klaas, T.; Pauscher, L.; Callies, D. LiDAR-mast deviations in complex terrain and their simulation using CFD. *Meteorol. Z.* **2015**, *24*, 591–603.

21. Hasager, C.; Stein, D.; Courtney, M.; Peña, A.; Mikkelsen, T.; Stickland, M.; Oldroyd, A. Hub Height Ocean Winds over the North Sea Observed by the NORSEWInD Lidar Array: Measuring Techniques, Quality Control and Data Management. *Remote Sens.* **2013**, *5*, 4280–4303.

22. Carbon Trust. *Carbon Trust Offshore Wind Accelerator Roadmap for the Commercial Acceptance of Floating Lidar Technology*; Technical Report CTC819; Carbon Trust: London, UK, 2013.

23. Fabre, S.; Stickland, M.; Scanlon, T.; Oldroyd, A.; Kindler, D.; Quail, F. Measurement and simulation of the flow field around the FINO 3 triangular lattice meteorological mast. *J. Wind Eng. Ind. Aerodyn.* **2014**, *130*, 99–107.

24. Bischoff, O.; Würth, I.; Gottschall, J.; Gribben, B.; Hughes, J.; Stein, D.; Verhoef, H. *IEA Wind RP 18. Floating Lidar Systems*; Technical Report RP 18; IEA Wind: Paris, France, 2017.

25. Gottschall, J.; Gribben, B.; Stein, D.; Würth, I. Floating lidar as an advanced offshore wind speed measurement technique: Current technology status and gap analysis in regard to full maturity. *Wiley Interdisciplin. Rev. Energy Environ.* **2017**, *6*, doi:10.1002/wene.250.

26. Wolken-Moehlmann, G.; Lange, B. Simulation of Motion-Induced Measurement Errors for Wind Measurements with LIDAR on Floating Platforms. In Proceedings of the Advancement of Boundary Layer Remote Sensing, ISARS, Paris, France, 28–30 June 2010.
27. Tiana-Alsina, J.; Gutiérrez, M.A.W.I.; Puigdefabregas, J.; Rocadenbosch, F. Motion Compensation Study for a Floating Doppler Wind Lidar. In Proceedings of the International Geoscience and Remote Sensing Symposium, IGARSS, Milan, Italy, 26–31 July 2015.
28. Borraccino, A.; Schlipf, D.; Haizmann, F.; Wagner, R. Wind field reconstruction from nacelle-mounted lidar short-range measurements. *Wind Energy Sci.* **2017**, *2*, 269–283.
29. Harris, M.; Bryce, D.; Coffey, A.; Smith, D.; Birkemeyer, J.; Knopf, U. Advance measurements of gusts by laser anemometry. *Wind Eng. Ind. Aerodyn.* **2007**, *95*, 1637–1647.
30. Harris, M.; Hand, M.; Wright, A. *Lidar for Turbine Control*; Technical Report NREL/TP-500-39154; National Renewable Energy Laboratory: Golden, CO, USA, 2006.
31. Bossanyi, E.; Kumar, A.; Hugues-Salas, O. Wind Turbine Control Applications of Turbine-Mounted Lidar. In Proceedings of the Science of Making Torque from Wind, Oldenburg, Germany, 9–11 October 2012.
32. Schlipf, D.; Kühn, M. Prospects of a Collective Pitch Control by Means of Predictive Disturbance Compensation Assisted by Wind Speed Measurements. In Proceedings of the German Wind Energy Conference (DEWEK), Bremen, Germany, 26–27 November 2008.
33. Schlipf, D.; Schuler, S.; Grau, P.; Allgöwer, F.; Kühn, M. Look-Ahead Cyclic Pitch Control Using LIDAR. In Proceedings of the Science of Making Torque from Wind, Heraklion, Greece, 28–30 June 2010.
34. Schlipf, D.; Fleming, P.; Kapp, S.; Scholbrock, A.; Haizmann, F.; Belen, F.; Wright, A.; Cheng, P.W. Direct Speed Control Using LIDAR and Turbine Data. In Proceedings of the American Control Conference, Boston, MA, USA, 19–22 August 2013.
35. Aho, J.; Pao, L.; Hauser, J. Optimal Trajectory Tracking Control for Wind Turbines During Operating Region Transitions. In Proceedings of the American Control Conference, Boston, MA, USA, 19–22 August 2013.
36. Schlipf, D. Prospects of Multivariable Feedforward Control of Wind Turbines Using Lidar. In Proceedings of the 2016 American Control Conference (ACC), Boston, MA, USA, 6–8 July 2016; pp. 1393–1398.
37. Fleming, P.A.; Scholbrock, A.K.; Jehu, A.; Davoust, S.; Osler, E.; Wright, A.D.; Clifton, A. Field-Test Results using a Nacelle-Mounted Lidar for Improving Wind Turbine Power Capture by Reducing Yaw Misalignment. In Proceedings of the Science of Making Torque from Wind, Copenhagen, Denmark, 18–20 June 2014.
38. Scholbrock, A.; Fleming, P.; Wright, A.; Slinger, C.; Medley, J.; Harris, M. Field Test Results from Lidar Measured Yaw Control for Improved Yaw Alignment with the NREL Controls Advanced Research Turbine. In Proceedings of the AIAA Aerospace Sciences Meeting, Kissimmee, FL, USA, 5–9 January 2015 2015.
39. Schlipf, D.; Fleming, P.; Haizmann, F.; Scholbrock, A.K.; Hofsäß, M.; Wright, A.; Cheng, P.W. Field Testing of Feedforward Collective Pitch Control on the CART2 Using a Nacelle-Based Lidar Scanner. In Proceedings of the Science of Making Torque from Wind, Oldenburg, Germany, 9–11 October 2012.
40. Scholbrock, A.; Fleming, P.; Fingersh, L.; Wright, A.; Schlipf, D.; Haizmann, F.; Belen, F. Field Testing LIDAR-Based Feed-Forward Controls on the NREL Controls Advanced Research Turbine. In Proceedings of the AIAA Aerospace Sciences Meeting, Grapevine, TX, USA, 7–10 January 2013.
41. Kumar, A.; Bossayni, E.; Scholbrock, A.; Fleming, P.; Boquet, M.; Krishnamurthy, R. Field Testing of LIDAR Assisted Feedforward Control Algorithms for Improved Speed Control and Fatigue Load Reduction on a 600 kW Wind Turbine. In Proceedings of the European Wind Energy Association Annual Event, Paris, France, 17–20 November 2015.
42. Schlipf, D. Lidar-Assisted Control Concepts for Wind Turbines. Ph.D. Thesis, University of Stuttgart, Stuttgart, Germany, 2016.
43. Simley, E. Wind Speed Preview Measurement and Estimation for Feedforward Control of Wind Turbines. Ph.D. Thesis, University of Colorado at Boulder, Boulder, CO, USA, 2015.
44. *Wind Turbines-Part 1: Design Requirements*, 3rd ed.; Technical Report IEC 61400-1; International Electrotechnical Commission: Geneva, Switzerland, 2005.
45. Sathe, A.; Banta, R.; Pauscher, L.; Vogstad, K.; Schlipf, D.; Wylie, S. *Estimating Turbulence Statistics and Parameters from Ground- and Nacelle-Based Lidar Measurements*; Technical Report; IEA Wind Task 32 Expert Report; IEA Wind: Paris, France, 2015.
46. Dimitrov, N.; Natarajan, A. Application of simulated lidar scanning patterns to constrained Gaussian turbulence fields for load validation. *Wind Energy* **2016**, *20*, 79–95.

47. Vasiljević, N.; Lea, G.; Courtney, M.; Cariou, J.P.; Mann, J.; Mikkelsen, T. Long-Range WindScanner System. *Remote Sens.* **2016**, *8*, 896.
48. Trabucchi, D.; Trujillo, J.J.; Kühn, M. Nacelle-based Lidar Measurements for the Calibration of a Wake Model at Different Offshore Operating Conditions. *Energy Procedia* **2017**, *137*, 77–88.
49. Bingöl, F.; Mann, J.; Foussekis, D. Conically scanning lidar error in complex terrain. *Meteorol. Z.* **2009**, *18*, 189–195.
50. Wagner, R.; Bejdic, J. *Windcube + FCR test at Hrgud, Bosnia and Herzegovina*; DTU Wind Energy: Copenhagen, Denmark, 2014; Volume E-0039.
51. Raach, S.; Schlipf, D.; Cheng, P.W. Lidar-based wake tracking for closed-loop wind farm control. *Wind Energy Sci.* **2017**, *2*, 257–267.
52. Aitken, M.L.; Banta, R.M.; Pichugina, Y.L.; Lundquist, J.K. Quantifying Wind Turbine Wake Characteristics from Scanning Remote Sensor Data. *J. Atmos. Ocean. Technol.* **2014**, *31*, 765–787.
53. Iungo, G.V.; Porté-Agel, F. Volumetric Lidar Scanning of Wind Turbine Wakes under Convective and Neutral Atmospheric Stability Regimes. *J. Atmos. Ocean. Technol.* **2014**, *31*, 2035–2048, doi:10.1175/JTECH-D-13-00252.1.
54. Trujillo, J.J.; Bingöl, F.; Larsen, G.C.; Mann, J.; Kühn, M. Light detection and ranging measurements of wake dynamics. Part II: two-dimensional scanning. *Wind Energy* **2011**, *14*, 61–75.
55. Leosphere. *Windcube FCR Measurements: Principles, Performance and Recommendations for Use of the Flow Complexity Recognition (FCR) Algorithm for the Windcube gRound-Based Lidar*; Technical Report; Leosphere: Orsay, France, 2017.
56. van Dooren, M.F.; Trabucchi, D.; Kühn, M. A Methodology for the Reconstruction of 2D Horizontal Wind Fields of Wind Turbine Wakes Based on Dual-Doppler Lidar Measurements. *Remote Sens.* **2016**, *8*, 809.
57. Astrup, P.; Mikkelsen, T.; van Dooren, M. *Wind Field Determination from Multiple Spinner-Lidar Line-of-Sight Measurements Using Linearized CFD*; Technical Report E-102; DTU Wind Energy: Copenhagen, Denmark, 2017.
58. Pauscher, L.; Vasiljevic, N.; Callies, D.; Lea, G.; Mann, J.; Klaas, T.; Hieronimus, J.; Gottschall, J.; Schwesig, A.; Kühn, M.; et al. An Inter-Comparison Study of Multi- and DBS Lidar Measurements in Complex Terrain. *Remote Sens.* **2016**, *8*, 782.
59. Churchfield, M.; Wang, Q.; Scholbrock, A.; Herges, T.; Mikkelsen, T.; Sjöholm, M. Using High-Fidelity Computational Fluid Dynamics to Help Design a Wind Turbine Wake Measurement Experiment. *J. Phys. Conf. Ser.* **2016**, *753*, 032009.
60. Doubrawa, P.; Barthelmie, R.J.; Wang, H.; Pryor, S.C.; Churchfield, M.J. Wind Turbine Wake Characterization from Temporally Disjunct 3-D Measurements. *Remote Sens.* **2016**, *8*, 939.
61. Trabucchi, D.; Trujillo, J.J.; Steinfeld, G.; Schneemann, J.; Kühn, M. Simulation of measurements of wake dynamics with nacelle and ground based lidar wind scanners. In *Book of Abstracts Wake Conference*; Gotland University: Visby, Sweden, 2011; pp. 170–174.
62. Joint Committee for Guides in Metrology (Working Group 1). *Evaluation of Measurement Data—Guide to the Expression of Uncertainty in Measurement*; Joint Committee for Guides in Metrology (JCGM): Sevres, France, 2008.
63. Vasiljevic, N. A Time-Space Synchronization of Coherent Doppler Scanning Lidars for 3D Measurements of Wind Fields. Ph.D. Thesis, Technical University of Denmark (DTU), Lyngby, Denmark, 2014.
64. Stawiarski, C.; Träumner, K.; Knigge, C.; Calhoun, R. Scopes and Challenges of Dual-Doppler Lidar Wind Measurements—An Error Analysis. *J. Atmos. Ocean. Technol.* **2013**, *30*, 2044–2062.
65. Frehlich, R. Simulation of Coherent Doppler Lidar Performance in the Weak-Signal Regime. *J. Atmos. Ocean. Technol.* **1996**, *13*, 646–658.
66. Vasiljevic, N. E-WindLidar Platform. Available online: http://e-windlidar.windenergy.dtu.dk (accessed on 9 February 2018).
67. Vasiljević, N.; Vignaroli, A.; Hasager, C.; Pauscher, L.; Klaas, T.; Lopes, J.; Bolstad, H.; Bardal, L. The Rise of Big Lidar Datasets and Need for Lidar Data Standardization, Contextualization and Dissemination. In Proceedings of the WindTech Conference, Boulder, CO, USA, 24–26 October 2017.
68. Wilkinson, M.D.; Dumontier, M.; Aalbersberg, I.J.; Appleton, G.; Axton, M.; Baak, A.; Blomberg, N.; Boiten, J.W.; da Silva Santos, L.B.; Bourne, P.E.; et al. The FAIR Guiding Principles for scientific data management and stewardship. *Sci. Data* **2016**, doi:10.1038/sdata.2016.18.

69. Wuerth, I.; Rettenmeier, A.; Schlipf, D.; Cheng, P.; Waechter, M.; Rinn, P.; Peinke, J. Determination of Stationary and Dynamical Power Curves Using a Nacelle-Based Lidar System. In Proceedings of the German Wind Energy Conference (DEWEK), Bremen, Germany, 7–8 November 2012.

remote sensing

MDPI

Article

Coherent Focused Lidars for Doppler Sensing of Aerosols and Wind

Chris Hill

Malvern Lidar Consultants, Great Malvern, Worcestershire WR14 1YE, UK; malvernlidar@gmail.com

Received: 16 January 2018; Accepted: 21 February 2018; Published: 16 March 2018

Abstract: Many coherent lidars are used today with aerosol targets for detailed studies of e.g., local wind speed and turbulence. Fibre-optic lidars operating near 1.5 μm dominate the wind energy market, with hundreds now installed worldwide. Here, we review some of the beam/target physics for these lidars and discuss practical problems. In a monostatic Doppler lidar with matched local oscillator and transmit beams, focusing of the beam gives rise to a spatial sensitivity along the beam direction that depends on the inverse of beam area; for Gaussian beams, this sensitivity follows a Lorentzian function. At short range, the associated probe volume can be extremely small and contain very few scatterers; we describe predictions and simulations for few-scatterer and multi-scatterer sensing. We review the single-particle mode (SPM) and volume mode (VM) modelling of Frehlich et al. and some numerical modelling of lidar detector time series and statistics. Interesting behaviour may be observed from a modern coherent lidar used at short ranges (e.g., in a wind tunnel) and/or with weak aerosol seeding. We also review some problems (and solutions) for Doppler-sign-insensitive lidars.

Keywords: coherent Doppler lidar; wind sensing; single-particle

1. Introduction

Near-infrared coherent lidars are familiar in anemometry and turbulence sensing. Their behaviour has been fairly well understood and modelled since the 1960s. Fibre-optic versions are increasingly used in the wind power industry for aerosol targets, and are also becoming popular for solid targets such as vibrating or rotating machinery, structural panels, and turbine blades. Several tutorials and reviews have recently been published, aimed at the growing number of readers—not necessarily optical or laser specialists—who need to understand the main features and limitations of these sensor tools. As they become widely used and extended to different technical areas, it is sometimes necessary to return to the fundamentals and check that older radar/lidar lessons are correctly translated and applied.

A recent review of modern fibre-optic lidars [1] discussed three main points:

(1) The most common references in the literature of "coherent continuous-wave focused monostatic lidars" are now some 25–40 years old but still worth reading. In particular, the carrier-to-noise analysis of Sonnenschein and Horrigan [2] agrees with alternative treatments based on the popular "antenna theorem" or "back-propagated local oscillator" (BPLO) approach. Their analysis applies to ZephIR and similar modern fibre-optic aerosol lidars.

(2) Such lidars work over a large range of conditions and spatial scales; for example, the so-called "probe volume" of a variable-focus lidar may easily vary over eight orders of magnitude. There may also be large variations in scattering particle density and average atmospheric backscatter. The assumption that the probe volume contains "many" scatterers can lead to simple mathematics (Gaussian statistics for long random walks where the central limit theorem

holds) and is extremely common but can be faulty (e.g., with very clean air or short measurement range).

(3) The desire for simple descriptions or "sensor performance metrics" conflicts with the complications of real lidar measurements. For example, the expressions "range resolution" and "bandwidth" have multiple meanings, and it is difficult (and often confusing) to characterise a lidar's performance by a single value. In a well-known sense, the axial resolution of a coherent CW focused monostatic lidar is a Lorentzian function with scale parameter equal to the beam Rayleigh range. In another important sense, the lidar can "resolve" scattering events with much finer range precision.

The present paper concentrates on aerosol scattering and a common commercial application: the "lidar Doppler" estimation of aerosol/lidar relative velocity and thus (by using several or many estimates) of wind flow and wind patterns. We start with a brief review of the beam geometry for a standard coherent lidar (Section 2). Sections 3 and 4 discuss the detector output for direct and heterodyne detection respectively. Section 5 has comments on the large preceding literature and reviews some disagreements about "lidar collection efficiency", which is one aspect of the dependence of carrier-to-noise (CNR) on beam geometry.

Section 6 discusses a computer simulation of multi-scatterer experiments. Section 7 describes the important practical constraint of sign ambiguity for moving targets and illustrates how, even if I&Q data or other indications of sign are not immediately available, that ambiguity can be removed for typical aerosol targets. For the common conical-scan or sector-scan geometry and its associated VAD (velocity-azimuth display) output, Section 8 discusses examples of measurement bias.

2. Lidar Geometry

Consider the sketches in Figure 1.

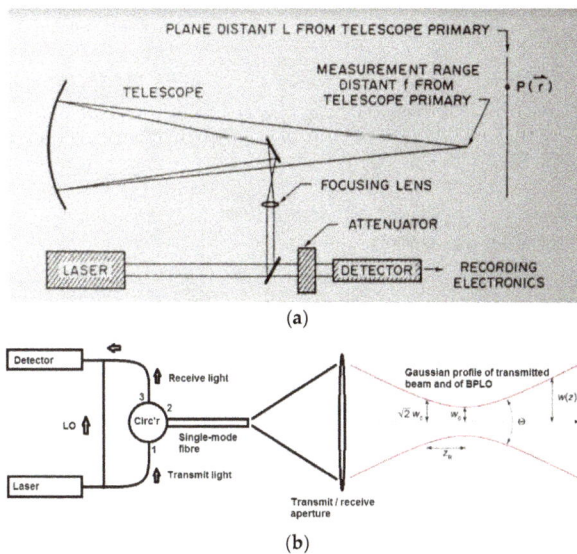

(a)

(b)

Figure 1. Schematics of laser Doppler systems for remote wind sensing. (**a**) Reproduced from Lawrence et al. [3] ("A laser velocimeter for remote wind sensing", Rev. Sci. Instrum. 1972, vol. 43, pp. 512–518) with the permission of AIP Publishing. This is concerned with Doppler measurements of wind-borne scatterers in the atmosphere at relatively short ranges, but is more widely relevant; (**b**) From Hill [1], concerned with modern fibre-optic lidars. See also below for Figure 2 and the original diagram in Lindelöw [4].

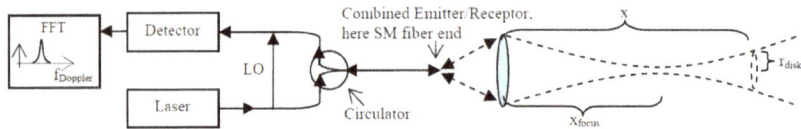

Figure 2. Schematic of focused monostatic CW coherent lidar based on fibre-optics components. Distance to focus is x_{focus}, and distance to target scatterer is x; these are called f and L in Figure 1a. Reproduced from Lindelöw [4].

The beam is brought to focus at a distance f from the telescope aperture (mirror or lens), and the scatterers are carried across the beam by the wind at a possibly different distance L, with a crosswind velocity component of V ms^{-1}.

Let the beam be a lowest-order Gaussian beam with a field described by

$$E(x,y,z) = \frac{E_0}{w(z)} \exp\left[\frac{-(x^2+y^2)}{w(z)^2}\right] \exp\left[jk\frac{(x^2+y^2)}{2R(z)^2}\right] \tag{1}$$

$$w(z) = w_0\sqrt{1 + (z/z_R)^2} \tag{2}$$

is the "beam $1/e^2$ intensity radius" at axial distance z from focus

$$R(z) = z + z_R^2/z \tag{3}$$

is the beam radius of phasefront curvature at z

$$z_R = \pi w_0{}^2/\pi \tag{4}$$

is the beam parameter or Rayleigh range (often notated b or b/2)

$$I(x,y,z) = \mid E(x,y,z) \mid^2 = \frac{E_0{}^2}{w(z)^2} \exp\left[-2\frac{(x^2+y^2)}{w(z)^2}\right] \tag{5}$$

Some constant factors and the extra on-axis (Gouy) phase shift have been omitted.

According to Equations (1) and (2), the beam extends indefinitely in any XY cross-section plane, but its field strength decays exponentially, with scale parameter w(z). The beam also extends indefinitely along the z axis, but its field strength decays according to $1/w(z)$, with scale parameter z_R. For a uniformly and densely seeded atmosphere, we are usually justified in neglecting the contributions from scatterers that are more than a few beam widths off-axis or more than a few Rayleigh ranges away from the beam focus—see below. For a single scatterer, of course, we are interested in the lidar response for any scatterer position if it is detectable.

For such a Gaussian beam of intensity $I(r) = I_0 \exp(-2(x^2 + y^2)/w^2) = I_0 \exp(-2r^2/w^2)$, illuminating one thin "slice" between ranges z and z + dz, some simple properties are

- $P_0 = I_0 (\pi w^2/2)$ is the total power, I_0 is the on-axis intensity, and w is the e^{-2} intensity radius of the beam.
- The power enclosed by a circle of radius h is $P_0 \cdot [1 - \exp(-2h^2/w^2)]$. If h equals 2w, about 99.97% of the total power is enclosed. This is effectively all the power for most practical purposes, and the mean intensity within the circle of radius h is nearly $P_0/(\pi h^2)$ if $h \geq 2w$.
- The probability distribution for the illumination intensity I is $p(I) = \frac{1}{I}(\frac{w^2}{2h^2})$ with normalisation to unit probability within the circle of radius h (i.e., the intensity lies between $I_0 \exp(-2h^2/w^2)$ and I_0).

Similar but z-averaged properties may be derived for a focused Gaussian beam whose width parameter w(z) varies along the beam according to Equation (2). However, the mean intensity and the p(I) then depend on the z integration limits, which may not be symmetric around the focus distance. For example, a "collimated" exit beam has a plane wavefront initially; the beam waist lies at the exit aperture (z = 0) and the beam expands as it propagates through positive ranges z; negative values of z are not relevant because they lie inside or behind the lidar.

Note that our Rayleigh range is defined here as $z_R = \pi w_0^2/\lambda$, so that the beam radius $w(z = z_R)$ is $\sqrt{2}$ times the beam waist radius w_0. Alternatively, the terms "beam parameter", "confocal distance", and "Rayleigh range" are sometimes defined as $2\pi w_0^2/\lambda$ to refer to the distance between the plane where $w(z = -z_R) = \sqrt{2}w_0$ and the corresponding plane on the other side of the waist where $w(z = z_R) = \sqrt{2}w_0$.

3. The Detector Output (Direct Detection)

Consider the average backscattered power reaching the receive aperture: a quantity proportional to the "photon count" due to scattering from all illuminated regions [5].

If there are N scatterers contributing, the instantaneous power is proportional to the magnitude-squared of the sum over N phasors (one from each scatterer). A time or ensemble average power, taken over all relative phases of the individual scatterer reflections (all phases being assumed equally probable), is proportional to the sum of N positive terms, one for each scatterer; each term depends on the scatterer's position in the Gaussian illuminating beam, and also has an inverse-square dependence of the strength of the spherical scattered wave as it propagates back to the receiver. Cross-terms average to zero, regardless of N; the averaging to zero results from the random uniform distribution of phases and does not require N >> 1. If we assume for simplicity that the fraction of scattered light received within the aperture (and falling on the detector) is the same for all scatterers at a particular range, that is usually a serious and restrictive assumption. The precise dependence on scatterer position may not be easy to express, if we wish to consider scatterers in the near field—for example, a thin slice so near the lidar aperture that the solid angle subtended by the aperture is not the same for all scatterers in the beam. We can only mention here a large literature on the detailed geometrical "form factors" and "collection efficiency" of direct-detection lidars [6].

This instantaneous power and this average power would be seen in direct-detection mode, where we assume that photoelectrons are generated independently from every small sub-area of the detector, and the total photoelectron current is (in the limit of large count) an accurate, "light-in-a-bucket" record (subject to some detector response bandwidth) of the time-varying optical power that falls on the whole detector surface. For the moment we can consider that a fixed (usually large) fraction of the backscattered laser light that falls within the receive aperture reaches the detector. This backscattered light forms an interference pattern due to its N randomly phased components (a Gaussian speckle pattern, in the limit N >> 1). Using a single-mode optical fibre in the receive channel of a direct-detection lidar has consequences. One advantage is that we know where we are looking (there is a well-defined "receive antenna" pattern—not to be confused with the phase-sensitive heterodyne-lidar antenna pattern below) but a disadvantage is that much light entering the receive aperture is lost because it does not match the fibre mode at the entrance facet; another consequence, often undesirable in direct detection, is that intensity fluctuations are not averaged out.

With this catch-all or "bucket" aperture, the average direct-detection backscattered power is proportional to the sum over the individual N illumination intensities—for any value of N or any seeding density; strictly, this holds without any lower limit, and we assume we never reach the other extreme where the scatterers are so dense that multiple scattering is significant.

In terms of the standard random-walk problem on a complex plane, we have at any one point (or any one sufficiently small detector element) the coherent summation of N randomly directed vectors [1]. If each vector has the same length a, the total "intensity"—the mean value of the square of the resulting distance from the origin—is $<I> = Na^2$. The second moment of intensity is $<I^2> = 2N(N-1)a^4 + Na^4$, which for large N approaches $2(Na^2)^2 = 2<I>^2$. The "normalised second moment" is thus $<I^2>/<I>^2 = 2 - (1/N)$. If the vector lengths are drawn from a random distribution [7], we can use the following result:

$$\frac{<I^2>}{<I>^2} = 2\left(1 - \frac{1}{N}\right) + \frac{1}{N}\frac{<a^4>}{<a^2>^2} \qquad (6)$$

Many further expressions for the intensity moments of various orders have been studied, often through the indirect mathematical methods of "generating functions". The probability distribution of intensity P(I) tends to be less tractable; few of the results, in terms of integrals of Bessel functions, reduce to useful analytical expressions. (Note that we use p(I) above to describe the spatial variations, and P(I) here to describe the time series at a detector point). For an illuminated collection of N non-fluctuating scatterers, with N "moderately large", Pusey et al. [8] expressed P(I) as the Gaussian-limit exponential exp(−I/<I>)/I plus a series of stated correction terms. Some of the integrals involved in random-walk analysis are difficult to compute with high precision, but this is not of major importance for lidar users.

A many-scatterer Gaussian-statistics limit of the random walk model should not be assumed without thought; also, the definition of "CNR" depends on whether we include times when no scatterers are present. In practice, perhaps with rather detailed "time-frequency" post-processing, we may increase the "CNR" by discarding data from such times and by discarding frequency bands which do not contain the (usually chirping) lidar returns.

Because of the Gaussian beam's circular symmetry, the backscattering from the full 3D scatterer-filled volume is usually treated by an integral over a set of thin discs (with radial scale w(z) and extending from z to z + dz) considered to fill the space on one side of the lidar aperture. We note that scattering from a collection of particles all confined to a very thin disk (or spherical shell etc.) may not satisfy the requirement of uniformly distributed phase [7].

4. The Detector Output (Heterodyne Detection)

Now consider the heterodyne operating mode by adding the LO (and BPLO) to the picture, with no other optical change; the backscattered light received at the detector is the same as before—say $P_s(N)$ for N scatterers—but in general, some fraction $(1 - \eta_{het})$ of this incident light will not be mode-matched with the LO and will not contribute to the heterodyne current, whose power is proportional to η_{het}, $P_s(N)$, and the local oscillator power P_{LO}. The overlap of incident light and LO light can be evaluated at the detector surface or (by a considerable extension of the original far-field "antenna theorem") at any other convenient plane. When we evaluate it in the target plane, the (time-averaged or ensemble-averaged) contribution to the heterodyne output power made by any scatterer is proportional to the intensity of the transmitted beam at the scatterer coordinates and also to the BPLO intensity there. Since we have assumed that transmit and BPLO beams are matched, we can describe the spatial variation equally by the product of transmit and BPLO intensities or by the square of the transmit intensity.

In direct detection, each small element of the detector produces its own photocurrent—which can be described by a time series of non-negative real numbers. In heterodyne detection, each small element produces a (usually dominant) shot noise photocurrent, plus a *modulation*—that is, an extra term (which oscillates at the heterodyne frequency, taking both positive and negative values as the interference fringes evolve). The net result of summing the outputs from the various detector elements is thus phase-sensitive: the total photocurrent is never negative, but the coherent (phase-sensitive) sum of the additional modulations—which is what interests us—can be positive or negative.

Once the lidar is shot-noise-limited, any further increase in LO power causes a proportionate increase in shot noise power and does not change the CNR; that is, any multiplying constant of LO power cancels top and bottom in a C/N expression.

The heterodyne current due to several scatterers is the vector sum of the individual currents, one for each scatterer; each term is proportional to the beam intensity and to the BPLO intensity at the scatterer. The mean square of the current (averaged over all relative phases as above) equals the sum of the individual squares; this is the same result as for the mean square of the length of a 2D random walk.

Consider a single scatterer blown transversely through the beam along a straight line at uniform speed V: its z and x values remain essentially constant, and its y value equals $(t - t_0)V$. As it traverses the beam it produces a detector output current with an intermediate-frequency (IF) component.

$$i(t) = 2R\sqrt{\eta_{het}P_{LO}P_s(t)}\ \cos{(\omega_c t + \theta(t))} \tag{7}$$

where R = detector responsivity, ω_c = offset radian frequency, and θ = phase shift. The time variation of i(t) includes variations in some or all of these parameters. Here we may take ω_c to be constant (there is no extra bulk target motion along the z-axis line of sight) and θ to express the time-varying Doppler phase shift. We may also take P_{LO} to be constant (the total local oscillator power at the detector) and P_s to be the backscattered power reaching the detector (not constant). The scalar η_{het} represents, for this single scatterer, the "heterodyne efficiency" with which the LO and backscattered light overlap at the detector. Note again that we write η_{het} proportional to the square of output current, i.e., to output power. In the literature on heterodyne lidars, there are various preferences for splitting the "efficiency" into several terms identified with different parts of the total system [1,9].

Usually, Equation (7) is taken to say that P_{LO} is a fixed quantity without spatial variation, and its value (in units of Watts)—the LO power that reaches the detector—is adjustable at our discretion; whereas, although the transmitted laser power is also a fixed number of Watts at our discretion, the value of $P_s(t)$ depends on other factors such as optical losses, atmospheric attenuation and the position of the scatterer in the beam. In practice, $P_s(t)$ will vary deterministically as the scatterer moves across the Gaussian intensity beam profile, reaching some peak value (see Section 5.4) when the scatterer makes its closest approach to the beam axis. The value of η_{het} for the particular distribution in space of scattered light at the detector (which will usually vary with time), and the particular (usually fixed) LO light distribution in space, is then a further matter of fact or calculation.

That is, in our notation, the spatial dependence in Equation (7), as the scatterer traverses the beam, brings a time dependence to P_s, and possibly to η_{het} and θ, but not to P_{LO}. This heterodyne current component i(t), due to a single scatterer at position y(t), is proportional to the transmitted beam intensity at the scatterer, i.e., the spatial variation is described by $\exp\left[\frac{-2(x^2+y^2)}{w(z)^2}\right]$. Note that the transverse variation of the transmitted beam is the Gaussian in Equation (1); the local oscillator (and hence the BPLO) will be assumed to keep a perfect copy of this Gaussian shape. Equation (14) in the early paper by Sonnenschein and Horrigan [2] describes "the square of the signal current produced by a single scatterer", and a corrected version is

$$|i_s|^2 \propto \frac{R^4}{\lambda^2 L^4\left[1 + \left(\frac{\pi R^2}{\lambda L}\right)^2\left(1 - \frac{L}{f}\right)^2\right]^2} \exp\left\{\frac{-4\left(\frac{\pi Rr}{\lambda L}\right)^2}{\left[1 + \left(\frac{\pi R^2}{\lambda L}\right)^2\left(1 - \frac{L}{f}\right)^2\right]}\right\} \tag{8}$$

L is the range to target, f is the range to beam focus, their R is the transmitted beam radius (our w(z) at the aperture where z = −f), and r is off-axis distance $(r^2 = x^2 + y^2)$.

In [1], we reviewed some of the differently notated but essentially similar versions of Equation (8) in the literature and illustrated the frequency-chirp behaviour of scatterers that traverse the beam. The product of chirp duration and chirp slope is roughly

$$\text{chirp excursion (Hz)} = \frac{\pi w(z)}{2V}\frac{2V^2}{\lambda R(z)} = \frac{\pi V w(z)}{\lambda R(z)} \tag{9}$$

The chirp changes sign if the wavefront curvature changes sign (i.e., if we consider events on one side of the beam focus and then the other). As we move away from the beam waist z = 0 (in either +ve or −ve direction), the absolute value of R(z) at first decreases, then reaches its minimum of R(|z| = z_R) = $2z_R$, then increases again. So, on each side of the beam focus, there are in principle two distances z corresponding with any given value of R greater than $2z_R$. These two distances are associated with the same chirp slope dθ(t)/dt but different beam widths w(z) and thus different envelope durations and different chirp excursions.

5. Previous Literature

5.1. Semiclassical Account of Laser Radar

There are large relevant literatures on coherent lidar, photon correlators, laser Doppler velocimeters (LDVs) and laser transit velocimeters (LTVs), electromagnetic scattering from small particles, interference effects in the presence of more than one scatterer, and so on (see for example [1,7,9], and their references). From an optical-radar viewpoint, the function of the scatterers is to provide, at the receiver, copies of the transmitted waveform [10]; when all the different copies are considered (with their various delays, attenuations, polarisations, frequency shifts etc.), we have a total field at any given detector element whose intensity at any instant is proportional to the mean rate of photoelectron production. That mean rate is vastly increased by the strong steady LO. There are typically very many photoelectrons, but each has a random (Poisson point process) time of origin. There is no one-to-one connection of individual photoelectrons with individual "photons" in this typical semiclassical account of a shot-noise-dominated coherent lidar. It is generally assumed that the photocurrents are so strong that, in line with our discussion above, the "full phase and intensity" information in the optical field is indeed transferred intact to the heterodyne detector output; for example, a very weak FM sideband (due to a faint micro-Doppler vibration) can still be isolated and examined and assigned a conventional SNR that is negligibly degraded by the (intermittent, discrete, Poisson) nature of the photoelectrons.

At power levels several orders of magnitude weaker, when we approach single-photon detection, this picture must change, but we retain it here. We neglect many complexities of scattering theory, vector wave effects, polarisation, and detector physics. But it is worth reviewing the detector output current (above) and referring now to some literature including frequently quoted "fundamentals of coherent lidar" papers.

The strength of the heterodyne current is usually judged by the average modulus-squared $<|i(t)|^2>$. (The average is taken over a time exceeding the longest fluctuation time, and there are usually several types of fluctuations present). This strength is determined by how well the scattered light and the LO overlap: the overlap integral, or antenna efficiency, includes both the magnitudes and the relative phases of the signal and LO terms. The need for transmit and receive antennas to be "matched" was familiar from earlier radar work, and was quickly imported and applied to lidar studies in the 1960s and 1970s. A large literature developed on various sub-topics such as:

- The "best" designs of telescopes and truncating apertures, according to several different metrics of efficiency;
- The benefits of angular selectivity or directionality of heterodyne antennas—and the corresponding requirement to establish and maintain precise alignment in practice;
- The statistics of detector outputs for various types of target: solid, liquid, or gas; few scatterers or many scatterers; concentrated or distributed in range; static or moving;
- The effects of one-way and two-way atmospheric turbulence;
- The differences between monostatic (shared apertures and collinear beams) and bistatic or multistatic lidars.

The literature inevitably swelled as different approaches and notations were developed and published by groups working on laser systems in private companies, universities, and government organisations.

5.2. Local Oscillator (LO) and BPLO

One important sub-topic is the "back-propagated local oscillator" or BPLO, which was presented by Siegman [11] as follows:

Consider the complex LO amplitude distribution falling on the photodevice surface (weighted by the quantum efficiency distribution if necessary). Reverse the direction of propagation of this LO distribution and allow the reversed wavefront to propagate back out through any optical elements that an incident signal wave would traverse. The resulting far-field or Fraunhofer diffraction pattern will be the antenna pattern of the optical heterodyne receiver.

The general principle is that we may reverse the direction of propagation of the wavefront and perform the overlap integral in any convenient place—not just the detector surface, but for example the telescope aperture or the plane(s) of the target(s). Two aspects of this BPLO approach may be especially relevant to modern wind lidars [1]. One is the advantageous choice of overlap plane. We are free to evaluate the overlap at different places: it may be that some, from a practical, computational point of view, are better than others. Zhao et al. [12] compare two expressions for a transmission function that relates a point in the scattering plane to a point in the detector plane. The light successively encounters a primary mirror, a secondary mirror, "a series of optical components, such as steering mirrors and polarizers", a detector lens, and the detector. The virtual BPLO encounters these in the reverse order, and the corresponding expression is derived by "changing the order of integrals and invoking the reciprocity theorem". Formally, the expressions are equivalent, but in the BPLO one, "the result of the first several steps of integration is common to all points in the scattering plane and needs only to be calculated once … In addition, if the system is well aligned and free from astigmatism, circular symmetric properties of the integrand further simplify the integration to a 1-D calculation … Thus the BPLO treatment greatly reduces the amount of computation".

The amount of computation needed for these integrals, and any "computational advantage" of the BPLO approach, typically decrease when we change to fibre-optic lidars—because they have fewer components and obstructions, and propagation within single-mode fibre needs negligible extra calculation. That is, the non-BPLO approach may be less tedious in fibre lidars than in free-space lidars, although still more tedious than the BPLO approach. We can calculate everything at the detector if we wish; nothing forces us into BPLO calculations.

The second point is a difference between fibre-based lidars and the more familiar free-space optical systems. Single-mode fibres act as spatial filters. They support only one spatial mode (transverse mode), so any light that arrives at a fibre entrance plane in other modes does not propagate any significant distance along the fibre and does not reach the fibre exit—it is lost. Similarly, any virtual BPLO light "arriving" at the fibre from the detector contributes nothing to the antenna pattern (and the overlap integral) unless it matches the fibre mode. The spatial form of the BPLO, once it leaves the fibre and continues through any transmit optics toward the target, is always that of the launched single transverse mode of the fibre. In particular, it is unaffected by apertures or obstructions between the fibre and the detector or by spatial variations in detector response. Nothing we do before the fibre can affect the antenna pattern after the fibre.

To the extent that detector output statistics are affected by the number and nature of the scatterers in the probe volume, this is an important difference between single-mode-fibre lidars and most free-space lidars. If we damage or partly block the fibre-lidar detector, then the probe volume (the physical extent and shape of the region contributing to the spatial overlap), and the carrier statistics, will not change. In free-space lidars there are often apertures, obstructions, and detector imperfections; these are often inaccessible and/or hard to adjust, yet they strongly influence the probe volume and statistics. Moreover, the practical difficulty of measuring detailed beam properties (such as BPLO shape and phase) causes uncertainty about what parts of a target are contributing, and how strongly. The ease of LO and BPLO alignment in fibre sensors, and the enforced limitation to one well-defined mode, reduce this uncertainty.

The situation is symmetrical; we could as easily have said "Nothing we do in free space can affect the illumination pattern at the detector—it is determined by the propagation of the single mode from the fibre end to the detector surface, through any distorting elements that may be present internally".

Zhao et al.'s caveats still apply: "The LO field at the detector should also be calculated very carefully ... the field at the waist is usually different from an ideal Gaussian distribution ... the criterion for ignoring the diffraction effect of the sharp-edged apertures for a Gaussian beam is quite stringent". The point is that the presence of a truly single-mode filter, somewhere in the optical chain between detector and target, defines the single mode that is relevant (e.g., in diffraction calculations) for both internal and external regions.

We neglect the possibilities of systems that are not reciprocal, e.g., the beams have significant frequency differences, and some components (such as modulators, amplifiers, and regenerative or self-aligning cavities) are frequency-selective or dispersive [3].

In practice, there are also specific, small, but possibly significant departures from ideal theory when fibres are involved:

- The fundamental mode of a single-mode fibre is normally modelled as a free-space Gaussian TEM_{00} in spatial profile, and we make this assumption here; in practice the match is very good but not perfect;
- Fibre-pigtailed collimators (FPCs) can improve the balance between lateral and angular misalignment effects by increasing the effective TEM_{00} radius at the coupling plane [13]. Lens focal length and lens-fibre distance will change accordingly, but otherwise, our efficiency calculations are not affected—the FPC/fibre combination behaves as a single-mode fibre with, at one end, a larger mode area;
- Higher-order terms in the mode field expressions imply that the wavefront in the fibre is slightly curved instead of plane. This curvature is familiar in hollow waveguide physics [14] and leads to measurable asymmetries in waveguide/beam coupling experiments, but we neglect it here.

There is a long history of cross-checks and calibrations of coherent lidars, with hard and distributed targets, and a sometimes baffling range of discrepancies. Experienced scientists, trying hard to account for all terms in the carrier-to-noise equations, still fall short by factors of around 2; indeed, Kavaya [15] noted that CNR theory and experiment often disagreed by nearly 3 dB, and not by a random variable factor ranging from say −3 dB to 3 dB. Note that this shortfall is in CNR (or sensitivity), and not necessarily in accuracy, probe length etc.

5.3. Collection Efficiency for Coherent Lidars

Evidently, then, in modern coherent lidar sensors the nature of the beam overlapping that determines optical efficiency and carrier-to-noise (and their variations as functions of range to target) is essential to lidar calibration and operation.

Some disagreements about this in the wind lidar literature were discussed within the UpWIND project [16–19]. First, Lindelöw [4] reconsidered the range weighting of wind lidars and proposed a redefinition of "probe length" based on the WPP ("wind peak profile") function; this is relevant to pulsed lidars that are range-gated (whether focused or not). These two issues of focus and range-gating are often present together, but it is possible to separate them and check the predicted discrepancies. For incoherent (randomly phased) returns from many aerosol scatterers, with the algorithms of signal processing and Doppler estimation that are most commonly used, the WPP proposal has not been widely accepted and has not altered the current standard treatments. It is mentioned here because "probe length" or "range resolution" is often presented as a figure of merit (that is, the shorter the better), and thus enters discussions of whether one lidar or another is "better", or whether pulsed or CW operation is "better". Readers should check which definition is being used.

A second topic is the "fibre lidar collection efficiency" relevant to focused monostatic coherent lidars, both pulsed and CW. Most published models use a Lorentzian for this efficiency or sensitivity function, whether or not the scaling with "1/area" (which is a Lorentzian for a Gaussian beam, with the conventions above) is recognised explicitly. Lindelöw [4] and then Lindelöw with Risø/Leosphere colleagues [17] preferred a function that is spatially narrower (tighter) than the Lorentzian. The discrepancy becomes larger at longer focus ranges, amounting to ~25% difference at typical large-turbine sensing heights of 100–150 m.

Lindelöw's model [4] "takes into account small receptor apertures and co-propagation of the local oscillator and the received backscatter in a single mode fiber ... A focused lidar will transmit a narrow beam of light with a waist at distance x_{focus}, typically at 20–200 m. A sketch of a fiber optic based focused monostatic coherent lidar is presented in Figure 31. The lidar has a combined emitter/receptor in the form of a fiber end positioned in front of the focusing lens". This "Sketch of system and principle of a focused monostatic cw coherent lidar based on fiber optic components" is our Figure 2 below, similar to Figure 1b above.

If we consider this as an imaging system, then an object or source at range x (in this case a thin illuminated disk of air) creates an image on the other side of the lens. In Lindelöw's treatment, the first stage of light propagation (from fibre through lens to this target object) and the last stage (just before recoupling to fibre) are described by standard Gaussian beam equations, but there is an intermediate stage (backscattering by the target, then formation of an image) under different assumptions and using thin-lens equations.

This mixing of Gaussian beams and rays has consequences illustrated in Figure 3.

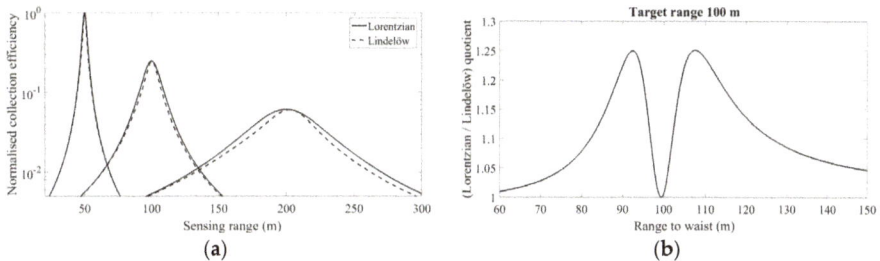

Figure 3. Two models for heterodyne lidar collection efficiency. (a) Normalised curves of range dependence for three different fixed focus ranges (50 m, 100 m, 200 m). Lorentzian (solid lines): efficiency scales as the reciprocal of beam area. Lindelöw (dashed lines): modified Gaussian optics/ray optics model proposed by Lindelöw [4]; (b) The quotient (Lorentzian efficiency)/(Lindelöw efficiency) for a fixed target range of 100 m and a varying focus range. Wavelength 1.575 μm, equivalent fibre core beam radius 4.5 μm, and lens focal length 200 mm (roughly representative of QinetiQ/ZephIR lidar).

The two "collection efficiency" functions are normalised so that their peak heights are equal for short focus range (where the differences between ray optics and Gaussian beam optics are negligible). This single normalisation factor is then applied for the other focus ranges. Any one curve in Figure 3a shows the efficiency for the given choice of focus (e.g., 100 m) as the target range, and only the target range, is varied. The Lorentzian curves peak at exactly the selected focus ranges (by construction, because our "focus range" means the distance where beam area is smallest); the curves for Lindelöw's model peak at slightly more distant ranges.

Note that the focus distance is assumed to be changed by varying the fibre-to-lens distance as mentioned above [20]. This means a slight but not always negligible dependence of beam size at the lens on x_{focus}. The exact form of the curves depends on this and on the choice of normalisation, but a comparison of the two approaches is not affected (because they use the same beam physics, and give identical results, for propagation to the target).

In Figure 3a, the efficiencies are plotted for three choices of x_{focus} (50 m, 100 m, 200 m). The quotient of the two efficiencies, for a fixed target range (100 m) and a varying focus range, is shown in Figure 3b.

Note that these are two different feasible experiments: the first considers how the fixed-focus lidar responds to scatterers placed at different ranges, and the second varies the fibre-to-lens distance to explore the response to scatterers placed at a fixed range.

Lindelöw [4] notes the disagreement between his approach and the early theory of Sonnenschein and Horrigan [2], which for our current geometry is equivalent to the BPLO approach. This "S&H" paper predates 1.5 μm fibre lidars but applies to them without modification because we assume untruncated pure TEM_{00} beams throughout. He notes that beams may overfill or underfill the detector, and that the older literature does not consider the modern layouts where received and LO beams co-propagate in optical fibre. But the transmit/BPLO approach remains valid, if the BPLO is correctly calculated (including any variations or truncations or damage of the "photodevice surface", and any spatial filtering in the fibre). Results may be expressed as a diffraction integral in the detector plane, the target plane or some other convenient plane. The standard approach is correct—or, at least, correctly developed within its stated approximations—and Lindelöw's, because of inconsistent shifting between ray and beam optics, is not. But the differences are small, of the same order as current errors in calibration, and so not easy to verify in practice—often less than 1 dB (in efficiency or power ratios) or a few % (in the range for peak efficiency). Lindelöw's −3 dB widths (FWHM) are 20–25% narrower than the conventional predictions (see also [21] and its Figure 2.9). No experimental checks are offered here, and few relevant ones are in the References; it would be good to see more.

Brewer et al. [22], before detailing a full diffraction-integral approach, describe a geometrical approach similar to Lindelöw's. They are more concerned with bistatic direct-detection lidars for imaging distant targets, but they are still treating single-mode fibres (which, although not carrying LO or BPLO beams, still reject any received light which is not mode-matched or, in geometric optics language, does not fall within the fibre numerical aperture). Some changes or extra details are needed before we apply their results to a monostatic focused coherent lidar: transmit and receive fibre/telescope optics are matched (in their worked example, the transmit aperture diameter must be taken as 10 cm), the far-field assumption is dropped, and care is needed when solving quadratic Gaussian-beam equations.

This last point is already familiar from Hill and Harris [20]. Brewer et al.'s worked example uses a nearly collimated transmitted beam that has a specified width (spot size) at a distant target. In their "far-field" solution a slightly diverging transmitted beam appears to have a virtual beam waist some 67 m behind the lidar. They ignore the slightly converging "near-field" solution with a real beam waist a similar distance in front of the lidar. The choice makes little difference to an ideal free-space imager, but more difference to a lidar used for near-field wind measurements.

Frehlich and Kavaya [23] develop useful general expressions and, after translation between various technical notations, show agreement with several other published accounts including Sonnenschein and Horrigan [2] where there are several misprints and, as occasionally happens, two cancelling factor-of-2 errors; Michael Kavaya has maintained and distributed a list of these and other errata.

Sonnenschein and Horrigan proceed to sum over a large collection N >> 1 of scatterers (representing a random diffuse target) and neither they nor Frehlich and Kavaya consider the single-particle or few-particle cases in detail.

Lawrence et al. [3] show good agreement between their CO_2 lidar and a conventional cup anemometer, but do not analyse the few-particle case.

5.4. Single-Particle Mode and Volume Mode

This interesting distinction in lidar processing should perhaps be better known. It gives an opportunity to recall the many friendships and the long productive career of Dr Rod Frehlich. Indeed, large parts of the present paper and of its References are comments on and restatements of his determined, thorough exploration of radar and lidar physics.

A NASA report by Kavaya et al. [24] includes Frehlich's comparison of single-particle mode (SPM) and volume mode (VM) measurements. The contributions from scattering particles can be defined in different ways. Essentially, VM adopts the theory and CNR definitions followed here and in most of the literature, while SPM measures a particle's peak contribution—that is, if a particle is detected in the probe volume, then we record the photocurrent contribution from the instant of closest approach to beam axis.

For decades, optical scientists have grumbled about the uncertainties in the shape(s) of laser beam(s), the resulting distortions in fringe patterns, and the positions and trajectories of scatterers. "Non-diametral traverses of particles across the scattering volume have also not been considered in the broadening formulae. There is no convenient way to account for these effects ... " [25]; see also Schulz-Dubois [26]. Our point here is that these two definitions give different results for range dependence, as shown in Figure 4. Errors may arise in calibration and interpretation unless a consistent choice is made.

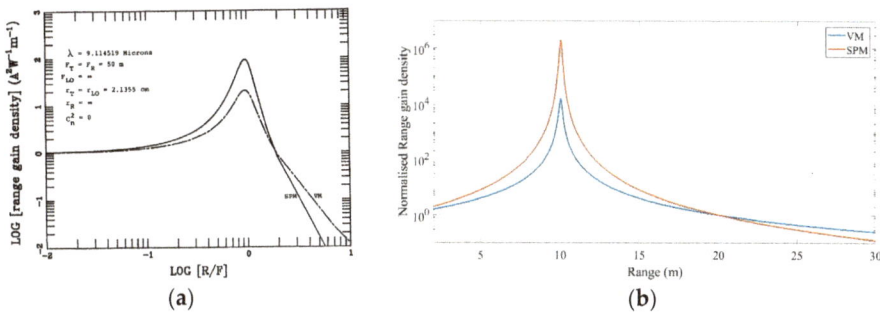

Figure 4. (a) "Normalized range gain density" from Kavaya et al. [24], for volume mode and single-particle mode ($\lambda = 9.11$ μm, beam radius to $1/e^2$ intensity points at the lens = 30.2 mm, focus distance = 50 m); (b) Same quantities calculated for a ZephIR-type lidar (1.57 μm, 25 mm, 10 m).

6. Simulation of Multiple Scatterers in a Lidar Beam

The simulated measurement volume is assumed here to have rectangular (usually square) cross section, with room for circles around the beam axis whose radii are at least $2w(z)$ for all relevant z. The decrease in average intensity as we move away from the focus along the z axis follows, as in Equations (1)–(5), a Lorentzian (rather than the steep Gaussian function for x and y), but for most practical purposes a choice of $(-3z_R, 3z_R)$ for the z-axis limits will include "almost all" the relevant scattering. Indeed, these choices are overkill in the sense that many of the simulated scatterers are illuminated so feebly that their contribution to the results is negligible; a conservative choice increases confidence in the simulations.

Thus, we choose a focused lidar beam geometry (a certain transceiver aperture and focused Gaussian beam parameters) and then simulate the movement of identical scattering particles which enter a measurement "box" at random times and at random positions on the x–z "face" of the box, the box being sufficiently large that any particles arriving outside it would not be "significantly" illuminated. Each scatterer has a random Poisson-process arrival time and crosses the beam at uniform speed, with a constant x and constant z each drawn from a uniform random distribution.

The scattering probe volume is shaped by our focused monostatic lidar [24], so the illumination is non-uniform, and the wavefronts encountered away from focus are non-planar. At any instant, $N_{tot}(t)$ scatterers are within the box, and the time or ensemble average of this number is proportional to the rate of the Poisson process and to the size of the box (the assumed measurement volume). For precise experimental checks, the measurement volume should be carefully specified. For example, in Figure 1b, the physically relevant volume lies to the right of the transmitting lens. Regions to the left are inside the lidar itself, and presumably not filled by scatterers; thus the "relevant" volume may not be symmetric around the beam focus.

At any instant, the number of scatterers "contributing", in the sense that they are significantly illuminated, is at most N_{tot} and may (if we size the box conservatively) be much smaller. During some time intervals, there may be no scatterer in the box or none that contributes significantly. The number of scatterers illuminated in a fixed measurement volume will generally change with time (as would happen in real life, because of random or wind-blown scatterer movement). The assumed measurement volume itself may be altered during a simulation run (notably because of a varying beam focus), but of course, consistency checks are advisable—runs for different measurement volumes may take very different amounts of computation, but their average results should be similar so long as the box and N_{tot} are sufficiently large.

There are many possible experimental scenarios, and many fluctuation processes and associated timescales. It is important to keep in mind what is meant by an "average"—whether over time, probability distribution, or ensemble of experiments.

For a collection of scatterers distributed (on average uniformly) across a slice of atmosphere at a certain range z, and for randomly distributed phase terms, the mean squared quantity $< |i(t)|^2 >$ contributed by that slice is proportional to the scatterer number density and inversely proportional to the illuminated area of the slice—so it peaks at the beam focus, where the area is least (see also Section 4 of [1] for this "1/area" relation). If we fix the number density and vary the beam area, the number of scatterers involved scales with beam area; but, because the average contribution per scatterer to the detector *power* scales as the intensity squared (i.e., as the inverse square of beam area), the net result is a power scaling with 1/area. We also see that the number of scatterers in a thin slice is proportional to beam area A, and the beam intensity at each scatterer is proportional to 1/A. So, the fractional backscattered power per slice is a constant, independent of beam size. So long as we continue to use a time-averaged definition of power, this conclusion applies when we vary the beam size for arbitrarily small seeding densities.

The heterodyne current due to several scatterers is the vector sum of the individual currents, one for each scatterer as in Equation (7); for matched transmit and BPLO beams, each term can be considered to vary as the beam intensity at the scatterer. The mean square of the current (averaged over all relative phases as above) equals the sum of the individual squares; this is the same result as for the mean square of the length of a 2D random walk.

We can illustrate some of the familiar results in the literature [1]. We assume a fixed transmit power, a fixed LO power, and the usual optical arrangement described above (focused monostatic coherent lidar, matched transmit and BPLO). First, Figure 5 shows the chirping behaviour of scatterers that cross the beam away from the focal plane.

This modelling approach was discussed in [1], and similar time-frequency plots are well known. Renard et al. [27] discuss airborne lidar options, and the general features of chirps are clear in their exemplary figure (reproduced here as Figure 6).

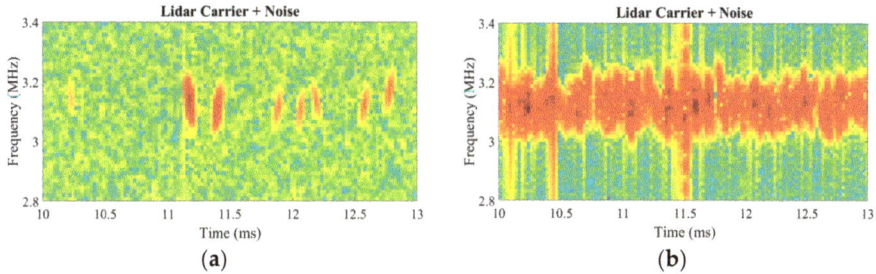

Figure 5. (a) Spectrogram of simulated lidar data for several scatterers crossing the beam. At any instant, during this short (~3 ms) simulation, the number of scatterers significantly illuminated within the probe volume is either 0 or 1; (b) Spectrogram of simulated lidar data (same duration ~3 ms) for a larger density of scatterers crossing the beam. At any instant, it is likely that several scatterers (N > 1) are significantly illuminated. The scatterers are given small random Doppler shifts to simulate slightly oblique paths across the beam. As usual with spectrograms there are smearing and rippling effects, and compromises are needed to choose filters and block lengths.

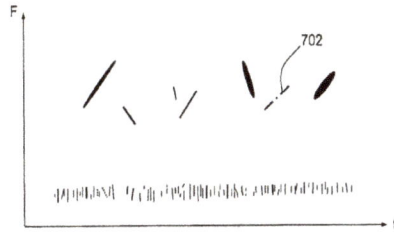

Figure 6. Example of time-frequency plot in Renard et al. [27]. See also Baral-Baron [28].

Second, we model an experiment where the beam waist position and radius are varied (for constant beam width at the lens); the target is modelled as randomly generated scatterers (of equal reflectance) within a box that encloses the waist as described above. The scatterer density (m^{-3}) is constant on average. We see in Figure 7 the variations in probe volume, CNR, and normalised second moment of intensity.

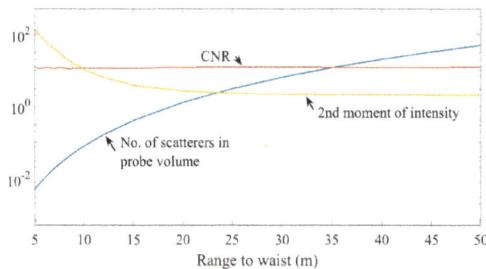

Figure 7. Results from one simulation of a lidar with uniform aerosol seeding density, fixed aperture, and varying focus range. Average CNR is almost independent of focus range, despite the variation of 4 orders of magnitude in probe volume. Normalised 2nd moment of intensity $<I^2>/<I>^2$ is large at short ranges but closely approaches 2 (characteristic of complex Gaussian statistics) if there are "many" randomly arranged scatterers (say a few tens or more) in the probe volume. Figure 1.4 in Banakh and Smalikho [9] is similar, illustrating their analytical expressions for "echo power" and the number of efficiently scattering particles, with specific model assumptions for a CO_2 atmospheric lidar ($\lambda = 10.6$ μm).

Here, the simulation assumes constant measurement time (120 ms) at each focus range. At the shortest ranges, there are so few scatterers contributing that the results for CNR and 2nd moment are noisy. (The "No. of scatterers in probe volume" plotted is a theoretical average number given by $(2\lambda^3/\pi^2) \times$ (focus distance/beam waist radius)$^4 \times$ (seeding density), so it is not noisy). As the range increases to tens of metres, the number of scatterers contributing rises quickly, because of this fourth-power dependence of probe volume on range [24]. The CNR is almost constant with range as we expect, and the 2nd moment of intensity falls to very nearly 2 when the probe volume contains tens of scatterers. This computer run uses $\lambda \sim 1.57$ μm and $w_0 \sim 15$ mm, with a seeding density of 500,000 m^{-3}, and with the scatterer reflectance and shot noise level set to give a CNR of about 10 in a bandwidth of 10 MHz. The 2nd moment is more than 100 at the shortest range of 5 m but within 1% of 2.00 for ranges above 45 m. Of course, more stable results at the shorter ranges can be obtained if (for example) we simulate a different experiment where the total number of contributing scatterers (rather than the measurement time) is fixed at each range.

Such a simulation is easily run for hundreds or thousands of scatterers on a PC. This was harder 40 years ago, for example when Mayo [29] modelled the "triply stochastic" physics of turbulent flow, particle arrivals, and photodetections.

Two more papers emphasising the practical side rather than simulations, but relevant to understanding the various models, are the following:

Jarzembski and Srivastava [30] discuss some interference effects for the case of two illuminated particles and treat some experimental time series at length. But they restrict their explanations to particles near the beam focus and use a corresponding simplification of the Sonnenschein and Horrigan equation. They neglect the information in the envelope of the two-scatterer lidar output, saying that "it does not contain phase properties of backscatter". For the linked amplitude and phase modulations in such FM interference experiments, see Hill et al. [31] and its references.

Harris et al. [32] draw fresh attention to the possibilities for single-particle lidar anemometry and to the need to consider different fluctuation statistics because the probe volume varies dramatically; they do not consider time series analysis, chirp behaviour, or spectral moment estimation in detail.

This section has only sketched a very wide topic of relevance to short-range and/or thinly seeded sensing (e.g., wind tunnel lidar and high-altitude anemometry); there are whole industries based on particle/suspension analysis through optical scattering.

7. Direction Sensing for VAD Lidars with Sign Ambiguity

7.1. Deciding the Sign

We mention two related issues in coherent lidar Doppler sensing. One is the requirement to attach signs (+ or −) as well as magnitudes to these estimates. The other is the estimation of the overall target motion vector after the estimation of several Doppler shifts. Apart from brief comments here and in Section 7.2 below, we cannot review the huge literature on estimating scatterer speed through the Doppler effect. For the moment we assume that the detector output yields a well-defined spectral peak and thus some reasonable mean-Doppler estimate.

Suppose for illustration that we are trying to estimate the magnitude and direction of the wind and using a coherent lidar and a "conical scan" or "velocity-azimuth display" (VAD) mode. With a system as described above, a change of sign (+/−) of the velocity component does not affect the detector output: that is, one lidar measurement (one brief "snapshot" at a single viewing angle) does not tell us whether the wind component along the line of sight is towards the lidar or away. Moreover, for a uniformly flowing wind, although a collection of several or many measurements at different viewing angles can tell us the wind speed (magnitude of wind vector), it still does not resolve this 180-degree ambiguity. For example, the ZephIR lidar collects approximately 50 separate estimates of wind velocity component per second, spaced evenly around its 360-degree conical scan, and the resulting "figure of 8" pattern allows an accurate estimate of wind speed. However, the same pattern would appear for a uniform wind flowing in the opposite (180 degrees different) sense (Figure 8).

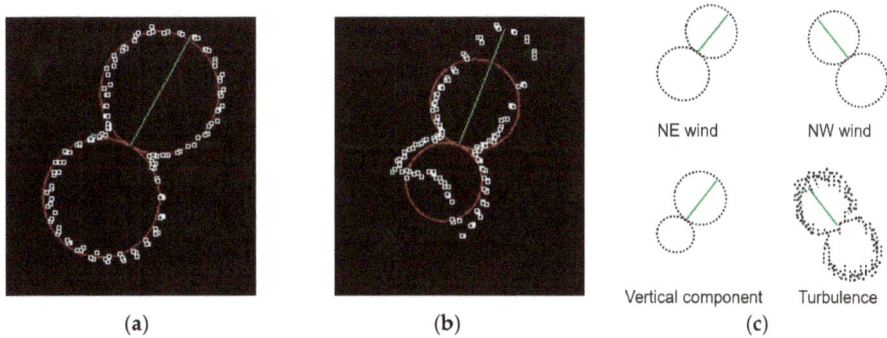

(a)　　　　　　　　　　(b)　　　　　　　　　　(c)

NE wind　　　NW wind

Vertical component　　Turbulence

Figure 8. The appearance of Doppler estimates in a polar plot for a conical-scan lidar without direction (+/−) discrimination. (**a**) "Figure-of-8" for a steady wind; (**b**) Perturbations soon after the passage overhead of an A320 jet. Data from trials with ZephIR at Birmingham airport; (**c**) Simplified guide to interpretation of "figures-of-8" (with thanks to Dr M Harris). The results from Hill et al. [33] here and in Figures 9a and 10 below are reproduced by permission of QinetiQ Limited.

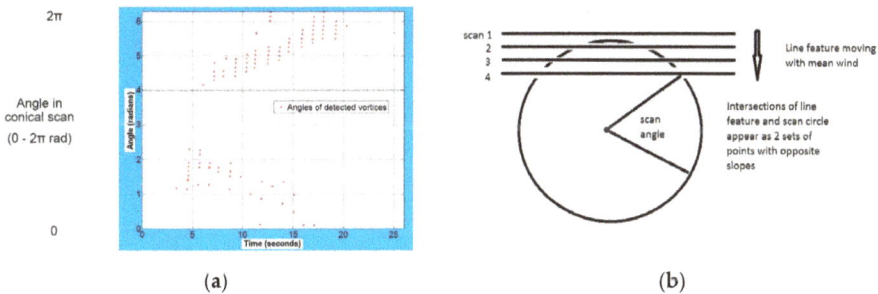

(a)　　　　　　　　　　　　　　　　(b)

Figure 9. A passing aircraft leaves a pair of trailing wake vortices which form (briefly and approximately) a stable "line feature" in the atmosphere. This line drifts with the mean wind across the scan circle defined by the measurement points of a conical-scan lidar, and the angles where the lidar detects a significant disturbance are recorded and plotted. (**a**) Example of data from trials with ZephIR at Birmingham airport, adapted from Hill et al. [33]; (**b**) Sketch of the principle. This plan view is simplified; in practice the lidar processing and scanning may take account of vertical as well as horizontal drifts and may have complicated algorithms for estimating the vortex and wind parameters.

Wind speeds are routinely estimated with fewer and/or less widely spread viewing angles if the CNR is sufficient and the wind is known (or assumed) to be steady so that we are confident of a sinusoidal variation of speed with scan angle or equivalently of the typical figure-of-8 polar plot of Figure 8. Schwiesow et al. [34] reported success with as little as 1/16 of a full scan (that is, less than half a radian), and Leosphere's Windcube lidars typically use 4 or 5 separate, fixed beam directions [35]. If we assume a steady wind and thus a figure-of-8 of known symmetric shape, far fewer than the 50 estimates per scan suffice for a good estimate of its size and orientation. But the same problem arises in all such cases: we need some extra information in order to determine the sign of the wind.

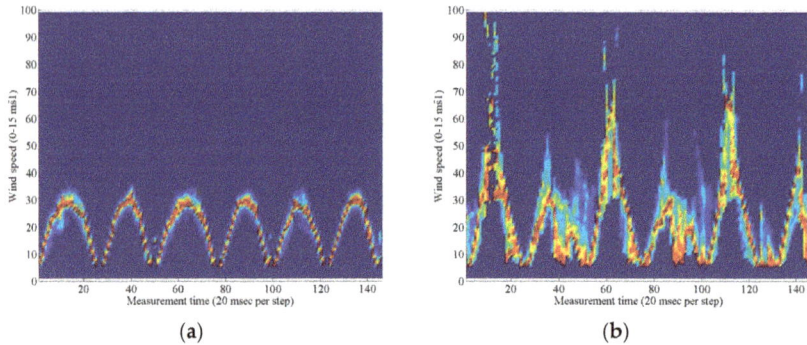

Figure 10. Spectrogram plots of ZephIR lidar data from airport trials. (**a**) Rectified cosine wave for a fairly steady airflow on the southern approach to Birmingham airport [33]. A polar plot would produce a "figure of 8" similar to Figure 8a; (**b**) Perturbed cosine wave after the passage of a Boeing 777. The vortex lines provide mainly positive Doppler contributions during one half-scan (vortex flow towards the lidar, faster than the mean wind) and mainly negative ones during the other. The scan angles at which the perturbation is judged "significant" according to some metric can be identified and added to a plot such as the one in Figure 9a.

Of course, a separate non-lidar instrument may be added to measure, at one point or more, the wind direction. In "ordinary" atmospheric conditions, it is sufficient to measure with a conventional anemometer (vane/cup/ultrasonic) near the surface, and trust that the wind direction does not veer sharply between measurements and between adjacent measurement heights; in this way, by assuming a reasonable continuity, we can assign the wind vector to the correct quadrant at successively greater heights [36]. But what if such a non-lidar instrument is unavailable or inappropriate?

There are several ways forward. An artificial frequency shift can be introduced, typically by an acousto-optic modulator, so that the zero-Doppler region appears shifted from zero frequency in the detector output. Or the in-phase and quadrature (I&Q) components of the complex carrier can be separately detected, so that their relative phase can be tracked. Both approaches tend to involve significant extra hardware and cost.

A third way, cruder but feasible, relies on the fact that real wind flows are not wholly "uniform"; they always have fluctuations, for example of density, velocity, and backscatter. Nonetheless we usually assume that they remain "uniform" or well-behaved over some extent and duration: that is, we adopt a "frozen flow hypothesis" whereby the spatial patterns containing these fluctuations are being transported with some overall mean wind motion. The patterns need only be sufficiently distinct, and evolve sufficiently slowly, for us to track them with adequate confidence: it is a matter of "sufficient" CNR and of timescales. If we look at the same small volume of atmosphere at intervals of hours or many minutes, we may see no significant correlations of the departures from uniformity; on a timescale of seconds or tenths of a second, we may have considerable success.

For example, Figure 9 shows unambiguous direction-sensing for a ZephIR lidar without frequency offset or I&Q processing [33]. In this case, the atmosphere is perturbed by aircraft wake vortices that typically last for many seconds and are (approximately) transported across the sky with the local mean horizontal wind component, while descending slowly towards the ground. During a series of one-second conical scans the lidar detects the perturbations caused by the vortex pair. For a typical single scan, there are several detections (marked by the small red symbols in Figure 9a); as time evolves, we see that these detections cluster in two groups that diverge. For one group the scan angles increase, and for the other they decrease, because the line of the vortex pair—which is approximately a straight line, considered on this scale of a few hundred metres—is entering the lidar's conical-scan pattern. For a line feature leaving the scan pattern, the two sets converge, and this distinction tells us the direction of the wind. For an ideal thin line, the two sets of detections coalesce when the line becomes tangent to the scan circle; in practice the region of lidar detection sensitivity is 10–30 m across, and so is the pair vortex structure from a typical medium aircraft. So, on this type of plot, the detections are clustered only loosely, and we expect scatter of several metres or more (or the corresponding spreads in angle/time), but after a few consecutive seconds of scanning, we are in no doubt of the overall tendency (divergence in angle of the two groups, or convergence?) and thus in no doubt of the $+/-$ sign decision.

A conventional X-Y time-frequency (rather than polar) plot shows a rectified cosine wave when the air is relatively unperturbed. When an aircraft passes overhead and the trailing vortices are sampled by the lidar scan, there are considerable additions to (and subtractions from) the cosine function (Figure 10).

The ZephIR example above, an early demonstration with eyesafe fibre-optic lidar, is deliberately simple: the airflow is strongly perturbed, and we also have prior expectations about the general form of the pair of counter-rotating vortices. But the principle [37–41] applies to the natural fluctuations of velocity, backscatter etc. in "ordinary" air, with no helpful jet aircraft passing, and hence was of interest for Malvern and Risø work with ZephIR-type lidars for wind applications. Evidently the "metrics" for correlation-tracking of this sort (typically based on the first few spectral moments viz. carrier strength, mean Doppler, and Doppler spread) will be adjusted to suit if we believe that particular geometrical features may be present; for example we may expect to see these linear vortices, or 2-D loops, or sheets of separated air flow, or 3-D volumes of turbulent air; but even a simple correlation plot without special assumptions may strongly indicate the likely sense of the overall wind motion. Many demonstrations are now in the literature with larger measurement sets, more detailed correlation and wavelet algorithms, and heftier processors [42,43].

7.2. Some Issues in Frequency Estimation

The estimation of a Doppler shift in wind lidar is usually treated as a "mean frequency" problem and attacked by forming power spectral estimates or autocorrelation functions. For light winds, there are obvious technical issues.

First, in the absence of a frequency-shifting scheme such as the modulator mentioned above, the informative carrier lies close to DC, where interferences such as local oscillator noise and laser relative intensity noise (RIN) are usually strongest. The influences of "intensity" noise and "phase" noise on frequency estimation depend on the estimator used, and we need to know the complex (intensity and phase) variations associated with laser instability; an estimate of the power spectral density attributed to RIN is insufficient. Evidently a two-point or "instantaneous frequency" algorithm, as it relies on the phases alone, should be unaffected by AM when the CNR is high, but other algorithms use more or less of the AM or envelope information—for good reasons—and will be correspondingly affected.

Second, how is a wind speed defined when the measurement volume—usually, as we have seen, a long thin volume extending through several or many metres—contains some scatterers approaching us and some scatterers receding? The coherent addition of complex components with different Doppler signs (clockwise and anticlockwise phasor rotations) does not necessarily give the same result as the incoherent addition of the sign-insensitive outputs.

This is an issue not just of "ambiguity" but of distinctly different results in the detector output. If we suppose for simplicity that there are only two scatterers present in the beam, one moving towards the lidar and one away, and that their individual contributions to the detector output have equal magnitudes and opposite signs, the net result in a frequency-shifted (direction-sensitive) coherent lidar is the sum of two oppositely rotating phasors.

Also, the common use of a blocking or highpass filter (to remove both carrier and dominant noise near DC) will in any case cause a bias in subsequent Doppler estimation. For example, the lowest few hundred kHz of ZephIR data (in practice, two or three frequency bins of the spectrogram display) are not used in Figure 10.

8. Errors and Bias in Doppler Lidars for Steady Winds

Finally, we discuss two examples of how (according to the manufacturer's publications) the extraction of wind velocity from the lidar detector output must be imperfect. Any error or bias is likely to be very small, but lidar users—including those involved in calibration and verification studies—are advised to check if and how any corrections are applied.

There is a practical issue in conical-scan or sector-scan lidars where lidar measurements are accumulated (usually in the form of averaged power spectral estimates or autocorrelations) during an appreciable interval and hence over an appreciable range of angles. For example, the standard ZephIR choice of 50 measurements per one-second scan means that each measurement represents a "wedge" or segment of angular extent $360/50 = 7.2$ degrees of scan. This, in turn, may mean a small but not entirely negligible error in estimates of the figure-of-8 size (wind speed), orientation (wind direction), and lobe symmetry (vertical wind component).

Suppose the lidar forms a spectral estimate at a given scan angle θ (not to be confused with the phase angle above) by accumulating short-term estimates over the range $\theta - \delta$ to $\theta + \delta$ and then finds a mean frequency (first spectral moment) from the accumulated estimate.

Denote the wind heading by θ_{true}, so that an ideal non-direction-sensing lidar would detect a maximum line-of-sight component at θ_{true} and $\theta_{true} + \pi$ radians, and a minimum line-of-sight component at $\theta_{true} + \pi/2$ and $\theta_{true} + 3\pi/2$ radians. Short-term estimates are incoherently accumulated at a steady rate (per radian of scan). The power per estimate will fluctuate around a steady mean (given assumptions of uniform operating conditions, backscatter etc.). But, neglecting the fluctuations, we may assume that each incremental scan segment $(\theta, \theta + d\theta)$ contributes a spectral increment of the same total power centred around a Doppler frequency proportional to $\mathrm{abs}(\cos(\theta))$.

The measured wind speed can thus be taken as proportional to the average of $\mathrm{abs}(\cos(\theta))$ over the interval $(\theta - \delta, \theta + \delta)$. Evidently, the wind speed may be underestimated near $\theta = \theta_{true}$ and $\theta_{true} + \pi$ and overestimated near $\theta_{true} + \pi/2$ and $\theta_{true} + 3\pi/2$. The direction-sensing lidar behaves slightly differently because of the signed function $\cos(\theta)$; the mean frequency is then proportional to $\cos(\theta)\cdot\sin(\delta/\delta)$.

First, we show results (as functions of θ) for 10 measurements per scan, that is $\delta = \pi/10$. It is assumed that scan segments are contiguous (without overlapping). In Figure 11a the scan origin $\theta = 0$ (the midpoint angle for the first segment) coincides with the true wind direction. In Figure 11b we assume the maximum possible "offset" so that the true wind heading coincides with the edge of a segment, that is, the angle where one segment finishes and the next one begins.

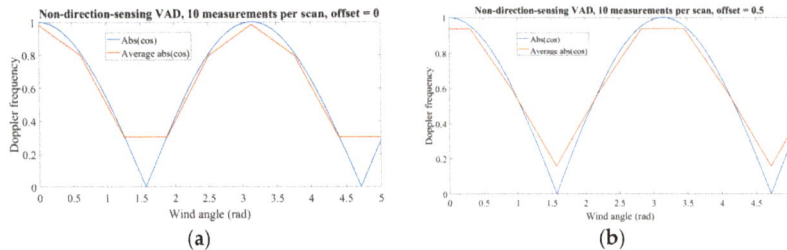

Figure 11. Rectified cosine function and its time-windowed average, representing the Doppler-estimating behaviour of a non-direction-sensing lidar; compare this with the typical ZephIR display in Figure 10a. For each scan rotation, the average of the cosine is obtained for each of 10 contiguous segments (arcs) of $\pi/5$ radians. (a) The true wind heading coincides with the midpoint angle of a segment (i.e., zero offset); (b) The true wind angle coincides with the edge of a segment (i.e., offset = 0.5 segment).

The "errors" shown here might be unacceptable. Figure 12 shows similar results for the ordinary ZephIR case of 50 measurements per scan, that is $\delta = \pi/50$ or $2\delta = 7.2$ degrees. Now the bias would almost certainly be negligible in practice. It is still noticeable very close to $\pi/2$ and $3\pi/2$ but, usually, these regions are not used—any measurements within them are discarded, because (as just mentioned) we do not trust the detector noise floor in the lowest spectral bin(s)—and the effect of this slight data loss on curve-fitting algorithms is unimportant.

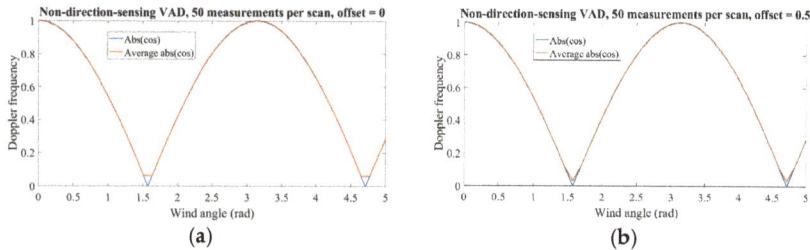

(a)

(b)

Figure 12. As for Figure 11, with 50 contiguous segments per scan rotation. The bias is now negligible. (a) The true wind heading coincides with the midpoint angle of a segment (i.e., zero offset); (b) The true wind angle coincides with the edge of a segment (i.e., offset = 0.5 segment).

Non-scanning lidars (such as standard WindCube) do not have this problem to the same extent, as their field of view is tighter—essentially, the beam angular width of at most a few milliradians rather than the wedge of a few degrees. But they have another issue if the speed and direction are estimated directly (as described in [19]) from a small number, typically 4, of fixed-angle measurements. When a horizontal wind speed is obtained as $V_h = \sqrt{u^2 + v^2}$, where its orthogonal components u and v are derived from lidar line-of-sight speed estimates, any small zero-mean random noises on these estimates will create small zero-mean random noises on u and v and (in general) non-zero-mean random noise on V_h. For zero-mean Gaussian noises on the initial estimates, this bias of V_h will be positive.

For each type of lidar, given our assumptions about uniform horizontal winds and our known beam parameters, a compensation function could reduce the bias of the individual measurements (before curve-fitting an equation for the wind vector). Or the biased measurements could be fitted to a modified equation that included terms to describe the bias.

9. Conclusions

This paper is a companion to [1] and other recent overviews of fibre-optic lidars; a common theme is the reexamination of basic equations and principles in radar or lidar. In [1], the emphasis was on the hundreds of fibre-optic lidars installed today for wind and especially wind turbine farm measurements; again we stressed what happens when the number of scatterers (components) changes through 0, 1, 2, . . . up to the "many" of near-Gaussian statistics. The present paper has considered the simulation of few-scatterer and multi-scatterer lidar experiments, and some problems (and solutions) for Doppler-sign-insensitive lidars.

We reviewed the BPLO description of the integrals for lidar CNR and efficiency. In a monostatic Doppler lidar with matched LO and transmit beams, "focusing of the beam gives rise to a spatial sensitivity along the beam direction that depends on the inverse of beam area; it follows that the sensitivity rises to a peak at the beam waist and falls symmetrically on either side" [20]. The sensitivity function Q used by Banakh, Smalikho et al. [44] is an exact Lorentzian, and (once we make the necessary adjustments in notation) it is the same Lorentzian used by Sonnenschein and Horrigan [2], Frehlich and Kavaya [23], Qi Hu et al. [45], and others. We contrasted this with Lindelöw's description of "collection efficiency", which departs from the "1/area" rule. A fair and separate question, though, is whether the usual calibration experiments are sufficiently precise to tell these two descriptions apart.

We reviewed the single-particle mode (SPM) and volume mode (VM) modelling of Frehlich et al. and some numerical modelling of lidar detector time series and statistics.

We have tried to highlight and explain behaviour that may be observed from a modern coherent lidar used at short ranges (e.g., in a wind tunnel) and/or with weak aerosol seeding. It is worth noting that non-Gaussian statistics (and opportunities for improved processing) may arise at focus distances of 10–20 m, even when there are 10^6–10^7 scatterers per cubic metre.

Finally, we reviewed some issues in direction sensing, sign ambiguity, and limited angular resolution for standard Doppler-sensing coherent lidars.

It has been interesting to watch these fibre-optic lidars grow into accurate reliable calibrated tools of the wind energy trade, with accompanying expectations and responsibilities.

Acknowledgments: I thank the two referees for helpful suggestions. For a better understanding of the sensing needs of the wind energy industry, I am indebted to many colleagues in the UpWIND consortium led by the Danish Technical University (DTU). I have been helped by technical discussions with Jean-Pierre Cariou (Leosphere), Mike Courtney, Mike Harris (ZephIR), Michael Kavaya, Petter Lindelöw, Jakob Mann, Guy Pearson (HALO Photonics), Anders Tegtmeier Pedersen, Kevin Ridley, and Michael Vaughan. These colleagues and their organisations are not responsible for any errors and omissions here.

Conflicts of Interest: The author declares no conflict of interest.

Glossary

BPLO	Back propagated local oscillator. A fictitious aid in analysing heterodyne lidars.
CNR	Carrier-to-noise ratio. Strictly this should mean a ratio C:N but it often appears as a quotient C/N or a decibel measure. In lidars with nonlinear processing steps such as frequency estimation, the CNR or carrier-to-noise (at the detector, before processing) is often distinguished from SNR or signal-to-noise (after processing); these two noises "N" may have quite different statistics.
CW	Continuous wave.
FM	Frequency modulation.
I&Q	In-phase and quadrature (the two parts of a complex quantity, usually the detector output current, giving a phase angle $\tan^{-1}(Q/I)$.
LO	Local oscillator.
SPM	Single-particle mode: lidar scattering events are individually registered and treated as follows: *if* a particle is detected in the beam, *then* the peak resulting detector output counts towards an overall metric. See VM.
TEM$_{00}$	Transverse electromagnetic mode of lowest order; the first in a nominally complete orthogonal set of "Gaussian beams" into which a propagating monochromatic lidar beam can be decomposed.
VAD	Velocity-azimuth display. Usually refers to a conical-scan lidar.
VM	Volume mode: the entire time series of a scattering event counts towards an overall metric. This is normal procedure in an atmospheric lidar illuminating a volume densely seeded with scatterers; individual scattering events are not usually isolated and examined even if that is technically feasible. But other metrics exist (see SPM) and may show different dependences on target range.
Windcube	A brand of pulsed lidars operating near 1.5 μm; based on research at ONERA, Palaiseau, and later developed and marketed by Leosphere.
ZephIR	A brand of CW lidars operating near 1.5 μm; based on research at QinetiQ, Malvern, and later developed and marketed by ZephIR Lidar.

References and Notes

1. Hill, C.A. Modern fibre-optic coherent lidars for remote sensing. *Proc. SPIE* **2015**, *9649*. [CrossRef]
2. Sonnenschein, C.M.; Horrigan, F.A. Signal-to-noise relationships for coaxial systems that heterodyne backscatter from the atmosphere. *Appl. Opt.* **1971**, *10*, 1600–1604. [CrossRef] [PubMed]
3. Lawrence, T.R.; Wilson, D.J.; Craven, C.E.; Jones, I.P.; Huffaker, R.M.; Thomson, J.A.L. A laser velocimeter for remote wind sensing. *Rev. Sci. Instrum.* **1972**, *43*, 512–518. In their lidar (our Figure 1a) the received light must make at least one extra reflection after the beamsplitter. In a common UK terminology this is an autodyne lidar, because the backscattered radiation "is collected by the telescope and allowed to reenter the laser cavity where it is amplified". The overlap of the reemerging radiation with the (BP)LO may or may not be stable and easily described—this also depends on the active cavity mode structure and the radiation frequencies. Most fibre-optic lidars, including the one in our Figure 1b, are not intended as "autodyne" in this sense. The word has various meanings.
4. Lindelöw, P. Fiber Based Coherent Lidars for Remote Wind Sensing. Ph.D. Thesis, Technical University of Denmark, Lyngby, Denmark, 2008.
5. Henderson, S.W. Review of fundamental characteristics of coherent and direct detection Doppler receivers and implications to wind lidar system design. In Proceedings of the 17th Coherent Laser Radar Conference, Barcelona, Spain, 17–20 June 2013.
6. Hao, C.H.; Guo, P.; Chen, H.; Zhang, Y.C.; Chen, S.Y. Determination of geometrical form factor in coaxial lidar system. *Proc. SPIE* **2013**, *8905*. [CrossRef]
7. Jakeman, E.; Ridley, K.D. *Modeling Fluctuations in Scattered Waves*; CRC Press: Boca Raton, FL, USA, 2006. See chapter 4's review of random-walk models and partly developed speckle.
8. Pusey, P.N.; Schaefer, D.W.; Koppel, D.E. Single-interval statistics of light scattered by identical independent scatterers. *J. Phys. A* **1974**, *7*, 530–540. [CrossRef]
9. Banakh, V.; Smalikho, I. *Coherent Doppler Wind Lidars in a Turbulent Atmosphere*; Artech House: Norwood, MA, USA, 2013.
10. Rye, B.J. Spectral correlation of atmospheric lidar returns with range-dependent backscatter. *J. Opt. Soc. Am.* **1990**, *7*, 2199–2207. Here Barry Rye thanks the late Ken Hulme for emphasising the superposition of scattered "copies".
11. Siegman, A.E. The antenna properties of optical heterodyne receivers. *Appl. Opt.* **1966**, *5*, 1588–1594. Siegman refers readers to his 1964 IEEE-MTT conference paper and to an unpublished memo by R E Brooks of the TRW company.
12. Zhao, Y.; Post, M.J.; Hardesty, R.M. Receiving efficiency of monostatic pulsed coherent lidars. 1: Theory. *Appl. Opt.* **1990**, *29*, 4111–4119. [CrossRef] [PubMed]
13. Wallner, O.; Winzer, P.J.; Leeb, W.R. Alignment tolerances for plane-wave to single-mode fiber coupling and their mitigation by use of pigtailed collimators. *Appl. Opt.* **2002**, *41*, 637–643. [CrossRef] [PubMed]
14. Tacke, M. The influence of losses of hollow dielectric waveguides on the mode shape. *IEEE J. Quantum Electron.* **1982**, *18*, 2022–2026. [CrossRef]
15. Kavaya, M.J. Coherent lidar: Factors of two among friends. In *NOAA Working Group on Space-Based Lidar Winds*; NASA/Langley Research Center: Hampton, VA, USA, 2002.
16. Hill, C.A. Calibration and optical efficiency of fibre-based coherent lidars. In *UpWIND Report*; Technical University of Denmark: Roskilde, Denmark, 2010.
17. Lindelöw, P.; Courtney, M.; Parmentier, R.; Cariou, J.P. Wind shear proportional errors in the horizontal wind speed sensed by focused, range gated lidars. *IOP Conf. Ser. Earth Environ. Sci.* **2008**, *1*. [CrossRef]
18. Cariou, J.P. *Pulsed Coherent Lidars for Remote Wind Sensing*; Remote Sensing Summer School at Risø-DTU: Roskilde, Denmark, 2010.
19. Cariou, J.P.; Boquet, M. LEOSPHERE pulsed lidar principles. In *Technical Report for UpWIND Work Package*; Technical University of Denmark: Roskilde, Denmark, 2010.
20. Hill, C.A.; Harris, M. Lidar measurement report. In *Technical Report for UpWIND Work Package 6*; Technical University of Denmark: Roskilde, Denmark, 2010.
21. Beuth, T. Analyse und Optimierung von Fokussierten LiDAR-Systemen für Windkraftanlagen (Analysis and Optimisation of Focused Lidar Systems for Wind Turbines). Ph.D. Thesis, Karlsruhe Institute of Technology, Karlsruhe, Germany, 2016.

22. Brewer, C.D.; Duncan, B.D.; Barnard, K.J.; Watson, E.A. Coupling efficiencies for general-target-illumination ladar systems incorporating single-mode optical fiber receivers. *J. Opt. Soc. Am.* **1998**, *15*, 736–747. [CrossRef]

23. Frehlich, R.G.; Kavaya, M.J. Coherent laser radar performance for general atmospheric refractive turbulence. *Appl. Opt.* **1991**, *30*, 5325–5352. [CrossRef] [PubMed]

24. Kavaya, M.J.; Henderson, S.W.; Frehlich, R.G. Theory of CW lidar aerosol backscatter measurements and development of a 2.1-μm solid-state pulsed laser radar for aerosol backscatter profiling. This is Contractor Report CR-4347; NASA: Washington, DC, USA, 1991. The "probe volume" for a reasonably tightly focused Gaussian beam is estimated as $\lambda^3 (F/D_0)^4$ where F is focal distance and D_0 is beam diameter at the lens. There are various acceptable versions of this in the literature (with different notations and factors of about π), such as $(8\lambda^3/\pi^2)(f/w_{lens})^4$, but our $8\lambda^3 R_{lens}^4/(\pi^2 w_0^4)$ in [1] seems misprinted.

25. Durst, F.; Melling, A.; Whitelaw, J.H. *Principles and Practice of Laser-Doppler Anemometry*, 2nd ed.; Academic Press: London, UK, 1981.

26. Schulz-Dubois, E.O. Photon correlation techniques in fluid mechanics. In Proceedings of the 5th International Conference, Kiel-Damp, Germany, 23–26 May 1982.

27. A Renard et al., French patent 2 948 459 (24 July 2009), US 2011/0181863 (23 July 2010); X Lacondemine et al., French patent FR 1202729 (12 October 2012), US 8,976,342 B2 (date of patent 10 March 2015). If we know the crosswind V and the chirp duration, we know the particle's distance from focus (though not whether it lies on the near or far side). If we do not know V, we need both duration and slope (measured, of course, at "sufficient" CNR). These papers correctly explain the beam/particle physics but there are minor problems: First, the radius of curvature is not shown decreasing (to a minimum at z_R) and then increasing again at longer ranges from the waist; second, there is a language or translation difficulty in the US version at: "In other words, at the Rayleigh distance z_R, the radius of curvature of the wavefronts is large and, as they move away from this point 205, their radius of curvature decreases". The French original is *En d'autres termes, au niveau de distance de Rayleigh z_R, les fronts d'onde sont tres incurvés, puis en s'éloignant de ce point 205, l'incurvation diminue.* But the curvature of the wavefronts, not their radius of curvature, is "large"—a local maximum—at z_R. I thank Xavier Lacondemine and Jean-Pierre Cariou for confirming that the English text is wrong here [1]. The French text is correct but *en s'éloignant de ce point 205* can be misinterpreted.

28. Baral-Baron, G. Traitements Avancés Pour L'augmentation de la Disponibilité et de L'intégrité de la Mesure de Vitesse 3D par LiDAR, dans le Domaine Aéronautique (Advanced Methods for Increasing the Availability and Integrity of Lidar Airborne 3D Velocimetry). Ph.D. Thesis, Supélec, Gif-sur-Yvette, France, 2014.

29. Mayo, W.T., Jr. Modeling laser velocimeter signals as triply stochastic Poisson processes. In Proceedings of the Minnesota Symposium on Laser Anemometry, Minneapolis, MN, USA, 22–24 October 1975.

30. Jarzembski, M.A.; Srivastava, V. Interference of backscatter from two droplets in a focused continuous-wave CO_2 Doppler lidar beam. *Appl. Opt.* **1999**, *38*, 3387–3393. [CrossRef] [PubMed]

31. Hill, C.A.; Harris, M.; Ridley, K.D.; Jakeman, E.; Lutzmann, P. Lidar frequency modulation vibrometry in the presence of speckle. *Appl. Opt.* **2003**, *42*, 1091–1100. [CrossRef] [PubMed]

32. Harris, M.; Pearson, G.N.; Ridley, K.D.; Karlsson, C.J.; Olsson, F.Å.; Letalick, D. Single-particle laser Doppler anemometry at 1.55 μm. *Appl. Opt.* **2001**, *40*, 969–973. [CrossRef] [PubMed]

33. Hill, C.A.; Bennett, J.; Smith, D. Airport trials with the Aviation ZephIR lidar. In Proceedings of the 15th Coherent Laser Radar Conference, Toulouse, France, 22–26 June 2009.

34. Schwiesow, R.L.; Köpp, P.; Werner, C. Comparison of cw-lidar-measured wind values by full conical scan, conical sector scan and two-point techniques. *J. Atmos. Ocean. Tech.* **1985**, *2*, 3–14. [CrossRef]

35. Leosphere. Available online: http://www.leosphere.com (accessed on 21 February 2018).

36. The assumption of "reasonable continuity" should be examined:

    The teeing area is protected rear and right by a thick stand of tall pine trees, so that, even in a stiff wind, you are often hitting from a pocket of comparative calm. The problem is that the trees end about halfway to the green, which is the point where any wind that happens to be around begins to affect the flight of the ball. Perhaps because this is the lowest section of the course, the further and larger problem is that the wind here swirls a lot. You can glance up at the tops of those big old pine trees to your right and note that they are blowing one way, then flick your eyes over to the green and watch the flag blowing in the exactly opposite direction. Picking the correct club and shot at this hole on anything but a dead calm day is, therefore, always something of a guessing game. (J Nicklaus and K Gorman, *My Story*, Random House, 1997).

The typical working ranges (50–250 m) and Lorentzian-halfwidth range resolutions of the current ZephIR lidar are well suited to modern par 3 holes and hard-to-judge approach shots. It is common for the wind direction on some areas of the author's home course to be quite different from that of the low grim clouds overhead; for example, a prevailing breeze from the west may swirl over the Malvern Hills, forming a roll or vortex structure that creates a westwards motion near ground level. The potentially large and rapid wind variations throughout such hilly or complex terrain are of interest to turbine farm companies as well as golf broadcasters.

37. Eloranta, E.W.; King, J.M.; Weinman, J.A. The determination of wind speeds in the boundary layer by monostatic lidar. *J. Appl. Meteorol.* **1975**, *14*, 1485–1489. [CrossRef]
38. Kolev, I.N.; Parvanov, O.P.; Kaprielov, B.K. Lidar determination of winds by aerosol inhomogeneities: Motion velocity in the planetary boundary layer. *Appl. Opt.* **1988**, *27*, 2524–2531. [CrossRef] [PubMed]
39. Zuev, V.E.; Matvienko, G.G.; Samokhvalov, I.V. Laser sensing of wind velocity by a correlation method. *Izv. Atmos. Ocean. Phys.* **1977**, *12*, 772–776.
40. Zuev, V.E.; Samokhvalov, I.V.; Matvienko, G.G.; Kolev, I.N.; Parvanov, O.P. Laser Sounding of Instantaneous and Mean Speed of Wind Using Correlation Method. Available online: https://ntrs.nasa.gov/search.jsp?R=19870000871 (accessed on 15 January 2018).
41. Samokhvalov, I.V. *Correlation Method of Laser Sounding Measurements of Wind Speed*; Nauka: Novosibirsk, Russia, 1985. (In Russian)
42. Dérian, P.; Mauzey, C.F.; Mayor, S.D. Wavelet-based optical flow for two-component wind field estimation from single aerosol lidar data. *J. Atmos. Ocean. Technol.* **2015**, *33*, 1751–1778. [CrossRef]
43. Hamada, M.; Dérian, P.; Mauzey, C.F.; Mayor, S.D. Optimization of the cross-correlation algorithm for two-component wind field estimation from single aerosol lidar data and comparison with Doppler lidar. *J. Atmos. Ocean. Technol.* **2016**, *33*, 81–101. [CrossRef]
44. Smalikho, I.N. On measurement of the dissipation rate of the turbulent energy with a CW Doppler lidar. *J. Atmos. Ocean. Opt.* **1995**, *8*, 788–793. See also: Mikkelsen, T. On mean wind and turbulence profile measurements from ground-based wind lidars: Limitations in time and space resolution with continuous wave and pulsed lidar systems. In Proceedings of the European Wind Energy Conference & Exhibition 2009, Marseille, France, 16–19 March 2009. This literature defines a probe scale $\Delta z = \lambda(f/w_{lens})^2$, roughly $\sim\pi$ times longer than the conventional Gaussian beam parameter or Rayleigh range.
45. Hu, Q.; Rodrigo, P.J.; Iversen, T.F.; Pedersen, C. Investigation of spherical aberration effects on coherent lidar performance. *Opt. Express* **2013**, *21*, 25670–25676. [CrossRef] [PubMed]

remote sensing

MDPI

Article

Optimizing Lidars for Wind Turbine Control Applications—Results from the IEA Wind Task 32 Workshop

Eric Simley [1,*], Holger Fürst [2], Florian Haizmann [2] and David Schlipf [2]

[1] Envision Energy USA Ltd., 1201 Louisiana St. Suite 500, Houston, TX 77002, USA
[2] Stuttgart Wind Energy, University of Stuttgart, Allmandring 5b, 70569 Stuttgart, Germany;
    fuerst@ifb.uni-stuttgart.de (H.F.); haizmann@ifb.uni-stuttgart.de (F.H.); schlipf@ifb.uni-stuttgart.de (D.S.)
*   Correspondence: eric.simley@envision-energy.com or esimley@gmail.com

Received: 20 April 2018; Accepted: 30 May 2018; Published: 1 June 2018

Abstract: IEA Wind Task 32 serves as an international platform for the research community and industry to identify and mitigate barriers to the use of lidars in wind energy applications. The workshop "Optimizing Lidar Design for Wind Energy Applications" was held in July 2016 to identify lidar system properties that are desirable for wind turbine control applications and help foster the widespread application of lidar-assisted control (LAC). One of the main barriers this workshop aimed to address is the multidisciplinary nature of LAC. Since lidar suppliers, wind turbine manufacturers, and researchers typically focus on their own areas of expertise, it is possible that current lidar systems are not optimal for control purposes. This paper summarizes the results of the workshop, addressing both practical and theoretical aspects, beginning with a review of the literature on lidar optimization for control applications. Next, barriers to the use of lidar for wind turbine control are identified, such as availability and reliability concerns, followed by practical suggestions for mitigating those barriers. From a theoretical perspective, the optimization of lidar scan patterns by minimizing the error between the measurements and the rotor effective wind speed of interest is discussed. Frequency domain methods for directly calculating measurement error using a stochastic wind field model are reviewed and applied to the optimization of several continuous wave and pulsed Doppler lidar scan patterns based on commercially-available systems. An overview of the design process for a lidar-assisted pitch controller for rotor speed regulation highlights design choices that can impact the usefulness of lidar measurements beyond scan pattern optimization. Finally, using measurements from an optimized scan pattern, it is shown that the rotor speed regulation achieved after optimizing the lidar-assisted control scenario via time domain simulations matches the performance predicted by the theoretical frequency domain model.

Keywords: wind energy; Doppler lidar; wind turbine controls; lidar-assisted control (LAC); IEA Wind Task 32

1. Introduction

In the past decade, lidar-assisted control (LAC) of wind turbines has become an important research topic in the wind energy community [1]. Whereas traditional wind turbine control systems rely on feedback measurements to control blade pitch, generator torque, and yaw direction, Light Detection and Ranging (lidar) allows preview information about the approaching wind to be used to improve wind turbine control. The shared goal of the various LAC research efforts is to enable a reduction in the levelized cost of energy (LCOE) of wind energy (i.e., the average cost per unit of energy over the lifetime of the turbine, including capital costs, operations and maintenance costs, and all other relevant expenses) through (a) an increase in energy production or (b) a decrease in turbine cost made possible by structural load reduction [1]. Although several successful field trials have been reported (as discussed in Section 2.3), LAC is generally still limited to research activities; there remain a variety of obstacles preventing the widespread use of LAC in the wind industry. The objective of this paper is to provide a summary of how lidar systems can be optimized specifically for control applications in order to overcome the barriers preventing the widespread deployment of LAC.

The results presented in this paper are based on the outcome of the IEA Wind Task 32 workshop "Optimizing Lidar Design for Wind Turbine Control Applications" held in Boston, MA in July 2016. IEA Wind Task 32: "Wind Lidar Systems for Wind Energy Deployment" is an international open platform with the objective of bringing together experts from the academic and industrial communities to identify and mitigate barriers to the use of lidar for wind energy applications. During the workshop, participants from academia, national research laboratories, as well as the lidar and wind turbine industries discussed the barriers preventing the widespread use of lidars for wind turbine control, strategies for overcoming those barriers, and ideas for maximizing the effectiveness of lidars for control applications. In this paper, both practical considerations for overcoming the obstacles to the use of lidars for control and theoretical approaches for optimizing lidar scan patterns are discussed.

The remainder of this paper is organized as follows. A review of lidar technology, LAC, and previous work on lidar optimization for control purposes is provided in Section 2. Sections 3 and 4 discuss the identified practical barriers to the widespread use of lidars for control and strategies for overcoming the barriers, respectively. Theoretical approaches to optimizing lidar scan patterns by minimizing either (a) the error between the lidar measurements and the rotor effective wind speed or (b) the deviation from the intended setpoint of a control variable of interest (e.g., rotor speed) with the use of LAC are described in Section 5. Section 6 outlines how a frequency domain model can be used to calculate measurement or controller setpoint error for different scan configurations. Optimal scan parameters are provided for a variety of scan scenarios. In Section 7, design choices required to maximize the utility of lidar measurements in a feedforward pitch control scenario are illustrated. Time domain simulations show that the rotor speed regulation (or equivalently, generator speed regulation) achieved by an optimized control scenario matches the performance predicted by the frequency domain model. Next, a technology outlook, discussing potential future directions of lidars for control purposes, is provided in Section 8. Finally, Section 9 concludes the paper.

2. Background

In this section, a summary of lidar systems developed for control purposes is provided, followed by a brief overview of the different categories of LAC that have been investigated and a discussion of successful field tests that have been reported. The remainder of the section provides a review of the subset of the LAC literature that focuses on lidar optimization for control purposes.

2.1. Lidar Systems for Control Applications

A variety of nacelle lidar systems have been developed for wind turbine control applications, both commercially and for research purposes. The first field demonstration of a nacelle-mounted lidar that is reported in the literature was made in 2003 by Harris et al. [2], where the authors tested a single-beam continuous wave (CW) lidar mounted atop a utility-scale wind turbine, measuring 200 m upstream of the turbine. CW lidars measure the wind velocity by focusing a laser beam at a specific range and detecting the Doppler shift of the light backscattered from aerosols at the focus point [3]. Note that, although CW lidars can only measure at a single distance at one time, the focus distance of the lidar can often be adjusted. The technology behind this initial CW lidar [2] eventually led to the creation of the ZephIR DM nacelle-mounted circularly-scanning CW lidar by ZephIR Lidar (Hollybush, Ledbury, UK). With a configurable focus distance between 10 m and 300 m, the ZephIR DM lidar completes a circular scan containing 50 measurement points once per second [4]. A similar ZephIR lidar system was placed in the hub, or spinner, of a MW-scale wind turbine, as discussed by Mikkelsen et al. [5], allowing continuous measurement of the wind inflow without periodic obstruction from blade passage. A more complex spinner-mounted lidar system developed by Sjöholm et al. [6], also based on a ZephIR, uses two rotating beam-redirecting prisms to scan 400 points per second, roughly evenly distributed over a 100 m-diameter disk, at a preview distance of 100 m, allowing two-dimensional turbulence structures to be examined. In contrast to research-oriented lidar systems relying on complex scan patterns, a nacelle-mounted CW lidar developed by Windar Photonics (Taastrup, Denmark) uses four fixed beam directions to measure the wind 80 m upstream of the turbine [7]. This simple design is meant to reduce the cost of nacelle lidars for control purposes.

Whereas CW lidars are limited to measuring one point at a time, but are capable of high sampling rates, pulsed lidar systems can measure the wind speed at multiple ranges along the beam simultaneously, but typically require longer sampling periods. Instead of tightly focusing the laser beam at a specific distance, pulsed lidars attribute the backscattered light to different measurement ranges (range gates) according to the elapsed time since the light was emitted by the lidar. Schlipf et al. [8] used a custom scanning pulsed lidar system based on the Leosphere Windcube V1 lidar to provide preview measurements for lidar-assisted pitch control field tests. The authors configured the lidar to scan a circular trajectory containing six evenly-spaced beam directions with five range gates from 42.7 m (1 rotor diameter (D)) to 85.3 m (2 D) upstream of the test turbine with a scan period of 1.33 s. Additionally, a five-beam pulsed nacelle lidar has been developed by Avent Lidar Technology (Orsay, France) for control applications. This lidar was used for LAC field field tests performed by Kumar et al. [9]. The five-beam Avent lidar provides measurements at up to 10 range gates per beam with a maximum range of 300 m and a scan period of approximately 1.25 s [10]. Note that, while CW lidars can typically measure the wind at a greater number of beam directions per scan than pulsed systems, pulsed lidars have the advantage of being able to measure the wind at multiple ranges per beam direction. Because each range gate corresponds to a different radial distance from the center of the rotor, pulsed lidars are capable of measuring a large portion of the rotor disk area with a small number of beam directions.

2.2. Lidar-Assisted Control Strategies

The study of LAC began in earnest with the 2005 simulation-based investigation of blade pitch control for rotor speed regulation and blade load reduction presented by Harris et al. [3]. Since then, LAC strategies have been explored using all three main types of wind turbine control actuation: yaw control, generator torque control, and blade pitch control. Furthermore, the use of lidar measurements for control has been investigated for both below-rated operation, particularly for optimizing power capture, as well as operation in above-rated wind speeds, where the objective is to regulate rotor speed and minimize structural loads.

In the below-rated control region, both lidar-assisted yaw control and generator torque control have been investigated. Traditionally, yaw control is performed using wind direction measurements from a nacelle wind vane. However, improper calibration and flow disturbance behind the rotor can produce inaccurate wind vane measurements and rotor misalignment. Field tests performed by Fleming et al. [11] show that a bias in the wind vane measurements can be detected and removed using measurements from a nacelle lidar, leading to an improvement in energy capture. Scholbrock et al. [12] discuss yaw control experiments where information from the wind vane is completely replaced by more accurate lidar measurements, similarly resulting in greater energy capture. However, several authors argue that most of the performance increase from full-time lidar-assisted yaw control could likely be achieved by simply removing the wind vane bias, which could be accomplished with a one-time lidar-based wind vane calibration [11,13,14].

During below-rated operation, the primary purpose of generator torque control is to regulate rotor speed. For a subset of the below-rated control region, the objective is traditionally to maintain the tip-speed ratio of the rotor at its optimal value in order to maximize power capture. The use of lidar preview measurements to improve optimal tip-speed ratio regulation via generator torque control has been shown to provide a small increase in energy capture [13–16]. However, as observed by Bossanyi et al. [13] and Schlipf et al. [16], the large generator torque fluctuations and drivetrain loads that result from this control strategy likely outweigh the modest benefits.

Blade pitch control has been recognized as one of the more promising types of LAC because of the significant rotor speed regulation improvement as well as structural load reduction that can be achieved. Both collective pitch control, requiring estimates of the "rotor effective wind speed" (e.g., rotor average wind speed) [13,17,18], and individual pitch control (IPC), in which each blade is controlled differently to account for spatial variations in the wind field [13,19–22], have been investigated. By mitigating the effect of spatially-varying wind speeds, using lidar measurements of either rotor effective quantities like wind shear or wind speeds local to the individual blades, IPC has the advantage of reducing blade loads as well as non-rotating loads transferred to the drivetrain and turbine structure beyond what can be achieved with collective pitch control. Note that, while most lidar-assisted IPC controller designs only use measurements of the wind *speed* variations over the rotor area, Wortmann et al. [23] have recently proposed a lidar-assisted IPC controller that mitigates the impact of yaw misalignment and vertical inflow in addition to wind shear on blade loads. A subset of lidar-assisted controllers combine pitch control and torque control to optimize the impact of the two types of actuation together [24–27], especially during the transition between below-rated operation (primarily torque control) and above-rated operation (primarily pitch control) [24].

When considering optimal lidar configurations for control applications, it should be noted that LAC applications typically require only a few seconds of preview time, as explained for a variety of lidar-assisted pitch control strategies; Laks et al. [28] find that a preview time of only 0.2 s is sufficient for reducing blade loads, Bottasso et al. [26] conclude that only 1 s of preview time is necessary to minimize tower fore-aft loads and generator speed error, and Dunne and Pao [29] find diminishing returns for generator speed regulation and pitch activity reduction with preview times beyond 2–3 s. As explained by Dunne et al. [21], less than one second is fundamentally required to overcome the delay from the pitch actuator, while up to a few additional seconds are required to overcome the delay from filtering the lidar measurements to remove unwanted noise.

2.3. Lidar-Assisted Control Field Tests

Despite the large LAC research literature from the past decade, only a small number of field tests have been reported. However, the results of these experiments show that the types of control improvements revealed by theory and simulation can be demonstrated in the field. As mentioned in Section 2.2, Scholbrock et al. [12] demonstrated a reduction in yaw error when replacing the wind vane measurements in the yaw controller of the three-bladed 600-kW CART 3 (Controls Advanced Research Turbine), located at the US National Renewable Energy Laboratory (NREL)'s National Wind Technology Center, with wind direction measurements from a nacelle lidar. These yaw control improvements were achieved using the circularly-scanning ZephIR DM CW lidar introduced in Section 2.1. Several lidar-assisted collective pitch control field tests were performed at NREL's National Wind Technology Center as well, all using simple feedforward controllers based on wind speed-to-pitch angle lookup tables. In 2012, Schlipf et al. [8] evaluated a feedforward pitch controller on the 600-kW, two-bladed CART 2 wind turbine using measurements from the custom circularly-scanning pulsed lidar system mentioned in Section 2.1. During the same year, a similar feedforward controller was tested on the CART 3 turbine using measurements from a BlueScout Optical Control System (OCS) three-beam pulsed lidar, as described by Scholbrock et al. [30]. Based on these two experiments, Schlipf et al. [8] report a reduction in rotor speed standard deviation during the above-rated operation when using LAC, while Scholbrock et al. [30] additionally reveal a decrease in tower fore-aft loads. Kumar et al. [9] discuss feedforward pitch control experiments on the CART 2 turbine using preview measurements provided by the five-beam pulsed Avent lidar system, described in Section 2.1 as well. By re-tuning the feedback pitch controller with which the feedforward controller is combined, the authors demonstrated the ability to maintain roughly the same rotor speed error as the original baseline feedback controller allowed, but with significant reductions in tower fatigue loads and pitch activity.

2.4. Lidar System Optimization for Control Applications

Although much of the literature on LAC focuses on controller design rather than the remote sensing aspect, a significant amount of research has been conducted specifically on the optimization of lidar systems for LAC. This section provides a review of existing research on the optimization of lidars for control applications, organized into four categories: (1) lidar optimization via the analysis of controller performance; (2) scan pattern optimization by minimizing measurement error using simulated lidar measurements; (3) scan pattern optimization by minimizing measurement error using computationally efficient frequency domain calculations; and (4) lidar configuration optimization to address practical considerations.

2.4.1. Optimizing Lidars by Assessing Controller Performance

While many lidar-assisted control studies rely on a single lidar scenario for simulations or field testing, some authors have directly analyzed the impact of different lidar configurations on controller performance. In an overview of the benefits of various LAC applications, Bossanyi et al. [13] arrive at the general conclusion that both pulsed and CW lidars are suitable for control purposes, as long as they can measure roughly 10 points distributed around the rotor area upstream of the turbine every second while providing a few seconds of preview time. They claim that additional measurement points and faster sampling provide diminishing returns, and that a circular scan pattern is nearly as useful as more advanced scan patterns. The lidar model used in this study includes the spatial averaging, or "range weighting", along the beam that is inherent to both pulsed and CW lidars as well as the limitation to line-of-sight (LOS) velocity measurements (i.e., the projection of the three-dimensional wind vector onto the beam direction). The authors also find that the optimal scan configuration depends on the wind field parameter being estimated; large cone angles are beneficial for measuring wind direction, but harm the accuracy of rotor effective wind speed and wind shear estimates. Finally, the authors suggest that spatial averaging along the beam is helpful for estimating rotor effective wind quantities for control because it resembles the spatial averaging of the wind field by the rotor.

Focusing on a circularly-scanning pulsed lidar, Koerber and Mehendale [31] discuss the potential for different LAC applications to reduce the cost of energy generated by a wind turbine. The authors investigate the impact of the number of beams (2, 4, 6, or 8) and the number of range gates (1–10) on rotor speed regulation and structural fatigue load reduction for a collective pitch feedforward controller. As expected, the study finds that the controller performance improves as the number of beams and range gates increases, with the largest improvement occurring when moving from two to four beams and diminishing returns beyond eight range gates. Although the authors do not reveal the specific range gate distances, for the maximum 10-range gate scenario, the range gates are roughly evenly spaced so that they span the entire rotor disk area when projected onto the rotor plane. It is concluded that coverage of the rotor plane area is the most important factor for the scan pattern. Finally, the authors show that sampling frequency has little impact on load reduction, as long as it is above 0.5 Hz.

After first analyzing the influence of the scan pattern, spatial averaging, and LOS velocity limitations on overall measurement quality, researchers began to introduce models of wind evolution into LAC simulations. Wind evolution describes the change in turbulent structures as they advect from the measurement location downstream towards the turbine. This represents a deviation from Taylor's frozen turbulence hypothesis [32], traditionally assumed during wind turbine simulation, which maintains that turbulent structures remain unchanged while traveling downstream at the mean wind speed. Wind evolution models are typically defined in terms of longitudinal spatial coherence (i.e., the correlation between wind speeds separated by different longitudinal distances as a function of frequency), with low frequency components of the turbulence remaining highly correlated as the wind travels downstream but high frequency components becoming decorrelated due to eddy decay [33].

Bossanyi [34] incorporated wind evolution into collective pitch LAC simulations using the theoretical Kristensen wind evolution model [33], and found that evolving turbulence does not significantly impact the load reduction potential of LAC. Laks et al. similarly included the Kristensen wind evolution model while simulating lidar-assisted IPC for a 600 kW wind turbine. Assuming a hub-mounted CW lidar with three rotating beams, the authors found that blade loads were reduced to nearly the same level as with frozen turbulence for a preview distance of 126 m (~3 *D*). At this preview distance, the spatial averaging along the lidar beam was found to effectively filter out the high-frequency structures in the wind that change the most due to wind evolution.

2.4.2. Optimizing Lidars by Assessing the Accuracy of Simulated Lidar Measurements

Another strategy for optimizing lidar scan patterns is the direct comparison between simulated lidar measurements and the true variables they are intended to represent, thereby removing the controller from the analysis entirely. Kragh et al. [35] examined how accurately different scan patterns are able to measure yaw misalignment (i.e., the relative wind direction) in stochastic turbulent wind fields with a constant yaw error imposed. By modeling a CW lidar with a focus distance of 100 m, the authors compared the accuracy of measurements from a linear scan pattern sweeping left and right at hub height, a circular scan pattern, and a 2D scan pattern that covers more of the rotor disk area, for cone angles of 15° and 30°. Using a 10-min averaging time, it was found that the circular scan results in the lowest error. For all scan patterns, the larger 30° cone angle improves accuracy. However, the authors showed that when horizontal shear is introduced to the wind fields, the accuracy of the wind direction estimates becomes worse; horizontal shear appears similar to yaw misalignment when measurements are constrained to LOS velocities.

Returning to longitudinal wind speed measurements, Simley et al. [36] simulated measurements from circularly-scanning CW and pulsed lidars rotating at the same speed as the 126 m-diameter rotor of the NREL 5 MW reference turbine model [37] in stochastic turbulent wind fields. The lidar optimization is made relevant for pitch controlled turbines by analyzing root mean square (RMS) measurement error for the "blade effective wind speed", approximated by averaging the longitudinal wind speed along a line representing the blade, weighted by the radially-dependent contribution to aerodynamic torque production. A scan radius near 44 m (70% of the rotor radius (*R*)) produced the most accurate measurements, in part because the aerodynamic torque production is concentrated near this region of the blade. For the CW lidar model, the optimal preview distance was found to be in the range of 150 to 180 m (1.2 to 1.4 *D*), depending on the amount of turbulence. Beyond the optimal preview distance, the amount of spatial averaging increases too much (the effective length that is averaged along the beam scales with the square of the focus distance). For shorter preview distances, the cone angle becomes too large and the measured LOS velocities contain too many contributions from the transverse and vertical wind components, instead of the longitudinal component of interest. For the pulsed lidar, however, the measurement error was found to always improve as preview distance increases because the amount of range weighting along the beam remains constant.

Whereas the previous two studies rely on stochastic turbulent wind fields, Simley et al. [38] simulated lidar measurements using a more physically realistic wind field generated using large-eddy simulation (LES), a computational fluid dynamics (CFD) technique. The LES wind field, with 8 m/s mean wind speed, includes the interaction with a model of an operating NREL 5 MW reference turbine, causing reduced velocities upstream of the rotor in the turbine's "induction zone". Wind evolution is also inherently included. Note that wind evolution places an additional penalty on long preview distances due to the decorrelation of the wind. By simulating a hub-mounted 3-beam circularly-scanning CW lidar at (a) the location of the operating turbine and (b) far upstream of the turbine, the impact of the induction zone on measurement quality was investigated. A model-based wind speed estimator, which relies on measured turbine variables to estimate the true wind disturbances, was used to determine the rotor effective wind speed as well as horizontal and vertical shear. With induction zone effects included, the optimal scan radii and preview distances for measuring rotor effective wind speed and shear were found to be 70% *R* (44 m) and 60% to 70% *D* (76–88 m), respectively. These values are only slightly smaller than the optimal parameters without induction zone effects.

2.4.3. Lidar Optimization Using Frequency Domain Techniques

As an alternative to performing lidar simulations in stochastic or CFD-based wind fields, metrics of lidar measurement quality can often be calculated directly in the frequency domain. This approach is possible because (a) stochastic wind fields are defined by their frequency domain statistics (turbulence power spectra and spatial coherence) and (b) many useful metrics for measurement quality are either directly based on frequency domain definitions, such as the coherence bandwidth, or can be easily calculated using frequency domain information, such as mean square measurement error. Measurement coherence (the correlation between the lidar measurements and the true wind speed at the rotor plane as a function of frequency) can be calculated using a frequency domain wind field model as long as the lidar measurements and true wind speed variables are defined as linear combinations of wind speeds. Because lidar measurements and rotor effective wind speeds can be modeled as weighted averages of the wind speeds along the lidar beam and along the blades or across the rotor plane, respectively, they are well-suited for frequency domain analysis.

There are two main advantages to determining measurement quality directly in the frequency domain rather than based on time domain simulations. First, generating stochastic or CFD-based wind fields and performing time domain lidar simulations is computationally expensive. Second, to arrive at statistically significant measurement quality statistics, results from many different time domain simulations must be averaged together. Frequency domain calculations, on the other hand, only need to be performed once for each lidar configuration and set of wind field parameters, and produce the exact measurement coherence.

Schlipf et al. [39,40] combine a simple exponential longitudinal coherence model of wind evolution, introduced by Pielke and Panofsky [41], with a standard wind field model based on frozen turbulence to form an evolving wind field model. The model is used to calculate the measurement coherence between pulsed lidar measurements and the rotor effective wind speed, defined as the spatial average of the longitudinal wind speeds across the rotor disk. Specifically, the authors calculate the −3 dB cutoff frequency of the transfer function from the lidar measurements to the rotor effective wind speed, describing the bandwidth where the measurements are correlated with the true wind disturbance.

In addition to lidar range weighting and LOS velocity limitations, the frequency domain measurement coherence model used in [40] includes the effects of sequential scanning (i.e., the use of a single laser to scan across all beam directions in a finite amount of time). In the pulsed lidar measurement model used, the velocity measurements at different range gates are delayed by the amount of time it takes them to reach the first range gate (determined by the mean wind speed according to Taylor's frozen turbulence hypothesis [32]). The delayed velocities, which should be roughly in phase with each other, are then averaged together to estimate the rotor effective wind speed. Schlipf et al. [39] describe the direct frequency domain calculation of lidar measurement coherence as a "semianalytic" approach because spatial averages are approximated using discretization along the lidar beams and across the rotor disk.

Building upon the theory described in [39], Schlipf et al. [40] optimize the scan trajectory of a nacelle-mounted circularly-scanning pulsed lidar with five range gates for a wind turbine with a 109 m rotor diameter by maximizing the −3 dB cutoff frequency of the rotor effective wind speed measurement transfer function. Subject to realistic lidar constraints, the cone angle, number of beam directions, the distance of the first range gate, and the spacing between range gates are optimized. Furthermore, the preview time provided by the measurement configuration must be large enough to overcome the time delay resulting from the measurement filter required to remove the uncorrelated high frequencies (discussed in [39,42]). Adhering to the constraints, the optimal scan pattern is found to consist of six beam directions along the scan circle (requiring a scan time of 4.8 s) with a cone angle of 21.8°, with the first range gate at 68 m (0.625 D) and a range gate spacing of 13.6 m.

This configuration allows the range gates to radially cover almost all of the rotor area. These parameters result in a measurement transfer function cutoff frequency of 0.03 rad/m (which can be converted to a temporal frequency in Hz by multiplying by $U/2\pi$, where U is the mean wind speed). More details about the semianalytic frequency domain model for computing measurement coherence used by Schlipf et al. [40] can be found in the dissertation of Schlipf [43].

Simley and Pao [44] use a frequency domain wind field model similar to that of Schlipf et al. [40], combining the Kristensen wind evolution model [33] with a standard wind field definition to semianalytically calculate measurement coherence and other metrics. However, instead of modeling the rotor disk average wind speed, the authors calculate measurement coherence for rotating "blade effective wind speeds", defined as the average of the longitudinal wind speeds along the length of the blade weighted by the local contribution to aerodynamic torque production. To determine measurement quality, the authors calculate the mean square error (MSE) between the lidar measurements and the blade effective wind speed, assuming an optimal minimum-MSE measurement filter is used (see [42]). Using the NREL 5 MW reference turbine model, Simley and Pao [44] optimize the scan parameters of a circularly-scanning hub-mounted CW lidar, finding that the measurement MSE is minimized using a scan radius of 44 m (70% R) and preview distance of 170 m (1.35 D).

In the dissertation of Simley [45], the blade effective wind speed model from Simley and Pao [44] is extended to calculate the "rotor effective" wind speed as well as linear horizontal and vertical shear that result from three rotating blades. The lidar scenario modeled in [45] consists of a CW lidar with three rotating beams, one for each blade. The measurements from the three beams are used to estimate the rotor effective wind speed and shear components. Wind evolution is included using an empirical exponential coherence formula based on the coherence measured in a variety of LES wind fields [46]. As described by Simley [45], for a mean wind speed of 13 m/s and a turbulence intensity (TI) of 10%, the optimal scan radius and preview distance for measuring the rotor effective wind speed for the NREL 5 MW reference turbine are found to be 38 m (60% R) and 100 m (0.8 D), respectively. The same parameters for measuring horizontal and vertical shear are found to be 44 m (70% R) and 113 m (0.9 D), respectively. By exploring different wind conditions, Simley [45] finds that the optimal lidar configuration changes very little as the mean wind speed varies. However, as TI increases, wind evolution becomes more severe, causing the optimal preview distances to become shorter.

2.4.4. Practical Lidar Considerations

Although the frequency domain measurement coherence models discussed in the previous section provide an efficient way to assess the accuracy of different scan patterns, they do not incorporate many practical lidar considerations. Several studies have focused on more practical aspects of lidar optimization, however.

In addition to their assessment of feedforward controller performance for different scan pattern parameters mentioned in Section 2.4.1, Koerber and Mehendale [31] discuss the need for high availability and reliability for LAC applications. For a collective pitch feedforward control system relying on measurements from a multiple-beam pulsed lidar system, the authors conclude that the load reduction and rotor speed regulation performance begins to suffer significantly when the availability drops below 50%. Here, availability is defined as the fraction of measurements that are considered acceptable during a given time series, with the majority of the unavailability arising from blockage by passing blades. Given that controller performance is acceptable with 50% availability, the authors suggest that nacelle mounting is a sufficient approach. Note that more specific definitions of availability for control applications, where even a few seconds of measurement unavailability can compromise control performance, are discussed by Davoust et al. [47]. Using a 5-beam pulsed lidar with 10 range gates, the authors identify the number of valid measurements out of all beam directions and range gates during the previous 3 s as a good indicator of usefulness, concluding that at least 40 valid measurements (out of a maximum of about 130) are needed for reliable control.

Davoust et al. [48] explore the impact of the scan parameters of a nacelle-mounted Avent pulsed lidar on availability, as well as the relationship between atmospheric conditions and availability. With respect to rotor blockage, the authors explain that one of the most important parameters that determines availability is the fraction of time that a lidar beam direction is unobstructed by a passing blade. This parameter is a function of the diameter of the blade root and the position where the beam intersects the rotor disk. The fraction of time when the lidar beam is unobstructed by the blades increases as the lidar cone angle increases, the height of the lidar above the rotor axis becomes larger, or the position of the lidar behind the rotor plane increases.

Davoust et al. [48] show that an additional parameter that impacts availability is the ratio between the averaging time used to measure a single LOS velocity and the time it takes for a blade to pass across the lidar beam direction. If this parameter is greater than 1, then the lidar beam will be unobstructed by the blade for part of the signal integration time. As long as the carrier-to-noise ratio (CNR) of the measurement remains above the necessary threshold, a reliable measurement can still be formed. The authors show that lidar availability can be described as a linear combination of the aforementioned geometrical blockage ratio and averaging time ratio parameters, improving as either parameter increases until the averaging time parameter reaches 1. Therefore, availability can be improved not only by adjustments to scan geometry and lidar position, but also by the measurement integration time.

3. Barriers to the Use of Lidar for Control

The following two sections recapitulate the outcome of the group discussions during the IEA Wind Task 32 workshop "Optimizing Lidar Design for Wind Turbine Control Applications". Furthermore, they summarize the main conclusions of the given presentations where wind turbine manufacturers as well as lidar vendors talked about their perspectives of the requirements and objectives of lidar systems for control purposes. The whole is augmented with the authors' point of view. In general, most of the raised issues are consented by a vast majority, but that does not necessarily represent a consensus of all the workshop participants. Finally, the barriers preventing the widespread use of lidar for control are structured as follows:

3.1. Lack of Clarity in Cost–Benefit Assessment

Based on the current state of the art, it is likely that LAC could contribute to lower LCOE either by (1) an increase in energy production, (2) a decrease in unit costs of a wind turbine by reduction of structural loads, or (3) an extension of the turbines' lifetime, which combines aspects of the aforementioned. However, the question is "how much exactly?". In particular, the turbine OEMs are interested in assessing the full lifetime costs of the lidar intended to be used for LAC. Besides the initial cost of the device (volume manufacturing already considered), this includes the costs for turbine integration, which consist of mechanical and electrical installation as well as alignment and calibration. Furthermore, all operational costs have to be taken into account. In the end, it all adds up to the need for appropriate cost models for LAC. They should reveal the link between the load reduction potential/increase in energy capture and all relevant aspects of the monetary impact on the wind turbine. Once a break-even point is exceeded, a business case can be established.

3.2. Absence of 100% Availability

It is generally accepted that a lidar device cannot provide uninterrupted measurement data throughout its operating time. Due to the passing of the blades, some portion of the measurements are not suitable for reconstructing the rotor effective wind speed or other wind characteristics. However, even when the lidar is mounted in the spinner, it is nearly impossible to receive 100% availability. Especially on sites with severe atmospheric conditions (clean air, heavy fog), it is challenging to get a high-quality signal all the time. The next difficulty is deciding when the quality should be considered as "good" and how trustworthy the data are. All in all, there is the need to identify a specific metric where the measurement availability can be regarded as acceptable for applicability to LAC actions.

3.3. Risk Assessment and Reliability Issues

On the basis of the preceding section, there are general concerns that come up when implementing LAC on a wind turbine. The question "what happens if there is no or a faulty signal from the lidar?" is ubiquitous among all workshop participants. Understandably, no one would risk harming the turbine or at worst putting people in danger, if the safe operation of the turbine depends on the reliability of the lidar itself. Because of failures in hardware components and/or in software applications, lidars have not always been able to provide service reliably in the past. It is claimed that the technology readiness level (TRL) of currently-available lidar systems for LAC is still too low. As a result, there are not only the aforementioned safety concerns but also the high costs of troubleshooting and repair. Overall, considerations must be made on how long the lidar's lifetime should be and how it should be designed to require the least maintenance effort.

3.4. Lack of Common Guidelines and Standards

As shown in the previous sections, there is a need for measures to evaluate the performance of a lidar device or of the whole LAC structure. These metrics, such as availability, have to be well defined and all their calculation constraints have to be fully revealed in order to ensure comparability. Another ambiguity is an appropriate threshold for the metrics. Depending on the application, it is not clear how to decide whether an availability value can be considered as "good" or when in terms of LAC an acceptable limit is reached. It would be desirable for all of this information to be provided in common sources to keep all stakeholders aligned. Ideally, a guideline or a standard specifically for LAC would exist.

3.5. Bringing Theoretical Knowhow and Practical Needs into Accordance

It has already been emphasized in Section 2 that many theoretical aspects of LAC have been covered quite well. Nevertheless, the technology has not advanced beyond prototype field testing on a wide scale yet. Significant long-term tests on multi-megawatt wind turbines have not been reported. On the one hand, this is mainly due to the aforementioned barriers, but it is also because the research community sometimes does not know the mechanisms of the industry as well as the needs of the end users. On the other hand, turbine manufacturers and lidar vendors are often not aware of what is theoretically possible and what is worth trying to implement. In summary, it can be stated that the level of collaboration between different stakeholders still has room to improve.

4. Suggestions for Mitigating the Barriers to the Use of Lidar for Control

The two questions posed during the second part of the workshop's roundtable discussion were: what are the design suggestions that can mitigate the identified barriers and what are the design suggestions to optimize lidars for control applications, taking into account the identified constraints? The following is an excerpt of the suggestions made by the workshop participants extended by some recommendations by the authors. There is no claim of completeness, neither should the impression arise that work has not already been done in some of the areas. The challenge for the LAC community is to bring all the different aspects together to adequately address the barriers.

4.1. Assess the Cost–Benefit Relationship by Using Methods of Systems Engineering

As mentioned in Section 3.1, the ultimate goal for LAC is to prove that the technology is able to reduce LCOE significantly. Because LAC is an interdisciplinary field of engineering where technically advanced systems interact with each other, a holistic approach is needed to fully manage complexity. Systems Engineering (SE) methods are designed to combine contributions and balance trade-offs among optimal performance, cost, and schedule while maintaining an acceptable level of risk covering the entire life cycle of a system. Thus, by combining technical and human-centered disciplines like optimization, reliability engineering, and risk management, SE is almost predestined to meet the challenges of LAC. However, there is probably no general answer to the question "how much?". It is only possible to investigate if turbine *A* with lidar *B* on site *C* is profitable, provided that the most important boundary conditions are known. A key factor for solving these large optimization problems is the existence of cost models for all systems involved. Thus, it is important to get the turbine manufacturers as well as the lidar suppliers involved. On the level of wind turbine optimization, there are already some promising approaches [49]. The integration of a lidar into the overall system is still missing though. Therefore, a collaboration with "IEA Wind Task 37: Systems Engineering" is highly recommended.

4.2. Improve Measurement Availability with Self-Adjusting Lidars

The mitigation of the availability problem is quite challenging, sometimes impossible. Even the best lidar cannot measure a LOS velocity if there are no aerosols in the air. In the case of finding a threshold when a single measurement can be considered as exploitable for further processing, the key is "adaptivity". By smartly adjusting e.g., CNR thresholds, sampling periods, lidar data processing parameters depending on wind evolution, or spacings of range gates, a significant increase in availability should be possible. Of course, the lidar should be capable of adjusting "by itself," without human intervention.

Another way to increase the percentage of availability is by redefinition. In terms of LAC, it would be more functional if rotor effective quantities were used as a metric. Because of averaging effects, it is not necessary for every LOS measurement to be available to calculate these measures and achieve 100% availability. A suitable metric to quantify the overall quality for LAC is the measurement coherence (see also Section 6.1).

Furthermore, one could address the availability matter by going a step backwards to the site assessment process. When evaluating if several areas are suitable for energy production through wind turbines, the sites could be checked for characteristics that are beneficial for LAC. Besides high aerosol concentration and practically no severe weather conditions, other measures have to be defined to characterize a site as "LAC compliant". This would probably limit the applicability of LAC to a small proportion of all explored sites, but once a location is found, the likelihood of good availability will be high. Especially for an emerging technology, it is important to reduce risk factors.

4.3. Improve Reliability and Gain Confidence in TRL by Fault Tolerant Control

Even under perfect environmental conditions, it is still possible to get no measurements from the lidar device because of hardware defects or software errors. These reliability issues were discussed a lot amongst the workshop participants. Generally, the more complexity that is added to a system, the more complicated the assurance of keeping technical products highly reliable becomes. A lidar with few but robust components and a manageable amount of features could probably be considered as less error prone. The challenge for the designer is to find a good trade-off between effort, time, and cost in particular. An important aspect to improving reliability in the near term is maintainability. First, it has to be ensured that the maintenance staff of a wind turbine operator is able to perform maintenance work during regular service intervals. Second, the subsystems of a lidar should be designed in a modular way so that they can be quickly replaced as entire units by plug-and play operations in case of a failure.

Finally, let us assume the worst case: a wind turbine is running in LAC mode and suddenly there are faulty measurements or no measurements at all because of availability or reliability problems. In a retrofit scenario for a fatigue load designed wind turbine, there is a simple but effective countermeasure. By switching back to a "normal operation mode", the wind turbine acts with its standard feedback controller for which it was originally designed. When availability exceeds the specified threshold or the lidar device is put back into operation, the feedforward mode takes over once again. Note that the challenge of smooth switching is still present.

However, the situation is more complicated when the margin of load reduction has already been taken into account during the design of the wind turbine. In that case, a lidar failure would immediately have a negative effect on the turbine's lifetime and on the safety of the whole system. On a fatigue load designed turbine, one could limit the damage by derating the turbine to a certain percentage of its rated power where the load margins are kept. In the end, it is a matter of the cost–benefit relationship again.

The by far riskiest scenario occurs when LAC is intended to be used in an extreme load driven wind turbine design to detect extreme events and counteract them. Here, even for an LAC compliant site with a self-adjusting highly reliable lidar, there is still some probability that an extreme operating gust could impact and destroy the turbine when the lidar is not in operation. This issue can only be bypassed by adding redundancy to the system.

4.4. Pave the Way towards International Standards

The first step of mitigating the barrier of missing standards has already been taken by publishing this article. This broad summary of lidar optimization for control purposes could serve as a basis for further input from relevant stakeholders. The medium-term goal should be to create a recommended practices document for LAC applications that is to be updated regularly.

Another important aspect regarding standards was revealed during the workshop. The question of how to carry out a type certification for a turbine with LAC came up. To address this question, an IEA Wind Task 32 workshop on the topic took place in Hamburg, Germany in early 2018.

4.5. Strengthen the Collaboration between Academia and Industry

The fact that 33 people attended the workshop clearly shows that there is already a strong motivation from universities (12 participants), national research laboratories (6), wind turbine manufacturers (10) and lidar vendors (5) in such a knowledge sharing event. Researchers and scientists presented their latest findings and representatives from industrial departments reported on their current developments and gave insights into their products and services. Furthermore, it was intended that the different stakeholders would share some of their experience with LAC, including the problems they face. In the future, the challenge for academia will be to summarize and condense its achievements in an easily understandable manner and point out the benefits clearly. To enhance the possibility of creating significant value, the industry should be encouraged to share more of their key knowledge (e.g., cost models)—of course in compliance with non-disclosure agreements between all parties.

4.6. Use a Standardized Simulation Environment in the Early Design Phase

When thinking about the optimization of lidars for control purposes, one should not be limited to the hardware device itself or the implemented software. In fact, there is still a huge potential for improving the preceding simulation process, more precisely the simulation tool chain. The better adjusted these techniques/tools are to the various subtasks of LAC, the more realistic and therefore more reliable the outcomes become. Consequently, establishing a kind of standardized LAC simulation environment would create added value. Furthermore, a modularized process with defined interfaces would ensure that every potential user is not limited to default features. One can optionally swap modules or extend them with needed functions. Wherever possible, new and useful developments should be shared within the community. The following modules are suggested as minimal requirements for a LAC simulation environment:

1. **Pre-Processors:** Tools for generating input files for the most common aeroelastic simulation tools (e.g., Bladed, FAST). Furthermore, lidar vendors could share scripts to convert the output data of their specific devices to a defined data format that is compatible with the other modules' interfaces.

2. **Wind Field Generator:** In the context of LAC, wind fields should be generated differently for realistic simulation. Several types of wind events (e.g., extreme operating gusts) have to be included in turbulent wind fields while ensuring acceptable computational effort [50].

3. **Lidar Simulator:** The lidar should be modeled with its most important properties, e.g., number of beams, scanning angles, number and spacing of range gates, update/measurement rate, range weighting function, probe length, etc.

4. **Wind Field Reconstruction:** Tools and algorithms for reconstruction of rotor effective wind characteristics from LOS data. A common data format with a defined interface is preferable for immediate application of new lidar devices.

5. **Controller Compilation Framework:** Here, the feedback as well as the feedforward controllers and all their subfeatures are included. It is not mandatory to use Matlab/Simulink (The MathWorks, Inc., Natick, MA, USA) like many control engineers, but it is beneficial when source code can be compiled to a dynamic link library (DLL) so the controllers can be shared within the community.

6. **Post-Processing:** A wide variety of tools for statistical evaluation and plotting of simulation results.

It is intended that, within one module class, there are several versions that can vary in complexity quite significantly. Depending on the stakeholder's needs, one can combine the submodules to create user-specific simulation environments. Wherever necessary, modules should also be executable independently to allow focusing on a specific problem.

5. Lidar Scan Pattern Optimization

While ideas for improving lidar availability and reliability, as well as other practical considerations, were presented in Section 4, this section focuses on the selection of lidar scan locations upstream of the turbine that result in measurements that best represent the actual wind variables of interest that arrive at the turbine. Lidar scan patterns can be optimized for different control strategies (e.g., collective pitch control for rotor speed regulation, IPC for asymmetric rotor load reduction, yaw control), which require measurements of different wind variables (e.g., rotor effective wind speed, wind shear, wind direction). Therefore, there is generally no single scan pattern that is optimal for all control objectives. Because of the great interest in collective pitch feedforward control for rotor speed regulation in the literature, however, the detailed investigations of scan pattern optimization presented in the rest of this paper will focus on measurements of the rotor effective wind speed.

For the purposes of the scan pattern optimization discussed here and in Section 6, rotor effective wind speed is defined as the time-varying mean value of the longitudinal u components of the wind over the rotor disk area. For simplicity, it is assumed that (1) all locations in the rotor disk have equal importance; (2) the response of the rotor is purely a function of the linear combination of wind speeds across the rotor disk; and (3) the transverse v and vertical w wind components can be neglected. Perfect alignment between the rotor and the mean wind direction is assumed (i.e., no yaw misalignment or vertical inflow angle).

The scan pattern optimization performed here is subject to the following potential sources of measurement error: (1) the evolution of the wind as it travels from the measurement location to the rotor plane; (2) the limitation to LOS measurements (i.e., the "corruption" of the wind speed measurements by the v and w components of the wind, when the u component of the wind is of interest); (3) range weighting, or the spatial averaging of the wind speeds along the beam direction; and (4) the fact that a scan pattern containing a finite number of points will not perfectly resemble a rotor disk average. All scan patterns are optimized for a rotor diameter of $D = 126$ m, the rotor diameter of the NREL 5 MW reference turbine [37].

5.1. Metrics for Assessing Measurement Quality

To compare the effectiveness of different scan patterns, an appropriate metric representing lidar measurement quality must be defined. One simple metric that describes measurement accuracy is the

MSE between the true rotor effective wind speed at the turbine u_{eff} and its estimated value based on the lidar measurement $\hat{u}_{eff}$:

$$\mathbf{E}\left[\left|u_{eff} - \hat{u}_{eff}\right|^2\right].\tag{1}$$

However, because lidar-based estimates of the rotor effective wind speed are not perfectly correlated with the wind at the rotor due to wind evolution and other sources of measurement error, it is beneficial to low-pass filter the measurements before they are used by the controller. As will be described in the next section, an optimal filter H_{opt} can be derived that minimizes the measurement MSE. Therefore, a useful metric for determining measurement accuracy is the MSE between the true rotor effective wind speed and its optimally-filtered lidar-based estimate:

$$\mathbf{E}\left[\left|u_{eff} - H_{opt}\hat{u}_{eff}\right|^2\right].\tag{2}$$

Metrics describing measurement quality can be made more meaningful by including the response of the turbine to the wind inflow, instead of concentrating on the wind variables alone. For example, the impact of the wind inflow on a turbine variable of interest y_{WT}, when the imperfect measurement $\hat{u}_{eff}$ is used as input to the feedforward controller, can be used to judge measurement quality. Assuming the objective of the lidar-assisted controller is to minimize the magnitude of a wind turbine variable, the variance of y_{WT} is an appropriate measurement quality metric. Specifically, for the control objective of minimizing the mean square error between the generator speed ω_{gen} and the rated generator speed $\omega_{gen,0}$, this metric becomes:

$$\mathbf{E}\left[\left|\omega_{gen} - \omega_{gen,0}\right|^2\right].\tag{3}$$

Finally, measurement quality can be judged by the highest frequency at which the measurement remains correlated with the true rotor effective wind speed. In this case, the objective of the scan pattern optimization is to maximize the correlation bandwidth, allowing the feedforward controller to mitigate the impact of as much of the frequency content of the wind as possible. The correlation between the rotor effective wind speed u_{eff} and its lidar-based estimate $\hat{u}_{eff}$ as a function of frequency is described by the magnitude-squared coherence between the two signals: $\gamma^2_{u_{eff}\hat{u}_{eff}}(f)$. The magnitude-squared coherence between two signals a and b is defined as

$$\gamma^2_{ab}(f) = \frac{|S_{ab}(f)|^2}{S_{aa}(f)\,S_{bb}(f)},\tag{4}$$

where $S_{ab}(f)$ is the cross-power spectral density (CPSD) between signals a and b, $S_{aa}(f)$ is the power spectral density (PSD) of a, and $S_{bb}(f)$ is the PSD of b. Values of coherence range from 0 to 1, where a value of 0 indicates that the two signals are completely uncorrelated while a value of 1 indicates perfect correlation. A reasonable value of coherence for representing the boundary between frequency components that are uncorrelated and correlated is 0.5 [51]. Therefore, a useful metric for indicating measurement quality is the frequency at which the measurement coherence crosses below 0.5:

$$f:\ \gamma^2_{u_{eff}\hat{u}_{eff}}(f) = 0.5.\tag{5}$$

5.2. Lidar System Configurations

The coordinate system and basic parameters used to describe the scan patterns analyzed in this paper are shown in Figure 1. A scan pattern may contain many different measurement points, but each point can be defined by its upstream preview distance d from the lidar unit in the $-x$ direction, its radial distance r in the yz plane from the point d m directly upstream of the lidar unit, and the azimuth angle ψ in the yz plane, where $\psi = 0$ indicates a measurement at the top of the scan circle defined by d and r. The measurement points can be equivalently described by their measurement distance F and cone angle θ. For simplicity, it is assumed that the lidar is located at the origin, which is defined as the rotor hub position.

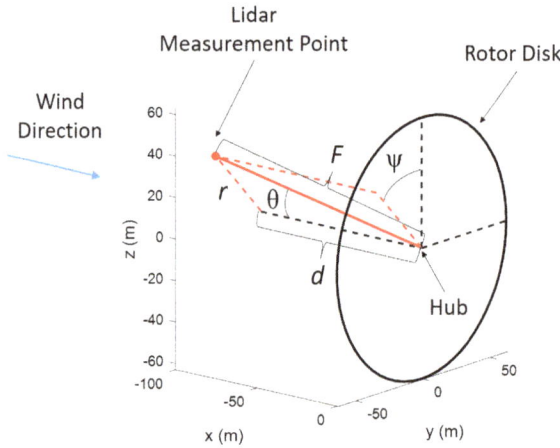

Figure 1. Scan pattern coordinate system and scan variables for a single lidar measurement point.

Both CW and pulsed lidars are examined here when analyzing scan patterns. Measurements from both types of lidars consist of wind speeds spatially averaged along the beam direction. However, the range weighting functions, which describe how heavily wind speeds at different radial distances are weighted, are different. The range weighting function for a CW lidar changes as the focus distance of the lidar increases; the full-width-at-half-maximum width of the weighting function is proportional to the square of the focus distance. Therefore, the farther a CW lidar is focused, the more spatial averaging occurs. The range weighting functions for pulsed lidars remain approximately constant regardless of the range gate distance. Figure 2 shows theoretical range weighting functions for a CW lidar with focus distances of 50 m, 100 m, and 150 m as well as for a pulsed lidar (shown without loss of generality at a measurement distance of 100 m), with parameters roughly based on the ZephIR 300 CW lidar [52,53] and the WindCube WLS7 pulsed lidar [53,54], respectively. More information about the range weighting functions can be found in Simley et al. [36].

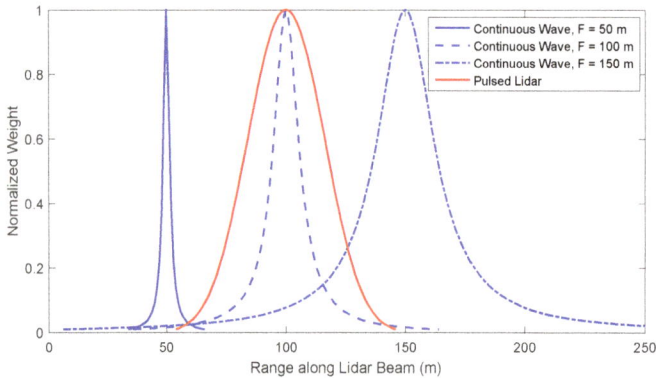

Figure 2. Normalized range weighting functions for a continuous wave (CW) lidar with focus distances of 50 m, 100 m, and 150 m as well as a pulsed lidar.

Several types of scan patterns are investigated in this paper for CW and pulsed lidars, all roughly based on lidar systems that have been developed commercially. The measurement quality achieved by each scan pattern is compared to a single point measurement upstream of the turbine at hub height, representing the most basic measurement scenario, as well as the longitudinal wind speed averaged over the rotor disk area upstream of the turbine, acting as an "ideal" preview measurement of the rotor effective wind speed. A preview distance of 50 m is used for the single-point and ideal measurements to match the closest range gate available for pulsed lidar measurements in this analysis.

After the point measurement, the simplest lidar scan pattern investigated is the two-beam lidar, with both beams measuring at hub height. This configuration is analyzed for a CW lidar, as shown in Figure 3b, and a pulsed lidar, illustrated in Figure 3c. Note that all pulsed lidar scan patterns modeled in this paper contain 10 evenly spaced range gates beginning at a range of 50 m and ending at the farthest range (no more than 300 m), parameters achievable with the Avent lidar system [9,10]. The next level of complexity is represented by a four-beam lidar, with CW and pulsed configurations shown in Figure 3d,e, respectively, based on the Windar lidar [7] and the four-beam version of the Avent system [10]. For the four-beam configuration, the beams are oriented at azimuth angles of ψ =45, 135, 225, and 315 degrees. The last realistic scan pattern investigated is the circular scan consisting of 50 points evenly spaced around a scan circle, based on the ZephIR DM lidar system [4], shown in Figure 3f. Because pulsed lidars typically require longer sampling times than CW lidars, only the CW configuration is analyzed for the circular scan scenario due to the high sampling rate required to complete the scan in a reasonable amount of time relative to the lidar preview time.

All of the commercial lidars that these scan patterns are inspired by can complete a full scan in approximately 1 s. Because of the relatively short scan times compared to typical feedforward control time scales, the calculations of measurement quality are simplified in this investigation by assuming each point in the scan pattern is measured simultaneously. Each of the scan patterns shown in Figure 3 is optimized by varying the preview distance d and scan radius r of the scan (or, equivalently, the measurement distance F and cone angle θ). For the pulsed lidar configurations, the 10 range gates measured are evenly spaced between 50 m and the maximum range given by the measurement distance parameter $F = \sqrt{d^2 + r^2}$.

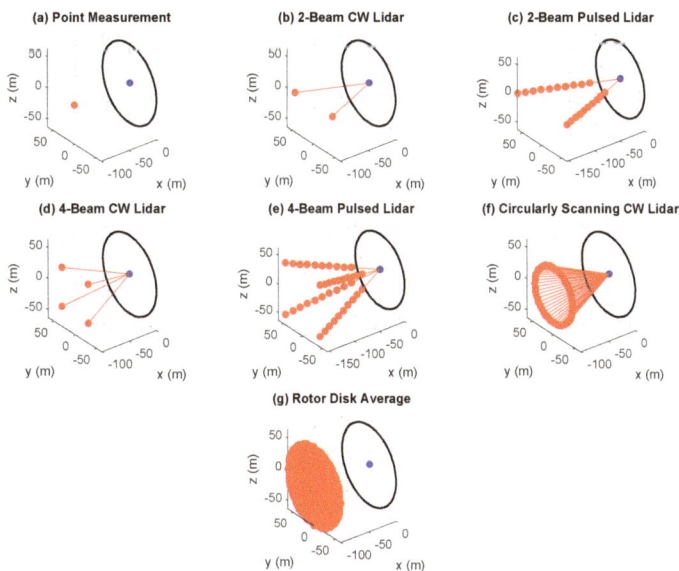

Figure 3. Lidar scan pattern scenarios investigated for scan pattern optimization.

5.3. Lidar-Based Wind Parameter Estimation

Line-of-sight velocity measurements from a lidar can be described as the opposite of the dot product between the unit vector in the direction that the beam is oriented ($\ell = [\ell_x, \ell_y, \ell_z]$) and the wind speed vector at the measurement point ($[u, v, w]$):

$$v_{LOS} = -\ell_x u - \ell_y v - \ell_z w. \tag{6}$$

Note that the true velocity measured by a lidar is a spatial average of the LOS velocities along the lidar beam, modeled as the integral of the LOS velocities along the beam weighted by the range weighting function. For control purposes, wind parameters such as the rotor average wind speed, wind shear, and wind direction are more useful than the LOS velocities provided directly by a lidar. Therefore, the wind parameters of interest must be estimated from the LOS velocities.

To estimate the wind parameters, an appropriate wind field model must first be defined. In general, it is not possible to distinguish between horizontal wind direction and horizontal shear (or vertical inflow angle and vertical shear) using LOS velocity measurements from a lidar scan pattern. Consequently, the two most common wind field models used for wind parameter estimation consist of (a) uniform wind speed across the rotor including horizontal and vertical wind direction, and (b) no transverse or vertical wind speeds, but linear horizontal and vertical shear [55].

Regardless of which wind field model is assumed, for a scan pattern with N unique beam directions, $\ell_1 \ldots \ell_N$, the solution for the estimate of the rotor average wind speed perpendicular to the rotor, used for optimizing CW lidars in this paper, is given by

$$\hat{u}_{eff} = -\frac{1}{N} \sum_{i=1}^{N} \frac{v_{LOS,i}}{\ell_{x,i}}. \tag{7}$$

However, for pulsed lidars, measurements at different range gates can be combined more effectively by delaying the measured velocity signals at each scan point by the estimated amount of time it takes the wind to travel to the plane where the closest range gate is located, according to Taylor's frozen turbulence hypothesis [32]. This effectively "stacks" the measurements from different range gates together before averaging by removing their relative time shifts (although the farther a range gate is from the rotor plane, the more the measurements will suffer from wind evolution). For pulsed lidars, an estimate of the rotor average wind speed, which is used for scan pattern optimization in this paper, can therefore be expressed as

$$\hat{u}_{eff}(t) = -\frac{1}{N} \sum_{i=1}^{N} \frac{v_{LOS,i}\left(t - \frac{x_0 - x_i}{U}\right)}{\ell_{x,i}}, \tag{8}$$

where x_i is the longitudinal position of measurement point i, x_0 is the longitudinal position of the range gate closest to the rotor, and U is the mean rotor effective wind speed. Note that, in practice, the estimation of the rotor effective wind speed could be improved by delaying measurements at the different range gates according to their height-dependent mean wind speed, accounting for wind shear. In this work, however, it is assumed that the wind advection speed is the same at all heights.

More information about wind parameter estimation using lidar measurements, including a discussion of nonlinear methods for estimating all five wind parameters described in this section simultaneously, can be found in Raach et al. [55].

6. Frequency Domain Optimization

This section begins with a discussion about how the lidar measurement quality metrics introduced in Section 5.1 can be directly computed via frequency domain calculations using a frequency domain wind field model. Next, the frequency domain calculation procedures are used to optimize the parameters for the scan patterns presented in Section 5.2 for each measurement quality metric.

6.1. Frequency Domain Calculations for Assessing Measurement Quality

Because lidar measurements and the rotor effective wind speed can be modeled as linear combinations of the wind speeds at different points in space, it is relatively straightforward to compute the measurement quality metrics in Section 5.1 using a frequency domain wind field model containing the power spectral densities (PSDs) of the three wind speed components and the spatial coherence between wind speeds at different points. As discussed in Section 2.4.3, direct frequency domain calculations of the metrics describing measurement quality are much more computationally efficient than calculations based on simulated lidar measurements and wind speeds.

The wind field model used here consists of the Kaimal turbulence spectrum and the spatial coherence model described in the International Standard IEC 61400-1 Ed. 3 [56]. Whereas the IEC standard only describes spatial coherence for the transverse y and vertical z directions, wind evolution is modeled here using the empirical longitudinal coherence formula based on LES data developed by Simley and Pao [46]. Note that the theoretical longitudinal spatial coherence model described by Kristensen [33] is often used as well. The spatial coherence model defined in the IEC standard [56] is modified further by extending the transverse and vertical spatial coherence formula used for the u component of the wind to the v and w components as well (the IEC standard defines the v and w components as having zero spatial correlation, but this could lead to unrealistically low error from LOS limitations).

Although Section 5.1 describes the possibility of assessing measurement quality using the raw lidar measurement MSE (Equation (1)), it is more appropriate to filter the lidar measurements to remove unwanted noise before using them in the controller (Equation (2)) [42]. Therefore, the following three metrics introduced in Section 5.1 will be used for scan pattern optimization in this section: (1) mean square measurement error with optimal filtering; (2) generator speed variance resulting from combined feedback-feedforward control with optimally filtered lidar measurements; and (3) the 0.5 measurement coherence bandwidth. As explained in Schlipf et al. [39] and Simley and Pao [42], the optimal filter that minimizes the MSE between the lidar measurement and the true rotor effective wind speed is given by the transfer function

$$H_{opt}(f) = \frac{S_{u_{eff}\hat{u}_{eff}}(f)}{S_{\hat{u}_{eff}\hat{u}_{eff}}(f)},\qquad(9)$$

where $S_{u_{eff}\hat{u}_{eff}}(f)$ is the CPSD between the rotor effective wind speed and its lidar-based estimate and $S_{\hat{u}_{eff}\hat{u}_{eff}}(f)$ is the PSD of the lidar measurement.

The simplest measurement quality metric to calculate is the correlation bandwidth, based on the measurement coherence: $\gamma^2_{u_{eff}\hat{u}_{eff}}(f)$. As derived by Simley and Pao [42], the measurement MSE using the optimal filter from Equation (9) can be calculated by integrating over the PSD of the measurement error:

$$\mathrm{E}\left[\left|u_{eff} - H_{opt}\hat{u}_{eff}\right|^2\right] = \int_0^\infty S_{u_{eff}u_{eff}}(f)\left(1 - \gamma^2_{u_{eff}\hat{u}_{eff}}(f)\right)df,\qquad(10)$$

where $S_{u_{eff}u_{eff}}(f)$ is the PSD of the true rotor effective wind speed.

Additionally, as shown by Simley and Pao [42], the generator speed MSE (letting ω_{gen} simply represent the deviation from the generator speed setpoint) resulting from the use of combined feedback-feedforward control with optimally-filtered lidar measurements can be expressed as

$$\mathrm{E}\left[\left|\omega_{gen}\right|^2\right] = \int_0^\infty \left|T_{\omega_{gen}u_{eff}}(f)\right|^2 S_{u_{eff}u_{eff}}(f)\left(1 - \gamma^2_{u_{eff}\hat{u}_{eff}}(f)\right)df,\qquad(11)$$

where $T_{\omega_{gen}u_{eff}}$ is the closed-loop transfer function from rotor effective wind speed to generator speed error. Equation (11) uses the assumption that the feedforward controller is designed to perfectly cancel the impact of the wind speed disturbance on the generator speed, as long as perfect preview information is provided. Following the pitch controller tuning recommendations in Hansen et al. [57],

71

the closed-loop transfer function $T_{\omega_{gen}u_{eff}}$ used in this paper is chosen to have a second order response with natural frequency $\omega_n = 0.6$ rad/s and damping ratio $\zeta = 0.65$. More information about the combined feedback-feedforward controller is provided in Section 7.1.

Detailed descriptions of how the PSDs $S_{u_{eff}u_{eff}}(f)$ and $S_{\hat{u}_{eff}\hat{u}_{eff}}(f)$ as well as the CPSD $S_{u_{eff}\hat{u}_{eff}}(f)$, used to compute all of the performance metrics discussed here, can be calculated from a frequency domain wind field model using Fourier properties are available in Schlipf et al. [40] and Simley [45].

6.2. Detailed Optimization Example: Circular Scan Pattern

All measurement quality calculations are performed for a wind field with mean wind speed $U = 13$ m/s (1.6 m/s above rated wind speed for the NREL 5 MW reference turbine) and IEC Class B normal turbulence model turbulence intensity [56]. Figure 4 reveals how the coherence bandwidth, optimally-filtered measurement MSE, and generator speed MSE with optimally filtered preview measurements depend on the scan radius and preview distance for the circular-scanning CW lidar scenario. The optimal scan radii that maximize the coherence bandwidth and minimize measurement MSE as well as generator speed variance are all around 40 m ($\sim$0.65 R) indicating that measurements at this radius yield the best approximation to a rotor disk average. Although the optimal preview distances are different for the three metrics ($\sim$70–90 m), they reveal the same general trend. Measurements closer to the rotor than the optimal preview distance, resulting in larger cone angles, suffer from LOS velocities that contain too many contributions from the transverse and vertical wind speeds. Measuring the wind farther away than the optimal preview distances causes wind evolution to become more severe, increasing measurement error as well. Note that the measurement quality is more sensitive to deviations in scan radius from the optimal point than to different preview distances. The specific optimal scan parameters for the different measurement quality metrics in Figure 4 depend on which frequencies are weighted most heavily in the particular metric.

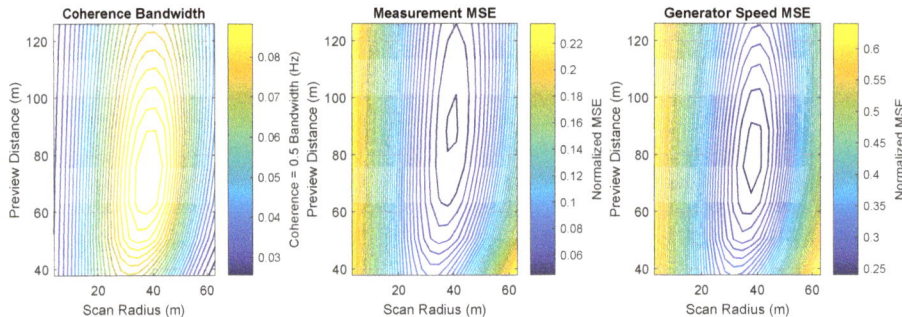

Figure 4. Dependence of coherence bandwidth, rotor effective wind speed measurement mean square error (MSE) (normalized by rotor effective wind speed variance), and linearized generator speed MSE (normalized by generator speed MSE with baseline feedback control) on scan radius r and preview distance d for the circular continuous wave (CW) lidar scan pattern.

6.3. Scan Pattern Optimization Results

For the seven lidar scan scenarios provided in Figure 3, the maximum achievable coherence bandwidth along with the minimum achievable measurement MSE and generator speed MSE corresponding to the optimal scan parameters with optimal filtering are compared in Figure 5. The provided measurement MSE values are normalized by the variance of the rotor effective wind speed, whereas the generator speed MSE values are normalized by the generator speed variance with feedback control only. All three measurement quality metrics reveal the same trends: (1) as the number of beams increases, the measurement accuracy increases as well and (2) for the same number of beams, the additional measurement ranges afforded by pulsed lidars improve measurement accuracy compared to a CW lidar. All lidar measurement scenarios improve upon a single point measurement of the wind speed at hub height. However, the measurement accuracy is much higher for the ideal rotor disk average wind speed measurements than for any realistic lidar scenario. Here, the only source of measurement error is wind evolution due to the 50 m preview distance between the measurement plane and the rotor disk.

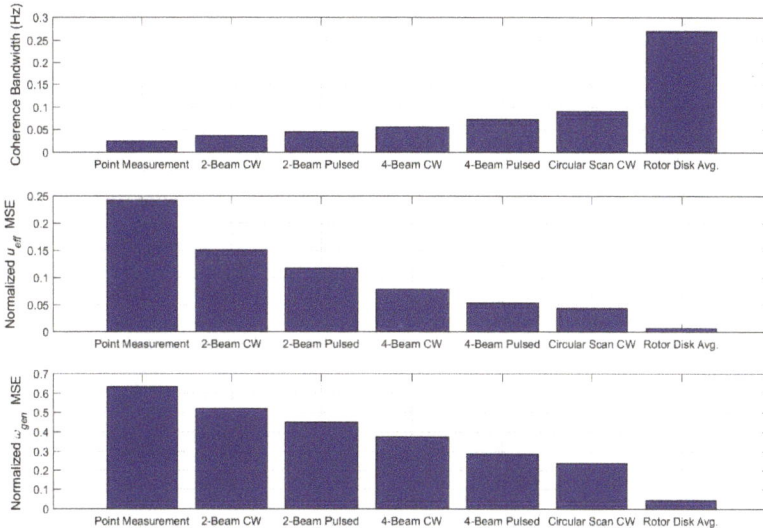

Figure 5. Maximum achievable coherence bandwidth and minimum achievable rotor effective wind speed measurement MSE (normalized by rotor effective wind speed variance) and linearized generator speed MSE (normalized by generator speed MSE with baseline feedback control) corresponding to the optimized scan parameters for each scan pattern category.

The optimal coherence bandwidth, measurement MSE, and generator speed MSE values shown in Figure 5 for the seven lidar measurement scenarios are listed in Table 1, along with the scan parameters required to achieve them. For the two pulsed lidar scenarios, the scan radius r, preview distance d, and focus distance F indicate the values corresponding to the farthest range gate measured. Table 1 reveals that the optimal scan parameters depend on the measurement quality metric that is being optimized. Note that the optimal scan parameters for maximizing coherence bandwidth and minimizing generator speed MSE are almost always the same. However, the optimal scan radius and preview distance values that minimize measurement MSE are typically greater than they are for the other two metrics.

The reasons why the optimal scan parameters that maximize coherence bandwidth or minimize generator speed MSE are shorter than for minimizing measurement MSE are related to the frequencies that are weighted most heavily when calculating each metric (see Equations (10) and (11)). For example, due to the frequency response of the closed-loop transfer function $T_{\omega_{gen}u_{eff}}$, the generator speed MSE is most sensitive to measurement error in the frequency band near 0.05 Hz, while the measurement MSE is more sensitive to measurement error at lower frequencies, where more of the energy in the wind is concentrated. This explains why generator speed MSE is minimized (and coherence bandwidth is maximized) using shorter preview distances, which prevent the coherence at higher frequencies from decaying too much from wind evolution. When minimizing measurement MSE, on the other hand, it is more important to measure farther away, thereby reducing LOS errors, which affect the low frequencies as much as the high frequencies.

Table 1. Optimal coherence bandwidth, rotor effective wind speed measurement mean square error (MSE) (normalized by rotor effective wind speed variance), and linearized generator speed MSE (normalized by generator speed MSE with baseline feedback control) along with optimal scan parameters for each scan pattern. Optimal scan radii and preview distances are expressed in meters and as fractions of the rotor radius and preview distance, respectively.

Optimal Parameters for Maximizing Coherence Bandwidth					
Scan Pattern	Coherence Bandwidth (Hz)	r	d	θ (deg.)	F (m)
Point Measurement	0.025	N/A	50 m (0.4 D)	N/A	N/A
2-Beam CW	0.037	22.1 m (0.35 R)	88.2 m (0.7 D)	14.0	90.9
2-Beam Pulsed	0.046	44.1 m (0.7 R)	163.8 m (1.3 D)	15.1	169.6
4-Beam CW	0.056	31.5 m (0.5 R)	75.6 m (0.6 D)	22.6	81.9
4-Beam Pulsed	0.073	53.6 m (0.85 R)	138.6 m (1.1 D)	21.1	148.6
Circular Scan CW	0.091	37.8 m (0.6 R)	75.6 m (0.6 D)	26.6	84.5
Rotor Disk Avg.	0.27	N/A	50 m (0.4 D)	N/A	N/A

Optimal Parameters for Minimizing Rotor Effective Wind Speed Measurement MSE					
Scan Pattern	Normalized u_{eff} MSE	r	d	θ (deg.)	F (m)
Point Measurement	0.242	N/A	50 m (0.4 D)	N/A	N/A
2-Beam CW	0.151	25.2 m (0.4 R)	100.8 m (0.8 D)	14.0	103.9
2-Beam Pulsed	0.118	50.4 m (0.8 R)	201.6 m (1.6 D)	14.0	207.8
4-Beam CW	0.079	37.8 m (0.6 R)	100.8 m (0.8 D)	20.6	107.7
4-Beam Pulsed	0.054	59.9 m (0.95 R)	151.2 m (1.2 D)	21.6	162.6
Circular Scan CW	0.045	41 m (0.65 R)	88.2 m (0.7 D)	24.9	97.2
Rotor Disk Avg.	0.007	N/A	50 m (0.4 D)	N/A	N/A

Optimal Parameters for Minimizing Generator Speed MSE					
Scan Pattern	Normalized ω_{gen} MSE	r	d	θ (deg.)	F (m)
Point Measurement	0.63	N/A	50 m (0.4 D)	N/A	N/A
2-Beam CW	0.52	22.1 m (0.35 R)	75.6 m (0.6 D)	16.3	78.8
2-Beam Pulsed	0.45	44.1 m (0.7 R)	163.8 m (1.3 D)	15.1	169.6
4-Beam CW	0.38	31.5 m (0.5 R)	75.6 m (0.6 D)	22.6	81.9
4-Beam Pulsed	0.29	53.6 m (0.85 R)	138.6 m (1.1 D)	21.1	148.6
Circular Scan CW	0.23	37.8 m (0.6 R)	75.6 m (0.6 D)	26.6	84.5
Rotor Disk Avg.	0.046	N/A	50 m (0.4 D)	N/A	N/A

Measuring farther away from the turbine adds the benefit of more volume averaging along the lidar beam due to the wider range weighting function for CW lidars. Longer preview distances also permit larger scan radii to be achieved with smaller cone angles, allowing greater radial coverage of the rotor disk area while reducing LOS errors. The additional volume averaging and greater rotor disk coverage help approximate the rotor effective wind speed, further explaining why longer preview distances and larger scan radii are optimal for minimizing MSE, where additional wind evolution at high frequencies can be tolerated.

It should be noted that the optimal scan radius and preview distance parameters listed in Table 1 are specific to a 126 m-rotor diameter turbine. However, when expressed in non-dimensional units of rotor diameters and rotor radii, the optimal scan parameters roughly translate to different rotor sizes. Furthermore, the optimal scan parameters presented in this section are only valid for measurements of rotor effective wind speed; when measuring other wind variables such as shear or wind direction, the optimal parameters are typically different. For example, as discussed by Simley [45], measurements of horizontal and vertical wind shear are more accurate with larger scan radii. Additionally, as presented by Kragh et al. [35], measurements of wind direction tend to improve with larger cone angles. Finally, the presence of yaw misalignment or vertical inflow may change the optimal scan parameters, likely favoring shorter preview distances so that the measured wind is more representative of what hits the rotor.

While this section presented a method for optimizing lidar scan patterns for measurement accuracy, it should be noted that in practice there may be other objectives that are as important or even more important than minimizing measurement error. For example, the load reduction offered by the lidar-assisted controller is likely a more useful metric for comparing different lidar scan patterns, although it is harder to calculate using frequency domain techniques. Additionally, the extra preview time provided by longer preview distances may be useful when attempting to detect extreme wind events and take necessary actions to protect the turbine. Finally, measuring the wind at multiple range gates with a pulsed lidar offers the advantage of being able to track wind speeds as they travel towards the turbine as well as allowing measurements at different preview distances to be combined to improve the simultaneous estimation of wind shear and direction, as described by Raach et al. [55].

7. Time Domain Optimization

In the two previous sections, the question of how lidar scan patterns can be optimized for control has been addressed using a few different metrics. This section now deals with the question "how can the lidar data processing and feedforward controller be optimized to improve the performance of wind turbine control?". To provide a basic understanding of LAC, the section focuses on the following two aspects:

- **Filtering**: Measurement filtering is crucial. No filtering or insufficient filtering of uncorrelated frequencies will cause unnecessary control actions, which will likely result in increased loads. By filtering out correlated frequencies, however, possible benefits of LAC algorithms will not reach their full potential.
- **Timing**: Temporal information about the inflowing wind field is necessary for control purposes. Because this information is available ahead in time, the lidar-assisted control action needs to be synchronized with the disturbance impact on the rotor.

The remainder of the section is organized as follows: Section 7.1 provides the basic background to understand the main objectives of the section. Thereafter, the feedforward collective pitch controller is introduced followed by a short description of the filter used for the lidar wind preview signal in Section 7.2. Finally, Section 7.3 summarizes the results from the time domain optimization.

7.1. Combined Feedback and Feedforward Control

This section first describes the design of a collective pitch controller with a combined feedback and feedforward controller assuming perfect wind preview. The section is based on the work of Schlipf [43], where more details can be found. Wind turbine control is designed to deal with variations in the wind to aim for the two most desirable goals: increasing the energy yield and reducing structural loads. Traditional feedback controllers are only able to react to impacts of wind changes on the turbine dynamics after these impacts have already occurred. The collective pitch feedforward controller is a promising approach to improve the control performance significantly over conventional feedback controllers due to its load reduction potential, robustness, and simplicity of implementation. This approach is based on a reduced nonlinear model and adds a feedforward update to the baseline collective pitch controller. It is designed for the entire full load region and it is able to almost perfectly cancel out the impact of the wind disturbance on the rotor speed assuming perfect wind preview. It is combined with an adaptive filter when used either with simulated lidar measurements or in real applications.

7.1.1. Reduced Wind Turbine Model for Controller Design

For controller design, a full aero-elastic model is too complex due to the required iterative calculation of the aerodynamics. Here, a SLOW (Simplified Low Order Wind turbine) model based on [58] is used considering only the rotor motion (see Figure 6 (left)). The model consists of a reduced servo-elastic and aerodynamic module (see Figure 6 (center)).

In the servo-elastic part, the motion of the rotor speed ω is described by

$$J\dot{\omega} = M_{aero} - M_{gen}, \tag{12}$$

where M_{aero} is the aerodynamic torque and M_{gen} is the generator torque. Further, J is the sum of the moments of inertia about the rotation axis.

In the aerodynamic part, the aerodynamic torque acting on the rotor with radius R is

$$M_{aero} = \frac{1}{2}\rho\pi R^3 \frac{c_P(\lambda,\beta)}{\lambda} \text{ with } \lambda = \frac{\omega R}{u_{eff}}, \tag{13}$$

where ρ is the air density, λ the tip-speed ratio, and c_P is the power coefficient. A two-dimensional look-up table shown in Figure 6 (right) is used, which is calculated from steady-state simulations with the full simulation model of the NREL 5 MW turbine [37].

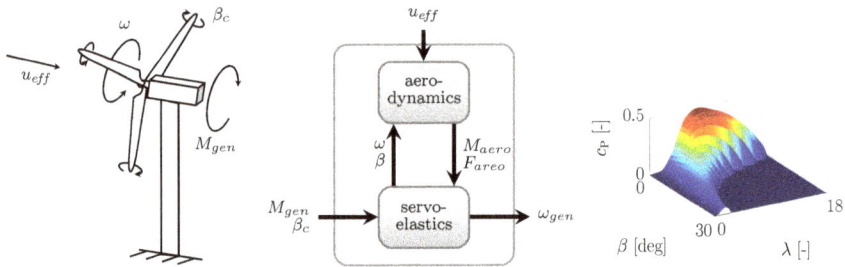

Figure 6. Degrees of freedom for the reduced nonlinear model (**left**). Control and disturbance inputs to the modules of the reduced model (**center**). Look-up tables of power coefficients (**right**).

7.1.2. Baseline Wind Turbine Controller

The baseline controller used in this paper is a simplified version of the NREL 5 MW controller described in [37] and combines an Indirect Speed Controller (ISC) and a collective blade pitch controller (CPC) (see Figure 7 (left)). The main control goal of the CPC is to maintain the rated rotational speed ω_{rated} in the presence of changing wind speed u_{eff} by adjusting the blade pitch angle β. The ISC additionally adjusts the generator torque M_{gen} to maintain constant electrical power above rated wind speed.

7.1.3. Collective Pitch Feedforward Controller

The basic idea of the "two-degrees-of-freedom-control" [59] is to complete the two main tasks of a controller (reference signal tracking and disturbance compensation) independently by a feedback and a feedforward controller, while the controllers should not obstruct each other. For example, it would be possible to calculate an update to the generator torque to cancel out wind speed changes and to maintain constant generator speed. However, this would yield changing electrical power and thus interfere with the control goal of the ISC.

Here, the feedforward controller is designed such that changes from the wind speed u_{eff} to the rotational speed ω are compensated by an additional blade pitch angle β_{FF}. In this case, the feedforward controller is not counteracting the control action of the CPC and ISC.

With the reduced nonlinear model introduced above, it can be shown that, if M_{gen} is used to control electrical power, then the aerodynamic torque needs to be held at its rated value to maintain constant rotor speed. This can be achieved for changing u_{eff} by adjusting the pitch angle using the static pitch curve $\beta_{ss}(u_{eff})$:

$$\beta_{FF} = \beta_{ss}(u_{eff}). \tag{14}$$

This curve is obtained by steady-state simulations (see Figure 7 (right)).

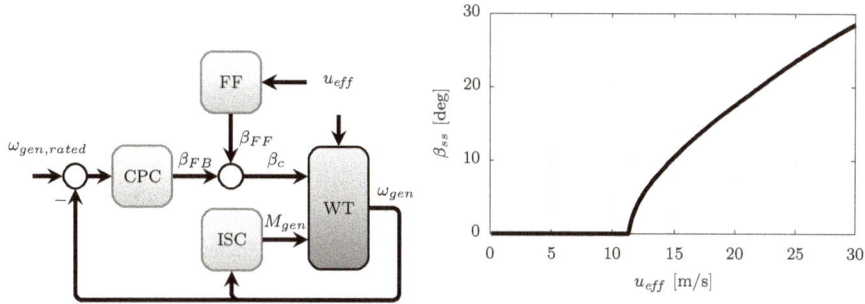

Figure 7. Feedback control loops and collective pitch feedforward control loop assuming perfect wind preview (**left**). Static pitch curve (**right**).

7.2. Design of an Adaptive Filter

As discussed in Sections 5 and 6, in reality, the rotor effective wind speed u_{eff} cannot be measured perfectly. While the lidar system measures in a three-dimensional wind field, it is only able to yield an estimate of the rotor effective wind speed $\hat{u}_{eff}$ due to limitations of the measurement principle. The wind field also evolves from the place of measurement to the rotor.

However, with an Adaptive Filter (AF), the lidar estimate $\hat{u}_{eff}$ can be matched in the best possible way to the rotor effective wind speed u_{eff}. The transfer function between the rotor effective wind speed measured by the lidar and that sensed by the turbine's rotor can be calculated by

$$G_{u_{eff}\hat{u}_{eff}}(f) = \frac{S_{u_{eff}\hat{u}_{eff}}(f)}{S_{\hat{u}_{eff}\hat{u}_{eff}}(f)}, \tag{15}$$

where $S_{u_{eff}\hat{u}_{eff}}(f)$ is the CPSD between both signals and $S_{\hat{u}_{eff}\hat{u}_{eff}}(f)$ is the PSD of the signal from the lidar. As introduced in Section 6 (see Equation (9)), a filter fitting to this transfer function will match the correlated part of $\hat{u}_{eff}$ to u_{eff} in an optimal minimum-MSE sense (see [42]). However, due to its simplicity in implementation, a linear low-pass filter can be fitted to the magnitude of the transfer function $G_{u_{eff}\hat{u}_{eff}}(f)$ for a given mean wind speed. Here, a first-order Butterworth filter with cutoff frequency f_{cutoff} is chosen. Because the correlation and thus the transfer function depends, among other factors, on the mean wind speed (e.g., see [8]), the cutoff frequency needs to be adjusted continuously. In this paper, constant correlation and mean wind speed is assumed.

In addition to the filtering, the timing of the feedforward control action is important. Due to the lidar measuring in front of the turbine, a certain preview time is obtained (depending mainly on the distance and the mean wind speed). A certain fraction of the preview will be consumed by the filtering. However, the feedforward control action β_{FF} needs to be synchronized with the rotor effective wind speed disturbing the system. Thus, an appropriate buffer time T_{buffer} needs to be found to buffer the filtered signal from the lidar.

The PSD of the rotor speed is a useful measure to evaluate the benefit of the combined feedback-feedforward controller over the conventional feedback controller. As described by Simley and Pao [42], the PSD for the feedback-only case can be calculated by multiplying the spectrum of the rotor effective wind speed $S_{u_{eff}u_{eff}}$ with the magnitude of the closed-loop transfer function $T_{\omega_{gen}u_{eff}}$ (introduced in Section 6.1):

$$S_{\omega,FB}(f) = |T_{\omega_{gen}u_{eff}}(f)|^2 S_{u_{eff}u_{eff}}(f).$$ (16)

As introduced in Equation (11), the PSD for the combined feedback-feedforward case with optimal filtering is determined by:

$$S_{\omega,FBFF}(f) = S_{\omega,FB}(f)(1 - \gamma^2_{u_{eff}\hat{u}_{eff}}(f)).$$ (17)

Thus, the measurement coherence $\gamma^2_{u_{eff}\hat{u}_{eff}}(f)$ has a direct impact on the optimal spectrum of the controlled variable. This spectrum can be used to evaluate the implementation of a feedforward controller.

7.3. Time Domain Optimization Results

Based on the PSDs and CPSD discussed in Section 6, wind time series with a length of 8192 s for the rotor effective wind speed u_{eff} and its lidar estimate $\hat{u}_{eff}$, using the optimal circularly-scanning CW lidar configuration for minimizing measurement MSE presented in Table 1, are generated. Next, simulations are carried out with the wind time series, the combined feedback-feedforward controller, as well as the wind turbine model described above. Here, 3×9 simulations are performed with nine different buffer times T_{buffer} and three different cutoff frequencies f_{cutoff}. For each simulation, the standard deviation of the rotor speed is calculated (see Figure 8).

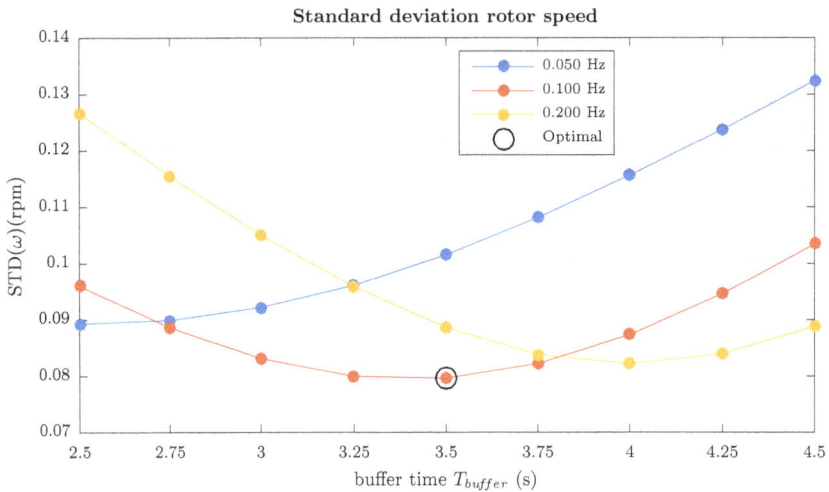

Figure 8. Standard deviation of rotor speed for different buffer times and cutoff frequencies using optimal circularly-scanning continuous wave (CW) lidar measurements.

In this case, the lowest value of 0.0796 RPM and thus the optimal result is achieved with a buffer time of 3.5 s and a cutoff frequency of 0.1 Hz. The standard deviation in the feedback-only case is 0.2078 RPM and, thus, an improvement of over 60% is obtained with LAC.

Finally, Figure 9 (left) compares the transfer function calculated using the frequency domain model described in Section 6 and the simplified filter used in the simulations. This confirms that the best filter in practice is close to the transfer function. Figure 9 (right) compares the PSDs of the rotor speed estimated from both the feedback-only and combined feedback-feedforward simulations to the theoretical values calculated using frequency domain methods presented in Section 6. The estimated spectra from the simulations fit well to the model. This confirms that the feedforward controller combined with the simplified filter is close to the optimal configuration predicted in Section 6.

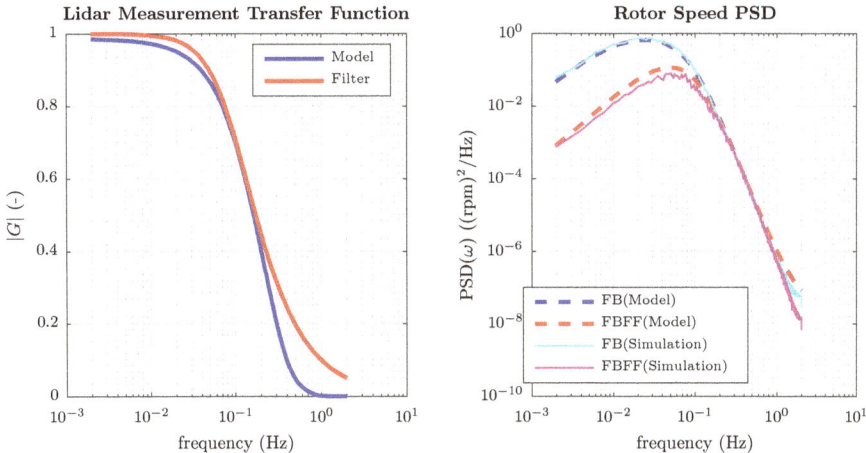

Figure 9. (**Left**) Transfer function from frequency domain model and fitted filter using optimal circularly-scanning CW lidar scan. (**Right**) Power spectral density of rotor speed for feedback-only (FB) and combined feedback-feedforward (FBFF) control from frequency domain model as well as simulations using optimal lidar scan.

8. Technology Outlook

There is a variety of technology that has the potential to influence the future development of LAC applications positively. Although the establishment is pushed forward with this work, there are various other promising control approaches, many of which are collected in Section 2. As a first step, these have to be implemented on multi-megawatt wind turbines in the field and long-term investigations have to be carried out. Another advantageous addition to conventional feedback controllers and CPC feedforward controllers is given by a multivariable approach where the transition between partial and full load operation is improved in a straightforward manner [27]. Further load reduction potential lies within the field of extreme event detection. A simulation environment has been developed that allows the modeling of realistic extreme events, aiding the design of controllers that can transition the turbine into a safety mode [60]. In addition to individual wind turbine control applications, preview measurements of the wind provided by lidar are beginning to be explored for measuring wakes as part of wind plant control strategies [61].

Wind field reconstruction methods are being improved by (1) applying nonlinear optimization methods for estimating shear and direction simultaneously [55] and (2) integrating lidar measurements into state estimation algorithms based on fluid dynamics models, allowing the individual components of the wind to be estimated at locations not limited to the measurement points, thereby partially overcoming LOS limitations and wind evolution [62]. Additionally, in the near future, it will be possible to study wind evolution phenomena by combining lidar measurements at several range gates with high resolution wind measurements performed by a swarm of unmanned aerial vehicles (UAVs) [63].

Finally, the primary question explored in this paper arises: "how should lidar hardware be designed in the future to fit control purposes optimally?". As outlined in the previous sections, there is no universal answer. Several boundary conditions like the control task (yaw, CPC, IPC, etc.), the turbine size, and the site characteristics, to list a few, determine optimal scanning angles and preview distances, etc. As a logical consequence, there is no "one" lidar for all LAC applications. In fact, one key to keeping a sustainable market position is to offer customizable lidar systems optimized for special purposes. This can even lead to varying designs for different turbines belonging to the same OEM. Of course, this does not imply developing dozens of lidar systems in parallel. A modularization approach could overcome the need for many different lidar configurations, by placing lidars in the context of a platform architecture. Starting from a common base, customer-specific modules could be assembled to create an entire system. Another advantage of modularization is that submodules could be optimized individually, potentially through outsourcing to specialized companies, which could increase reliability and save costs at the same time. In addition, maintenance services would likely be simplified. However, pointing out this scenario does not mean that every lidar should be trimmed to fulfill a specific task. On the contrary, there is still a need for more complex high-tech lidar systems in the future. They are typically more costly and over-designed for certain applications, but the flexibility to address many different questions with a single device can compensate for these disadvantages. Especially for the research community, this will continue to be of inestimable value.

9. Conclusions

The IEA Wind Task 32 workshop "Optimizing Lidar Design for Wind Turbine Control Applications" made it clear that there is no single metric that can determine whether a lidar is optimal for control purposes. Indeed, workshop participants had different views of what it means to optimize a lidar for control purposes. Instead, several objectives must be achieved to arrive at an optimal lidar for control purposes, including availability, reliability, measurement accuracy, and system cost. A review of the literature on lidar optimization for control applications revealed a large body of work on the optimization of lidar scan patterns via the resulting controller performance, time domain assessments of measurement accuracy, and direct frequency domain calculations of measurement coherence and measurement error. More practical aspects of lidar optimization have been treated as well, including the determination of availability requirements through feedforward control simulations as well as analyses of how the lidar beam angle, mounting position on the nacelle, and sampling time affect measurement availability.

Presentations and group discussions at the workshop, augmented by additional input from the authors, reveal a number of practical barriers preventing the widespread use of lidars for control. Some of the main barriers related to the lidar system are:

- High lidar measurement availability is necessary to maximize the value creation of lidar-assisted control (LAC). However, due to weather and atmospheric conditions, 100% availability cannot be achieved.
- Lidar reliability is seen as a potential problem. Again, the value creation depends heavily on the ability of the lidar to function properly when needed.
- Difficulty in assessing the cost–benefit relationship for LAC.
- The lack of common guidelines for assessing lidar applicability to LAC.
- The need to bring theoretical knowledge from the research community and practical requirements from industry into accordance.

Proposed solutions to these barriers include:

- Lidars that can adapt to changing measurement quality to increase availability. For example, a lidar could switch to shorter measurement ranges when poor atmospheric conditions make measuring at the ideal preview distance impossible.
- Reducing the amount of lidar downtime due to reliability issues by developing a regular maintenance plan involving modular lidar components.
- Using a systems engineering approach to improve the understanding of the cost–benefit relationship of LAC, where the complexity of the lidar system can be optimized.

- The creation of common documents providing guidelines for evaluating LAC performance and defining variables such as availability. The recent IEA Wind Task 32 workshop"Certification of Lidar-Assisted Control Applications" served as a first step in this direction.
- Perhaps most importantly, further collaboration between the research community and industry is needed to ensure that common goals are being targeted and the latest methods are being employed.

Scan pattern optimization was analyzed from a theoretical perspective by using frequency domain methods from the literature to directly calculate mean square rotor effective wind speed measurement error and measurement coherence bandwidth for various continuous wave and pulsed lidar scenarios. The models were extended in this paper to include the direct calculation of mean square generator speed error for a simple feedforward control scenario. This paper only considered scan pattern optimization for rotor effective wind speed. In general, the optimal configurations for measuring shear or wind direction are different. The main conclusions from the frequency domain optimization are:

- In general, measurement locations should approximate a rotor disk average. Line-of-sight errors penalize large cone angles and errors from wind evolution penalize long preview distances.
- As the number of beams increases, measurement/generator speed error decreases.
- For a fixed number of beams, measurement/generator speed error is lower with a pulsed lidar, measuring at multiple range gates.
- Of all the realistic scan patterns investigated, the circularly-scanning continuous wave lidar yields the lowest measurement/generator speed error. For minimizing generator speed error, the optimal scan parameters for this scan pattern consist of a scan radius of 60% of the rotor radius and preview distance of 60% of the rotor diameter of the 126 m-rotor diameter NREL 5 MW reference turbine (resulting in a focus distance of 84.5 m and a cone angle of 26.6°).
- The optimal scan parameters for the 4-beam pulsed lidar configuration, combining measurements from multiple ranges, consist of a scan radius of 85% rotor radius and preview distance of 110% rotor diameter for the farthest range gate, yielding a cone angle of 21.1°.

It should be noted, however, that, while minimizing measurement error is an important objective, it is not the only objective that should be considered. For example, a lidar with a suboptimal scan pattern but very high availability and reliability may be more useful than a lidar with an optimal scan pattern but lower availability. This further highlights the need for a systems engineering approach to lidar optimization for control applications.

The steps involved in the design of a feedforward pitch controller for rotor speed regulation were described, illustrating how lidar measurements should be filtered and delayed properly to maximize the controller effectiveness. Through time domain simulations, it was shown that the rotor speed regulation achieved by the controller matches the value predicted by the theoretical frequency domain model.

Finally, there are many promising directions that lidars for control applications could take in the future. For example, the design of LAC scenarios for extreme event detection is gaining importance. Additionally, new approaches to wind field estimation, such as the coupling of raw lidar measurements with fluid dynamics models, are being explored. Finally, it is anticipated that a modular lidar design philosophy will be adopted to address the variety of lidar system requirements for different control applications.

Author Contributions: E.S. organized the writing of this paper and wrote the Abstract, Introduction, Sections 2, 5 and 6, as well as the Conclusions. H.F. wrote Sections 3 and 4, in part using information from the Workshop meeting minutes prepared by F.H., as well as Section 8. Section 7 was written jointly by D.S., H.F. and F.H. The authors all contributed to the planning, execution, and followup activities of the IEA Wind Task 32 Workshop "Optimizing Lidar Design for Wind Turbine Control Applications," which was led by E.S. E.S. organized the first exercise of the workshop, on which much of Section 6 is based. D.S. and H.F. prepared and led the second exercise of the workshop, from which much of the material in Section 7 is drawn.

Funding: The work performed by Holger Fürst and Florian Haizmann was partially funded by the German Federal Ministry for Economic Affairs and Energy (BMWi) in the framework of the German joint research project

ANWIND. David Schlipf acknowledges support from IEA Wind Task 32. Additionally, IEA Wind Task 32 is acknowledged for supporting the workshop "Optimizing Lidar Design for Wind Turbine Control Applications" in Boston, MA (July 2016), which served as the foundation for this work. Finally, the publication of this work was supported by Envision Energy USA, Ltd.

Acknowledgments: The authors thank the participants of IEA Wind Task 32 Workshop "Optimizing Lidar Design for Wind Turbine Control Applications" for their contributions to the workshop, including presentations and input during group discussions, which were incorporated into this paper. Thirty-three participants attended the workshop, representing lidar suppliers, wind turbine manufacturers, academia, and the research community.

Conflicts of Interest: The authors declare no conflict of interest.

References

1. Scholbrock, A.; Fleming, P.; Schlipf, D.; Wright, A.; Johnson, K.; Wang, N. Lidar-Enhanced Wind Turbine Control: Past, Present, and Future. In Proceedings of the American Control Conference, Boston, MA, USA, 6–8 July 2016.
2. Harris, M.; Bryce, D.; Coffey, A.; Smith, D.; Birkemeyer, J.; Knopf, U. Advance Measurements of Gusts by Laser Anemometry. *Wind Eng. Ind. Aerodyn.* **2007**, *95*, 1637–1647. [CrossRef]
3. Harris, M.; Hand, M.; Wright, A. *Lidar for Turbine Control*; Technical Report, NREL/TP-500-39154; National Renewable Energy Laboratory: Golden, CO, USA, 2006.
4. Medley, J.; Barker, W.; Harris, M.; Pitter, M.; Slinger, C.; Mikkelsen, T.; Sjöholm, M. Evaluation of Wind Flow with a Nacelle-Mounted, Continuous Wave Wind Lidar. In Proceedings of the European Wind Energy Association Annual Event, Barcelona, Spain, 10–13 March 2014.
5. Mikkelsen, T.; Angelou, N.; Hansen, K.; Sjöholm, M.; Harris, M.; Slinger, C.; Hadley, P.; Scullion, R.; Ellis, G.; Vives, G. A Spinner-Integrated Wind Lidar for Enhanced Wind Turbine Control. *Wind Energy* **2013**, *16*, 625–643. [CrossRef]
6. Sjöholm, M.; Pedersen, A.T.; Angelou, N.; Abari, F.F.; Mikkelsen, T.; Harris, M.; Slinger, C.; Kapp, S. Full Two-Dimensional Rotor Plane Inflow Measurements by a Spinner-Integrated Wind Lidar. In Proceedings of the European Wind Energy Association Annual Event, Vienna, Austria, 4–7 February 2013.
7. Windar Photonics. Available online: http://www.windarphotonics.com/ (accessed on 19 April 2018).
8. Schlipf, D.; Fleming, P.; Haizmann, F.; Scholbrock, A.K.; Hofsäß, M.; Wright, A.; Cheng, P.W. Field Testing of Feedforward Collective Pitch Control on the CART2 Using a Nacelle-Based Lidar Scanner. In Proceedings of the Science of Making Torque from Wind, Oldenburg, Germany, 9–11 October 2012.
9. Kumar, A.; Bossanyi, E.; Scholbrock, A.; Fleming, P.; Boquet, M.; Krishnamurthy, R. Field Testing of LIDAR Assisted Feedforward Control Algorithms for Improved Speed Control and Fatigue Load Reduction on a 600 kW Wind Turbine. In Proceedings of the European Wind Energy Association Annual Event, Paris, France, 17–20 November 2015.
10. Borraccino, A.; Courtney, M. *Calibration Report for Avent 5-Beam Demonstrator Lidar*; Technical Report, DTU Wind Energy E-0087; DTU Wind Energy: Roskilde, Denmark, 2016.
11. Fleming, P.A.; Scholbrock, A.K.; Jehu, A.; Davoust, S.; Osler, E.; Wright, A.D.; Clifton, A. Field-Test Results using a Nacelle-Mounted Lidar for Improving Wind Turbine Power Capture by Reducing Yaw Misalignment. In Proceedings of the Science of Making Torque from Wind, Lyngby, Denmark, 17–20 June 2014.
12. Scholbrock, A.; Fleming, P.; Wright, A.; Slinger, C.; Medley, J.; Harris, M. Field Test Results from Lidar Measured Yaw Control for Improved Yaw Alignment with the NREL Controls Advanced Research Turbine. In Proceedings of the AIAA Aerospace Sciences Meeting, Kissimmee, FL, USA, 5–9 January 2015.
13. Bossanyi, E.A.; Kumar, A.; Hugues-Salas, O. Wind Turbine Control Applications of Turbine-Mounted Lidar. In Proceedings of the Science of Making Torque from Wind, Oldenburg, Germany, 9–11 October 2012.
14. Schlipf, D.; Kapp, S.; Anger, J.; Bischoff, O.; Hofsäß, M.; Rettenmeier, A.; Smolka, U.; Kühn, M. Prospects of Optimization of Energy Production by LiDAR Assisted Control of Wind Turbines. In Proceedings of the European Wind Energy Association Annual Event, Brussels, Belgium, 14–17 March 2011.
15. Wang, N.; Johnson, K.; Wright, A. Combined Feedforward and Feedback Controllers for Turbine Power Capture Enhancement and Fatigue Loads Mitigation with Pulsed Lidar. In Proceedings of the AIAA Aerospace Sciences Meeting, Nashville, TN, USA, 9–12 January 2012.

16. Schlipf, D.; Fleming, P.; Kapp, S.; Scholbrock, A.; Haizmann, F.; Belen, F.; Wright, A.; Cheng, P.W. Direct Speed Control Using LIDAR and Turbine Data. In Proceedings of the American Control Conference, Washington, DC, USA, 17–19 June 2013.

17. Schlipf, D.; Kühn, M. Prospects of a Collective Pitch Control by Means of Predictive Disturbance Compensation Assisted by Wind Speed Measurements. In Proceedings of the German Wind Energy Conference (DEWEK), Bremen, Germany, 26–27 November 2008.

18. Schlipf, D.; Schlipf, D.J.; Kühn, M. Nonlinear Model Predictive Control of Wind Turbines using LIDAR. *Wind Energy* **2013**, *16*, 1107–1129. [CrossRef]

19. Schlipf, D.; Schuler, S.; Grau, P.; Allgöwer, F.; Kühn, M. Look-Ahead Cyclic Pitch Control using LIDAR. In Proceedings of the Science of Making Torque from Wind, Heraklion, Greece, 28–30 June 2010.

20. Laks, J.; Pao, L.; Wright, A.; Kelley, N.; Jonkman, B. The Use of Preview Wind Measurements for Blade Pitch Control. *IFAC J. Mechatron.* **2011**, *21*, 668–681. [CrossRef]

21. Dunne, F.; Schlipf, D.; Pao, L.Y.; Wright, A.D.; Jonkman, B.; Kelley, N.; Simley, E. Comparison of Two Independent Lidar-Based Pitch Control Designs. In Proceedings of the AIAA Aerospace Sciences Meeting, Nashville, TN, USA, 9–12 January 2012.

22. Kragh, K.A.; Hansen, M.H.; Henriksen, L.C. Sensor Comparison Study for Load Alleviating Wind Turbine Pitch Control. *Wind Energy* **2014**, *17*, 1891–1904. [CrossRef]

23. Wortmann, S.; Geisler, J.; Konigorski, U. Lidar-Assisted Feedforward Individual Pitch Control to Compensate Wind Shear and Yawed Inflow. In Proceedings of the Science of Making Torque from Wind, Munich, Germany, 5–7 October 2016.

24. Aho, J.; Pao, L.; Hauser, J. Optimal Trajectory Tracking Control for Wind Turbines during Operating Region Transitions. In Proceedings of the American Control Conference, Washington, DC, USA, 17–19 June 2013.

25. Mirzaei, M.; Soltani, M.; Poulsen, N.K.; Niemann, H.H. Model Predictive Control of Wind Turbines using Uncertain Lidar Measurements. In Proceedings of the American Control Conference, Washington, DC, USA, 17–19 June 2013.

26. Bottasso, C.L.; Pizzinelli, P.; Riboldi, C.E.D.; Tasca, L. LiDAR-Enabled Model Predictive Control of Wind Turbines with Real-Time Capabilities. *Renew. Energy* **2014**, *71*, 442–452. [CrossRef]

27. Schlipf, D. Prospects of Multivariable Feedforward Control of Wind Turbines Using Lidar. In Proceedings of the American Control Conference, Boston, MA, USA, 6–8 July 2016.

28. Laks, J.; Simley, E.; Pao, L.Y. A Spectral Model for Evaluating the Effect of Wind Evolution on Wind Turbine Preview Control. In Proceedings of the American Control Conference, Washington, DC, USA, 17–19 June 2013.

29. Dunne, F.; Pao, L.Y. Optimal Blade Pitch Control with Realistic Preview Wind Measurements. *Wind Energy* **2016**, *19*, 2153–2169. [CrossRef]

30. Scholbrock, A.; Fleming, P.; Fingersh, L.; Wright, A.; Schlipf, D.; Haizmann, F.; Belen, F. Field Testing LIDAR-Based Feed-Forward Controls on the NREL Controls Advanced Research Turbine. In Proceedings of the AIAA Aerospace Sciences Meeting, Grapevine, TX, USA, 7–10 January 2013.

31. Koerber, A.; Mehendale, C. Lidar Assisted Turbine Control... An Industrial Perspective. In Proceedings of the American Wind Energy Association WINDPOWER Conference, Chicago, IL, USA, 5–8 May 2013.

32. Taylor, G. The Spectrum of Turbulence. *Proc. R. Soc. Lond. Ser. A Math. Phys. Sci.* **1938**, *164*, 476–490. [CrossRef]

33. Kristensen, L. On Longitudinal Spectral Coherence. *Bound.-Layer Meteorol.* **1979**, *16*, 145–153. [CrossRef]

34. Bossanyi, E.A. Un-Freezing the Wind: Improved Wind Field Modelling for Investigating Lidar-Assisted Wind Turbine Control. In Proceedings of the European Wind Energy Association Annual Event, Copenhagen, Denmark, 16–19 April 2012.

35. Kragh, K.A.; Hansen, M.H.; Mikkelsen, T. Precision and Shortcomings of Yaw Error Estimation using Spinner-Based Light Detection and Ranging. *Wind Energy* **2013**, *16*, 353–366. [CrossRef]

36. Simley, E.; Pao, L.Y.; Frehlich, R.; Jonkman, B.; Kelley, N. Analysis of Light Detection and Ranging Wind Speed Measurements for Wind Turbine Control. *Wind Energy* **2014**, *17*, 413–433. [CrossRef]

37. Jonkman, J.; Butterfield, S.; Musial, W.; Scott, G. *Definition of a 5-MW Reference Wind Turbine for Offshore System Development*; Technical Report, NREL/TP-500-38060; National Renewable Energy Laboratory: Golden, CO, USA, 2009.

38. Simley, E.; Pao, L.Y.; Gebraad, P.; Churchfield, M. Investigation of the Impact of the Upstream Induction Zone on LIDAR Measurement Accuracy for Wind Turbine Control Applications using Large-Eddy Simulation. In Proceedings of the Science of Making Torque from Wind, Lyngby, Denmark, 17–20 June 2014.

39. Schlipf, D.; Mann, J.; Cheng, P.W. Model of the Correlation between Lidar Systems and Wind Turbines for Lidar-Assisted Control. *J. Atmos. Ocean. Technol.* **2013**, *30*, 2233–2240. [CrossRef]

40. Schlipf, D.; Haizmann, F.; Cosack, N.; Siebers, T.; Cheng, P. Detection of Wind Evolution and Lidar Trajectory Optimization for Lidar-Assisted Wind Turbine Control. *Meteorol. Z.* **2015**, *24*, 565–579. [CrossRef]

41. Pielke, R.A.; Panofsky, H.A. Turbulence Characteristics along Several Towers. *Bound.-Layer Meteorol.* **1970**, *1*, 115–130. [CrossRef]

42. Simley, E.; Pao, L.Y. Reducing LIDAR Wind Speed Measurement Error with Optimal Filtering. In Proceedings of the American Control Conference, Washington, DC, USA, 17–19 June 2013.

43. Schlipf, D. Lidar-Assisted Control Concepts for Wind Turbines. Ph.D. Thesis, University of Stuttgart, Stuttgart, Germany, 2016.

44. Simley, E.; Pao, L.Y. Correlation between Rotating LIDAR Measurements and Blade Effective Wind Speed. In Proceedings of the AIAA Aerospace Sciences Meeting, Grapevine, TX, USA, 7–10 January 2013.

45. Simley, E. Wind Speed Preview Measurement and Estimation for Feedforward Control of Wind Turbines. Ph.D. Thesis, University of Colorado at Boulder, Boulder, CO, USA, 2015.

46. Simley, E.; Pao, L.Y. A Longitudinal Spatial Coherence Model for Wind Evolution based on Large-Eddy Simulation. In Proceedings of the American Control Conference, Chicago, IL, USA, 1–3 July 2015.

47. Davoust, S.; Mashtare, D.; Markham, T.; Shane, C.; Stinson, K.; Velociter, T.; Krishna Murthy, R. Evaluation of LiDAR Performance for Practical Turbine Control Implementation. In Proceedings of the European Wind Energy Association Annual Event, Paris, France, 17–20 November 2015.

48. Davoust, S.; Jehu, A.; Bouillet, M.; Bardon, M.; Vercherin, B.; Scholbrock, A.; Fleming, P.; Wright, A. Assessment and Optimization of Lidar Measurement Availability for Wind Turbine Control. In Proceedings of the European Wind Energy Association Annual Event, Barcelona, Spain, 10–13 March 2014.

49. Dykes, K.; Ning, A.; King, R.; Graf, P.; Scott, G.; Veers, P. Sensitivity Analysis of Wind Plant Performance to Key Turbine Design Parameters: A Systems Engineering Approach. In Proceedings of the 32nd ASME Wind Energy Symposium National Harbor, MD, USA, 13–17 January 2014. doi:10.2514/6.2014-1087. [CrossRef]

50. Schlipf, D.; Raach, S. Turbulent Extreme Event Simulations for Lidar-Assisted Wind Turbine Control. *J. Phys. Conf. Ser.* **2016**, *753*, 052011. [CrossRef]

51. Dunne, F.; Pao, L.Y.; Schlipf, D.; Scholbrock, A.K. Importance of Lidar Measurement Timing Accuracy for Wind Turbine Control. In Proceedings of the American Control Conference, Portland, OR, USA, 4–6 June 2014.

52. Pitter, M.; Slinger, C.; Harris, M. *Introduction to Continuous-Wave Doppler LIDAR, Chapter 4 in Remote Sensing for Wind Energy*; Technical Report, DTU Wind Energy-E-Report-0029(EN); DTU Wind Energy: Roskilde, Denmark, 2013.

53. Mikkelsen, T. On Mean Wind and Turbulence Profile Measurements from Ground-Based Wind Lidars: Limitations in Time and Space Resolution with Continuous Wave and Pulsed Lidar Systems—A Review. In Proceedings of the European Wind Energy Conference, Stockholm, Sweden, 16–19 March 2009.

54. Cariou, J.P. *Pulsed Lidars, Chapter 5 in Remote Sensing for Wind Energy*; Technical Report, DTU Wind Energy-E-Report-0029(EN); DTU Wind Energy: Roskilde, Denmark, 2013.

55. Raach, S.; Schlipf, D.; Haizmann, F.; Cheng, P.W. Three Dimensional Dynamic Model Based Wind Field Reconstruction from Lidar Data. In Proceedings of the Science of Making Torque from Wind, Lyngby, Denmark, 17–20 June 2014.

56. IEC 61400-1 *"Wind Turbines-Part 1: Design Requirements"*, 3rd ed.; Technical Report; International Electrotechnical Commission: Geneva, Switzerland, 2005.

57. Hansen, M.; Hansen, A.; Larsen, T.; Øye, S.; Sørensen, P.; Fuglsang, P. *Control Design for a Pitch-Regulated, Variable-Speed Wind Turbine*; Technical Report, Risø-R-1500(EN); Risø National Laboratory: Roskilde, Denmark, 2005.

58. Bottasso, C.L.; Croce, A.; Savini, B.; Sirchi, W.; Trainelli, L. Aero-Servo-Elastic Modelling and Control of Wind Turbines using Finite-Element Multibody Procedures. *Multibody Syst. Dyn.* **2006**, *16*, 291–308. [CrossRef]

59. Horowitz, I.M. *Synthesis of Feedback Systems*; Academic Press Inc.: New York, NY, USA, 1963.

60. Hagemann, T.; Haizmann, F.; Schlipf, D.; Cheng, P.W. Realistic simulations of extreme load cases with lidar-based feedforward control. In Proceedings of the German Wind Energy Conference (DEWEK), Bremen, Germany, 17–18 October 2017.
61. Raach, S.; Schlipf, D.; Borisade, F.; Cheng, P.W. Wake Redirecting using Feedback Control to Improve the Power Output of Wind Farms. In Proceedings of the American Control Conference, Boston, MA, USA, 6–8 July 2016.
62. Towers, P.; Jones, B.L. Real-time Wind Feld Reconstruction from LiDAR Measurements using a Dynamic Wind Model and State Estimation. *Wind Energy* **2016**, *19*, 133–150. [CrossRef]
63. Molter, C.; Cheng, P.W. Optimal Placement of an Airflow Probe at a Multirotor UAV for Airborne Wind Measurements. In Proceedings of the European Rotorcraft Forum (ERF), Milan, Italy, 12–15 September 2017.

remote sensing

MDPI

Article

Wind Turbine Wake Characterization with Nacelle-Mounted Wind Lidars for Analytical Wake Model Validation

Fernando Carbajo Fuertes [1], Corey D. Markfort [2,3] and Fernando Porté-Agel [1,*]

[1] Wind Engineering and Renewable Energy laboratory (WiRE)-École polytechnique fedérale de Lausanne (EPFL), Lausanne 1015, Switzerland; fernando.carbajo@epfl.ch

[2] IIHR-Hydroscience & Engineering, The University of Iowa, Iowa City, IA 52242, USA; corey-markfort@uiowa.edu

[3] Civil and Environmental Engineering, The University of Iowa, Iowa City, IA 52242, USA

* Correspondence: fernando.porte-agel@epfl.ch; Tel.: +41-(0)-21-693-61-38

Received: 31 March 2018; Accepted: 20 April 2018; Published: 25 April 2018

check for updates

Abstract: This study presents the setup, methodology and results from a measurement campaign dedicated to the characterization of full-scale wind turbine wakes under different inflow conditions. The measurements have been obtained from two pulsed scanning Doppler lidars mounted on the nacelle of a 2.5 MW wind turbine. The first lidar is upstream oriented and dedicated to the characterization of the inflow with a variety of scanning patterns, while the second one is downstream oriented and performs horizontal planar scans of the wake. The calculated velocity deficit profiles exhibit self-similarity in the far wake region and they can be fitted accurately to Gaussian functions. This allows for the study of the growth rate of the wake width and the recovery of the wind speed, as well as the extent of the near-wake region. The results show that a higher incoming turbulence intensity enhances the entrainment and flow mixing in the wake region, resulting in a shorter near-wake length, a faster growth rate of the wake width and a faster recovery of the velocity deficit. The relationships obtained are compared to analytical models for wind turbine wakes and allow to correct the parameters prescribed until now, which were obtained from wind-tunnel measurements and large-eddy simulations (LES), with new, more accurate values directly derived from full-scale experiments.

Keywords: wind energy; atmospheric boundary layer; wind turbine wake; wind lidar; turbulence; wake modeling; field experiments

1. Introduction

The wind flow around the rotating blades of a wind turbine creates aerodynamic forces that result in a torque on the rotor axis, which ultimately generates electrical energy, and an axial thrust force, which pushes back the rotor. Following Newton's third law, these actions are compensated with reactions on the wind flow, altering its characteristics within a volume downstream of the wind turbine that is called the wake region [1]. The reaction force of the thrust creates an axial induction opposite to the air motion direction which reduces the kinetic energy of the flow, causing a reduction in velocity. The reaction torque, instead, creates a tangential induction which causes the flow to spin in the opposite sense of the rotation of the blades. Since the reaction aerodynamic forces have a dynamic nature and they generate important shear locally in the flow, they result as well in increased levels of turbulence. A wind turbine wake has two main negative effects on surrounding wind turbines within its area of influence. First, the kinetic energy deficit results in a decrease in energy production [2,3], and second, the higher turbulence levels result in higher fatigue loads and a potential life time reduction [4].

The correct understanding, characterization, and accurate modeling of wind turbine wakes is of utmost importance for accurate power prediction of wind farms [5,6] as well as layout optimization [7,8]. Wind turbine wake models may also play a key role in the control of wind farms [9,10].

Wind turbine wake models can be analytical [11], numerical [12], empirical, or a mixture of them [13]. In all cases the wake models need to be validated with experimental data. Wind tunnel experiments present some advantages for validation purposes (e.g., repeatability, flow control, wind turbine control, wind farm layout, etc.) [14–20] but it is very challenging to ensure complete flow similarity for scaled tests [21]. Ideally, the validation would include a comparison of model prediction under different conditions of the atmospheric boundary layer (ABL) with measurements of full scale wind turbine wakes. The measurement technique best suited to measurements of the wake is the wind lidar (Light Detection and Ranging), which is a remote sensing measurement technique based on the Doppler effect of reflected laser light from aerosol. A pulsed wind lidar, in particular, is able to measure wind speed with relatively high spatial and temporal resolutions (around 20 m and 10 Hz can be easily achieved with state-of-the-art systems under normal atmospheric conditions) up to distances of a few kilometers.

Comparing full-scale measurements of wind turbine wakes and model predictions is particularly challenging and presents a number of difficulties, especially given the limited amount of data available during the experiments. The discussion presented by Barthelmie et al. [22] is particularly interesting and relevant for this manuscript. They address the issues related to correctly establishing the free stream flow characteristics (i.e., horizontal wind speed, wind direction, nacelle orientation and yaw misalignment, turbulence intensity and atmospheric stability), to the accuracy of the site specific power curve and thrust coefficients, and to ensuring equivalent time averaging in models and measurements. These difficulties often arise from two important sources, which are the horizontal inhomogeneity and the non-stationarity of the atmospheric flow (i.e., the horizontal gradients and the natural fluctuations in the wind speed and direction in any period).

Recently, an increasing number of studies have investigated wind turbine wakes either via planar or volumetric scans with ground-based scanning lidars [23–31] using different scanning strategies and post-processing algorithms. Nevertheless, nacelle-based lidar experiments [32–36] have inherent advantages when measuring wakes. Some of these advantages are: the lidar always has the same alignment with the rotor, this alignment is independent from the wind direction, the errors due to the assumption of unidirectional average flow are smaller, and it can perform horizontal planar scans of the wake.

The objective of this study is to present an experimental setup and data post processing methodology for the characterization of single wind turbine wakes under different atmospheric conditions based on two nacelle-mounted lidars. The first lidar is upstream-looking and it is dedicated to the characterization of the inflow conditions in terms of average wind speed, turbulence intensity, yaw and vertical wind shear. The second lidar is downstream-looking and executes horizontal plan position indicator (PPI) scans of the wake. This allows accurate measurements of the velocity deficit in the wake and is ideal for the comparison with the predictions from wake models.

2. Methodology

This section describes the characteristics of the site where the tests were performed, as well as the meteorological tower, the wind turbine, the lidar setup, and the methodology used to analyze the data.

2.1. Test Site

The selected test site is located at the Kirkwood Community College campus, in the state of Iowa. The wind turbine studied is a 2.5 MW Liberty C96 model, manufactured by Clipper Windpower. It is equipped with a Supervisory Control And Data Acquisition (SCADA) system which continuously collects data at 10 min intervals about the wind turbine operation. The main characteristics of the wind turbine are detailed in Table 1.

Table 1. Wind turbine main characteristics.

Clipper Windpower-Liberty C96	
Rated power	2.5 MW (at 15 m/s)
Rotor diameter (D)	96 m
Tower height	80 m
Minimum rotor speed	9.5 rpm
Maximum rotor speed	15.5 rpm
Cut-in wind speed	4 m/s
Cut-off wind speed	25 m/s

Figure 1 shows the power coefficient of the wind turbine and the blade pitch angle imposed by the control system as function of the incoming wind speed as registered by the SCADA system. This information helps determining the range of usable wind speeds for the analysis of the wake. The power coefficient C_P is quasi-constant and close to a value of 0.37 for velocities between 5 and 10 m/s. This is the range of velocities where the wind turbine operates optimally and therefore the aerodynamic forces are most important. Once the wind velocity reaches 10 m/s the control system changes the pitch angle of the blades, reducing the aerodynamic efficiency of the rotor and effectively decreasing the thrust forces. We consider that the thrust coefficient C_T is close to constant within the same range of wind speeds, which is common in most wind turbines, and limit the study of the wake to this range. In the absence of manufacturer's data, and given the similarity in terms of thrust coefficient from commercial wind turbines, a value of $C_T = 0.82$ is estimated for the 5–10 m/s wind speed range.

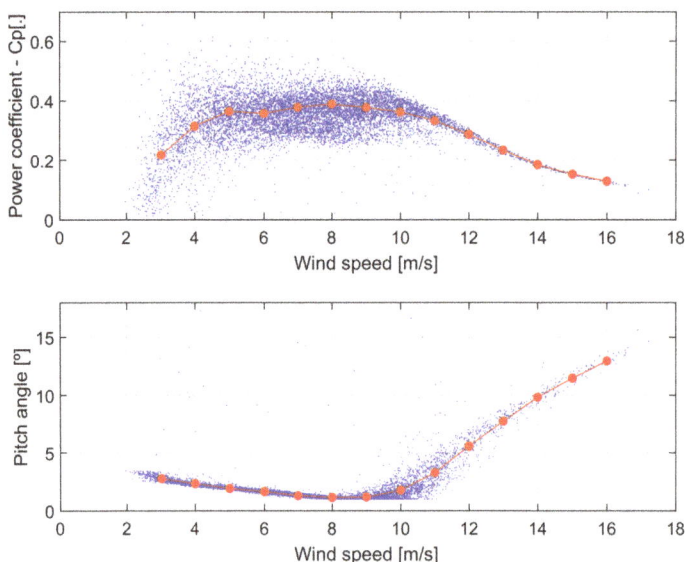

Figure 1. Power coefficient and blade pitch angle as a function of the wind speed for the 2.5 MW Liberty C96 wind turbine obtained from SCADA data. The 10 min values are shown in blue dots and binned averages in red. A quasi constant power coefficient is observed for wind speeds from 5 to 10 m/s.

Figure 2 shows a satellite image of the Kirkwood Community College campus and its surroundings with the location of the wind turbine and the meteorological tower. The predominant wind directions in the area are NW and SSE as shown in Figure 3. The wind turbine is situated at an elevation of 246 m and the surrounding terrain can be considered as rolling terrain. In an area of

3 km around the turbine the maximum elevation difference does not exceed 30 m and the terrain slope rarely exceeds 1%. The surface roughness, which plays a role in the ambient turbulence of the ABL, changes from higher roughness lengths associated to the suburban area of the city of Cedar Rapids (W to N directions) to lower values associated to agricultural fields (NE to SW directions). The campus is equipped with a 106 m tall meteorological tower situated at a distance of approximately 900 m from the wind turbine towards the SSW direction. The tower is equipped with sonic anemometers, cup anemometers and wind vanes—among other instruments—situated at heights of 10, 32, 80 and 106 m for the characterization of the ABL.

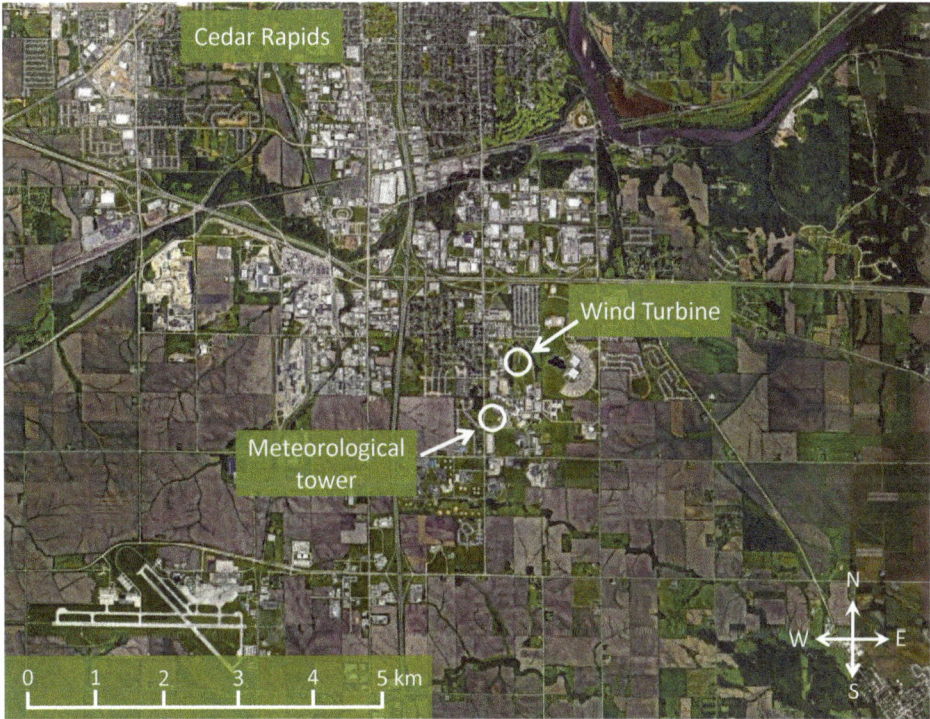

Figure 2. Location of the wind turbine and the meteorological tower inside Kirkwood's campus as well as the outskirts of Cedar Rapids. Map data: Google, Image NASA.

Figure 3 presents the wind rose for the last ten years obtained from measurements at the Eastern Iowa Airport, shown in the lower left corner of Figure 2. The airport is situated approximately 5 km to the SW of the test site. The wind presents two main directions: NW are affected by relatively high surface roughness of the suburban area of Cedar Rapids, while SSE are affected by the lower roughness of the agricultural fields.

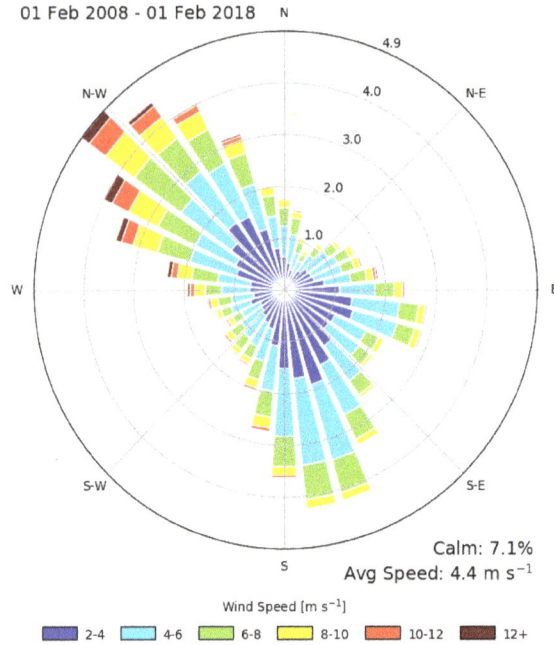

Figure 3. Wind rose measured at the Eastern Iowa Airport for 10 years, situated approximately 5 km to the SW of the wind turbine and visible in Figure 2. The wind presents two main directions: NW and SSE. Data obtained from Iowa State University, Iowa Environmental Mesonet. The length of the bars indicate frequency in percentage units.

2.2. Lidar Setup

The two lidar units used in the experiment are StreamLine models manufactured by Halo-Photonics. These instruments are infrared Doppler pulsed wind lidars which emit 1.5 μm wavelength pulses at a frequency of 10 kHz. They are scanning lidars, which means that the laser beam can be oriented towards any direction thanks to a steerable head. The units are able to provide measurements of the radial velocity with a resolution of 3.82 cm/s at intervals of 18 m along the laser beam direction or Line-of-Sight (LoS) and their measurement range extends from 63 m to more than 1000 m under most atmospheric conditions.

Both lidars are mounted on level platforms installed on the nacelle of the wind turbine. The first unit is an upstream-looking lidar dedicated to the characterization of the incoming flow conditions of the ABL and the second one is a downstream-looking lidar dedicated to the characterization of the wind turbine wake. A sketch of the lidar setup measurement configuration is presented in Figure 4. The forward-looking lidar laser beam is commonly blocked by the passage of the wind turbine blades and a quality check algorithm is implemented in order to filter out blocked measurements. Between 5 and 20% of the measurements are rejected under normal operating conditions, although this rarely compromises the analysis of the inflow data.

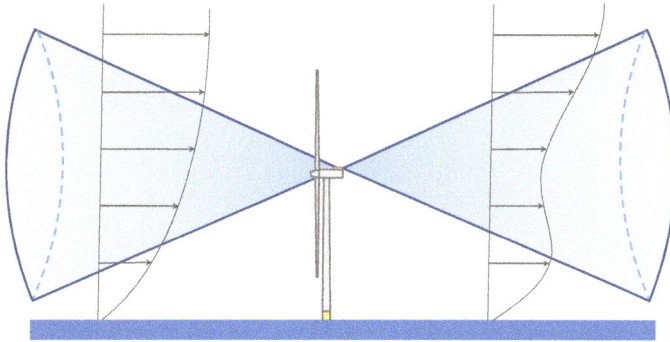

Figure 4. Sketch of the lidar setup with one lidar dedicated to the measurement of the incoming flow conditions and another one dedicated to the characterization of the wake.

2.3. Upstream Scanning

Both the inflow and the wake scans are grouped into synchronized periods of 30 min. The length of the periods is a compromise between the need for multiple samples in order to decrease the statistical error of the measurements and the requirement of stationarity of the flow. Each period of upstream scanning is divided into four successive scans that quantify different variables of the incoming flow. The calculations assume horizontally homogeneous flow in the region 250–600 m (2.6–6.25D) upstream of the rotor. Only measurements within this range are taken into consideration. Some of the measurements are redundant, such as the wind speed at hub height U_{hub}, yaw angle γ and longitudinal turbulence intensity TI_x, and they help to understand the degree of stationarity of the atmospheric conditions during the 30 min periods. The sub-indexes *ppi*, *rhi* and *st* indicate parameters obtained from Range Height Indicator (RHI), Plan Position Indicator (PPI) and staring-mode scans respectively.

An example of the characterization of the inflow conditions for the period between 22h30 and 23h00 (GMT-6) of 15 September 2017 is provided in the following subsections.

2.3.1. Yaw

The determination of the yaw angle γ_{ppi} and the wind speed at hub height $U_{hub,ppi}$ is done with upstream horizontal PPI scans of a $\pm60°$ range around the rotor axis direction. The scans are performed at an angular resolution of 4° and a measurement frequency of 3 Hz during 5 min. The yaw angle and the wind speed at hub height are determined by fitting a cosine function to the radial velocities as a function of the azimuth angle φ as detailed in the following equation:

$$Vr(\varphi) = U_{hub,ppi} \cos(\varphi - \gamma_{ppi}).\qquad(1)$$

The azimuth angle is defined as the horizontal angle between the laser beam orientation and the downstream axis of the rotor. An example of the wind speed is provided in Figure 5 where one can see all the radial velocity measurements taken in the upstream range previously described and plotted against the azimuth angle, the cosine fit and the calculated yaw angle.

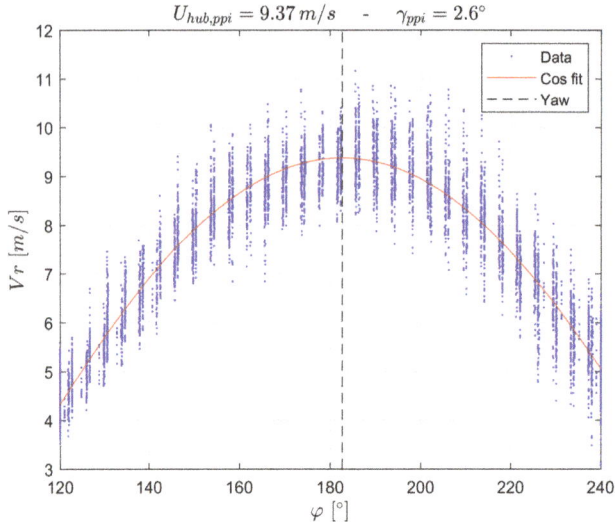

Figure 5. Example of the estimation of the yaw angle and the horizontal wind speed at hub height with plan position indicator (PPI) scans. Data corresponding to the period 22h30–22h35, 15 September 2017.

2.3.2. Vertical Profile of the Horizontal Velocity

The determination of the vertical profile of horizontal velocity is done with upstream vertical Range Height Indicator (RHI) scans aligned with the rotor axis direction of a ±15° around the horizontal plane. The scans are performed with an angular resolution of 1° in the elevation angle θ and a frequency of 3 Hz during 5 min. The measurements of radial velocity are corrected with the elevation angle and the yaw angle previously quantified in Section 2.3.1 in order to obtain an estimation of the undisturbed horizontal velocity at different heights:

$$u_\infty(z) = Vr(z)/(\cos(\theta)\cos(\gamma_{ppi})) . \tag{2}$$

The measurements of the horizontal component of the wind velocity are divided into blocks of 10 m in the vertical direction, and the average $U_\infty(z)$ and standard deviation $\sigma_{u,rhi}(z)$ are calculated. From these vertical profiles it is possible to extract the values at hub height of the mean horizontal velocity $U_{hub,rhi}$ and the longitudinal turbulence intensity $TI_{x,rhi} = \sigma_{u,rhi}(z_{hub})/U_{hub,rhi}$. An example is provided in Figure 6. Figure 6a shows all the corrected horizontal wind speed measurements taken in the upstream range previously described and plotted against height, while Figure 6b shows the binned statistics for each block of 10 m in the vertical direction.

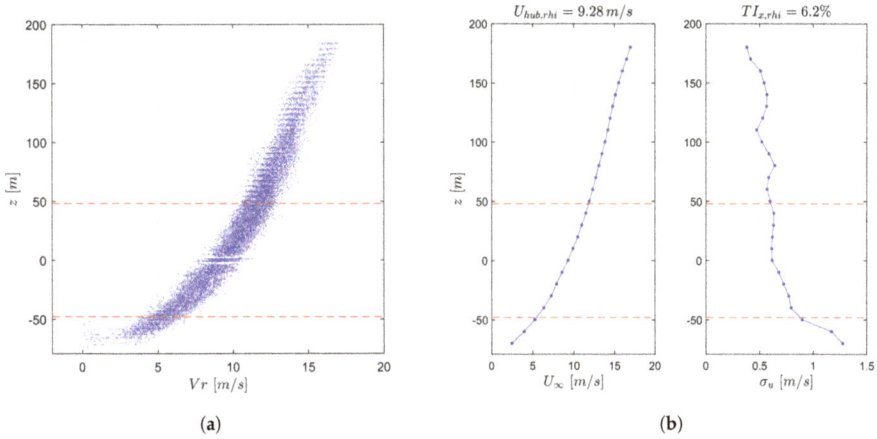

Figure 6. Example of the characterization of the vertical profile of horizontal velocity with range height indicator (RHI) scans. The red dashed lines indicate the upper and lower limits of the rotor. (**a**) All measurements; (**b**) Binned statistics. 22h35–22h40, 15 September 2017.

2.3.3. Turbulence Intensity

The determination of the longitudinal turbulence intensity $TI_{x,st}$ and a supplementary calculation of the wind speed at hub height $U_{hub,st}$ is done with a staring-mode scan aligned with the rotor axis direction for a duration of 10 min. Measurements are taken at a frequency of 1.5 Hz. Only the measurements of radial velocity in the mentioned upstream range are considered and their average and standard deviation calculated. The wind speed at hub height is corrected with the yaw angle previously calculated in Section 2.3.1:

$$U_{hub,st} = \overline{Vr}_{(\varphi=0°)} / \cos(\gamma_{ppi}). \tag{3}$$

The standard deviations of the longitudinal component of the wind speed and the radial velocity are equated:

$$\sigma_u = \sigma_{Vr(\varphi=0°)}, \tag{4}$$

and the longitudinal turbulence intensity is calculated as:

$$TI_{x,st} = \sigma_u / U_{hub,st}. \tag{5}$$

The transversal turbulence intensity calculation uses a horizontal staring-mode scan of the same duration at an angle of 90° with the rotor axis direction. The same relationship between the standard deviations of the longitudinal component of the wind speed and the radial velocity is used:

$$\sigma_v = \sigma_{Vr(\varphi=90°)}, \tag{6}$$

Additionally, the transversal turbulence intensity is calculated as:

$$TI_{y,st} = \sigma_v / U_{hub,st}. \tag{7}$$

A supplementary estimation of the yaw angle can be obtained by:

$$\gamma_{st} = \tan^{-1}\left(\overline{Vr}_{(\varphi=90°)} / U_{hub,st}\right) \tag{8}$$

The calculated values of the wind speed and longitudinal turbulence intensity for the period 22h40–22h50, 15 September 2017, are $U_{hub,st} = 9.12$ m/s and $TI_{x,st} = 5.7\%$. The values of the longitudinal turbulence intensity and yaw angle for the period 22h50–23h00 are $TI_{y,st} = 4.2\%$ and $\gamma_{st} = 2.17°$.

2.4. Downstream Scanning and Reconstruction of Planar Velocity Fields

The wake scanning consists of consecutive downstream horizontal PPI scans with a range of $\pm 20°$ around the rotor axis direction. The scans are performed at an angular resolution of 2° and a frequency of 2 Hz during 30 min.

The reconstruction of the longitudinal velocity fields in terms of its average and standard deviation follows these steps:

1. The average radial velocity $\overline{Vr}(\varphi, r)$ and its standard deviation $\sigma_{Vr}(\varphi, r)$ for all PPI scans are calculated at each point in space separated 2° in the azimuth φ and 18 m in the radial direction r, conforming a regular polar grid.
2. The average radial velocity is corrected with the calculated yaw angle γ_{ppi} in order to estimate the longitudinal velocity component:

$$\overline{u}(\varphi, r) = \overline{Vr}(\varphi, r) / \cos(\gamma_{ppi} - \varphi) . \tag{9}$$

3. The standard deviation of the streamwise velocity component and the radial velocity are assumed to be the same and are directly equated:

$$\sigma_u(\varphi, r) = \sigma_{Vr}(\varphi, r) . \tag{10}$$

4. The values in the polar grid are interpolated linearly into a Cartesian grid of 10 m resolution obtaining $\overline{u}(x, y)$ and $\sigma_u(x, y)$, more suitable to the post-processing of the data and the comparison with wake models.

An example of the average and standard deviation fields is provided in Figure 7, where it is possible to see the effect of the interpolation from polar coordinates to Cartesian. The interpolation from a polar grid with a resolution of 2° and 18 m into a Cartesian one with a resolution of 10 m means that the data will be slightly oversampled overall except closer to the rotor, where it will be slightly downsampled in the transversal direction. This should not affect the results obtained from the post-processing of the data. Although the turbulence fields of the wind turbine wake are not used in this study, the authors consider illustrative to show it as proof of the potential of the simple reconstruction technique used from PPI scans.

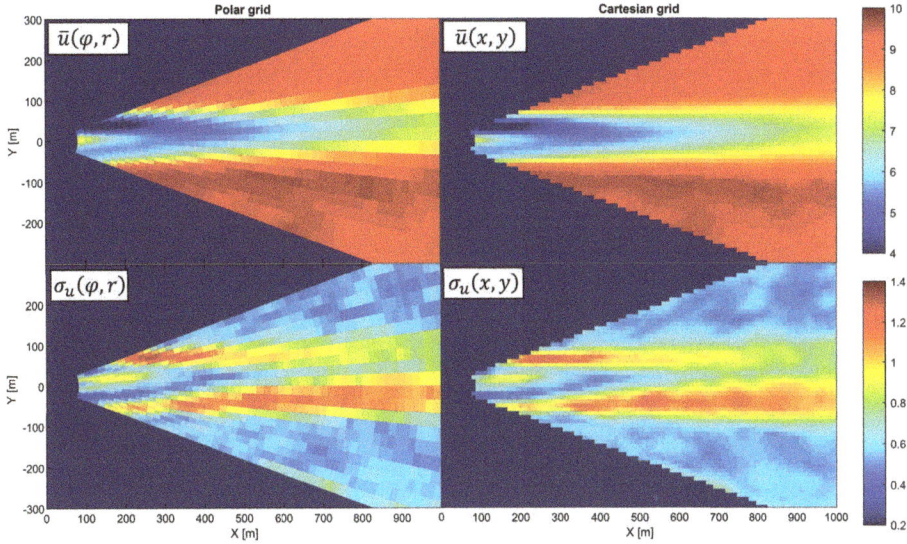

Figure 7. Example of the average and standard deviation of the wake in polar and Cartesian coordinates. 22h30–23h00, 15 September 2017. Units in m/s.

2.5. Wake Analysis

The analysis of the measurement of the wake of the wind turbine will vary depending on the objective of the study (e.g., wake meandering, yaw and skew angles, near/far wake determination, etc.). In this case the objective is the comparison with the predictions of the analytical model by Bastankhah and Porté-Agel in [11], who assumed Gaussian velocity deficit profiles in the far wake region and used mass and momentum conservation to link the growth rate of the wake and the recovery of the velocity deficit for different inflow conditions.

The local velocity deficit is defined as the difference between the local wind speed and the undisturbed wind speed at hub height:

$$\Delta \bar{u}(x,y) = U_{hub} - \bar{u}(x,y).$$

(11)

The far wake of the wind turbine is defined as the region of the wake that exhibits self-similar velocity deficit transversal profiles, which are well approximated by a Gaussian function of the form:

$$\Delta \bar{u}(x,y) = C(x)e^{-\frac{(y-y_C(x))^2}{2\sigma_y(x)^2}},$$

(12)

where C is the amplitude in m/s and corresponds to the velocity deficit along the centerline, y_c corresponds to the deviation of the center of the wake in meters from the longitudinal rotor axis and, finally, σ_y is the standard deviation in meters and corresponds to the wake width. In order to avoid contamination from data outside the area of interest, such as horizontal inhomogeneities of the free stream wind flow, the Gaussian fit to measured data uses a weighted nonlinear least squares regression. The weighting function is the resulting Gaussian function, but 50% wider. The goodness of the fit has been estimated by calculating the correlation ρ between measured and Gaussian fitted velocity profiles. It gives an indication of the beginning of the far wake region, or, conversely, the length of the near wake ℓ_{nw}. A value of $\rho = 0.99$ has been selected in [37] as the threshold to determine this distance.

An example of the analysis of a wind turbine wake is presented in Figure 8. In Figure 8a it is possible to see the velocity deficit vectors measured and the fitted Gaussian functions. It is noticeable the characteristic bimodal velocity deficit profile in the near wake, where a Gaussian function is not a good representation. In contrast, in the far wake the velocity deficit profile is self similar and shows an almost perfect fit to a Gaussian function. Figure 8b shows the longitudinal evolution of the Gaussian parameters of the fit. The first plot from the top shows the correlation coefficient between the measured velocity deficit and the Gaussian fit, together with the threshold $\rho = 0.99$ that indicates the beginning of the far wake [37]. In the case depicted it occurs at a distance of 3.9D from the rotor. The second plot shows the growth of the wake width in the longitudinal direction. Several wake models assume a linear expansion of the wake and it is possible to observe that it is a good assumption for the far wake region. The figure shows the coefficients of the linear fit in the form:

$$\sigma_y(x) = \left(k^* \frac{x}{D} + \varepsilon\right) D, \tag{13}$$

where D is the rotor diameter, k^* is the longitudinal growth rate of the wake width and ε is the wake width at the rotor plane.

The third plot of Figure 8b shows the velocity deficit along the centerline C and its decrease in the longitudinal direction, corresponding to the recovery of the wake velocity. Based on conservation of mass and momentum, this parameter has been linked to the wake width in the far wake by the following relationship [11]:

$$\frac{C(x)}{U_{hub}} = 1 - \sqrt{1 - \frac{C_T}{8(\sigma_y(x)/D)^2}} \ . \tag{14}$$

The plot also shows in a continuous red line the prediction of the C parameter by the analytical model, which shows good agreement for the far wake using the calculated wake expansion from the previous quadrant. It is also possible to observe that the prediction deviates substantially from the measured values when the Gaussian profile is not a good representation of the velocity deficit profile ($\rho < 0.99$). Finally, the last plot shows the deviation of the center of the Gaussian profile from the axis of rotation. The deviation follows a linear trend for the far wake as well and it is possible to calculate the skew angle of the wake by:

$$\chi = \tan^{-1}(\partial y_c / \partial x), \tag{15}$$

which, in the case presented, is 1.3°. This parameter could be useful to study the relationship between the yaw angle and the skew angle.

The analysis procedure described above is applied to all the 30 min periods in which the experiment has been divided and results are presented in Section 3.

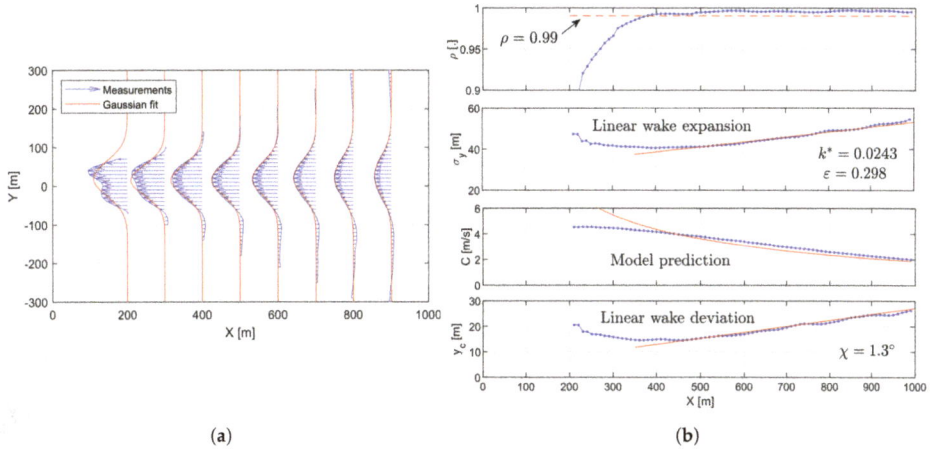

(a)   (b)

Figure 8. Example of the analysis of the wake of the wind turbine. (**a**) Velocity deficit and Gaussian fits; (**b**) Downstream evolution of the fit parameters. 22h30–23h00, 15 September 2017.

3. Results

This section describes the first measurements used for the analysis of the wake of the wind turbine as well as the conditions which entail a removal of those periods which are not suitable. The aggregated results are then presented and compared to data used to further validate and calibrate the analytical model discussed in Section 2.5.

The experiment data consists of a series of 30 min periods obtained between 20 August and 16 October 2017. One 30 min period is obtained every two hours for a total number of approximately 700. The inflow conditions for each period have been thoroughly studied in order to filter out those not suitable for the analysis of the wake. Criteria that were used to filter out measurement periods include (although are not limited to):

- Down times of the wind turbine.
- Wind speed outside the 5–10 m/s range.
- Low signal-to-noise ratio of the lidar measurements due to precipitation.
- Non-stationary undisturbed wind speed at hub height (comparison of $U_{hub,st}$, $U_{hub,ppi}$, $U_{hub,rhi}$).
- Non-stationary undisturbed wind direction at hub height (comparison of γ_{st}, γ_{ppi}).
- Non-stationary undisturbed turbulence intensity at hub height (comparison of $TI_{x,st}$, $TI_{x,rhi}$).
- Horizontal inhomogeneity of the wind speed in the surroundings of the wind turbine (this horizontal inhomogeneity can be easily observed when reconstructing the average of the longitudinal velocity field $\bar{u}(x,y)$ and observing the regions not affected by the wake. It can be seen that the case presented in Figure 7 shows a horizontally homogeneous flow outside the area of influence of the wake).
- Changing orientation of the rotor by the control system of the wind turbine.
- Measured yaw angles above $\pm 10°$.
- Disagreements among the inflow measured by the nacelle-mounted lidar, the data from the meteorological tower, and the SCADA data.

From all the collected data, only 44 periods have been selected as suitable for analysis, which yields a validity rate of around 6%. Three of the selected cases are shown for illustrative purposes in Figure 9. They are ordered in increasing inflow turbulence intensity at hub height from left to right. The top quadrants present the adimensional velocity deficit in the horizontal plane at hub height and it is observed that the wind speed recovery occurs significantly faster for higher turbulence conditions. It is also observed that lower turbulence conditions retard the occurrence of a self-similar Gaussian velocity deficit profile, indicating there is a significantly longer near wake region. The bottom quadrants present the longitudinal turbulence intensity in the horizontal plane at hub height. It is observed that the background turbulence intensity levels in those areas not affected by the wind turbine wake as well as the turbulence generated by the shear, which is greatest in the mixing layers at the edge of the wake. The values of the incoming turbulence intensity, wake growth rate, wake width at the origin, and length of the near wake for these three cases are highlighted in Figures 10–12.

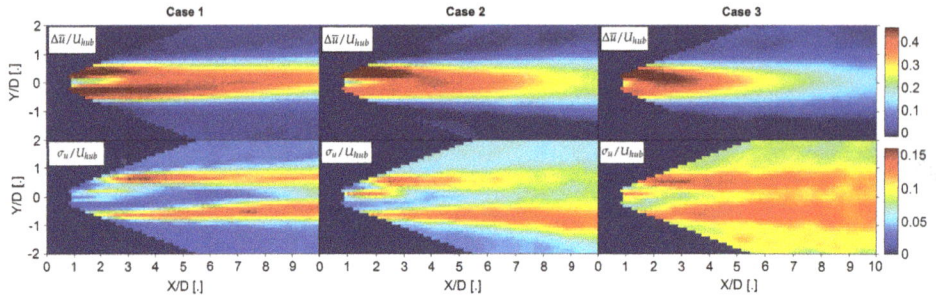

Figure 9. Example of three cases of wakes for increasing different longitudinal turbulence intensity conditions at hub height (Case 1–2.2%; Case 2–4.6%; Case 3–9.9%). The top quadrants present the adimensional velocity deficit while the bottom quadrants present the longitudinal turbulence intensity in the horizontal plane at hub height. These three example cases are further referenced in Figures 10–12.

The relationship found between the wake width growth rate k^* and the longitudinal turbulence intensity TI_x is presented in Figure 10. The data from the full scale field experiment (blue stars) as well as data from validated LES simulations (black squares) and wind tunnel experiment (black circle), presented in [11], agree well, taking into account the significant variability of the data. The data show clearly that the rate of growth of the wake width increases with the turbulence intensity. The growth of the wake is linked to the velocity recovery by mass and momentum conservation as already discussed. This implies a faster recovery of the velocity deficit for higher background turbulence since turbulence enhances flow mixing and the transfer of momentum from the undisturbed flow region into the wake. The linear fit to the full-scale field data is presented as a dashed red line in Figure 10, and it can be expressed as:

$$k^* = 0.35 \, TI_x \, . \tag{16}$$

The linear relationship of Equation (16) is similar to the one used in [38] by fitting a straight line to the data presented in [11], $k^* = 0.383 \, TI_x + 0.0037$ (presented as a dashed black line in Figure 10). When using these relationships, it is important to take into account the variability of the data, which indicates that it is not uncommon to find wake growths that differ by a factor of two or three for very similar conditions of longitudinal turbulence intensity. This suggests that further experiments should be addressed to understand the role of other variables that could also play an important role on the development of the wind turbine wake.

One particularly interesting effect to study is the occurrence of different wake growths in the vertical and horizontal directions, which makes the velocity deficit profiles not self-similar in the radial direction, leading to an elliptical profile instead of a circular one. Only considering the horizontal growth rate violates mass and momentum conservation and could be a reason for the variability of the results. Volumetric downstream scans with a similar setup have been performed during the same dates to further study this fact and preliminary analysis proves the occurrence of non circular velocity deficit profiles under certain atmospheric conditions. A more complete analysis can be provided in the future.

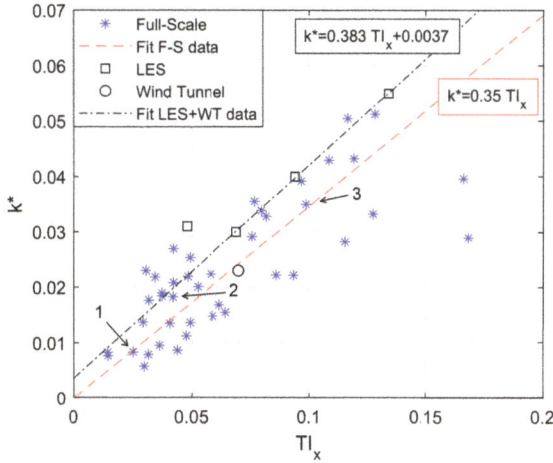

Figure 10. Relationship between wake growth k^* and longitudinal turbulence intensity TI_x. In blue all the data collected during the experiment, in dashed red the linear fit to the full-scale field data presented in Equation (16). In black the data obtained from [11] and the linear fit used in [38]. Numbers 1 to 3 indicate the cases presented in Figure 9.

The relationship between the growth rate of the wake and the wake width at the rotor plane ε is presented in Figure 11. ε is not physical, but rather a theoretical parameter, which indicates the hypothetical wake width at the rotor plane when considering a Gaussian wake from its origin (i.e., no existence of a near wake region). The correlation is negative, which means that for a higher growth rate, the width at the rotor plane is smaller. The figure presents the data obtained during this experiment (blue stars) together with its linear fit, which allows for the calculation of the wake width at the origin for a particular wake growth rate is presented as a dashed red line:

$$c = 1.91k^* | 0.34 . \tag{17}$$

The agreement between the full-scale field data and the data from the validated LES simulations (black squares) is good, but it is poor for the wind tunnel measurements (black circle).

Finally, the relationship between the length of the near wake ℓ_{nw} and the longitudinal turbulence intensity is presented in Figure 12. The setup used, and the range of the downstream PPI scans (see Section 2.4) does not allow for scans of the full width of the wake for shorter distances than approximately 200 m and, therefore, shorter near wake lengths are not included in the analysis. This is represented by the gray shaded area at the bottom of the figure. Similarly to the effect of the growth rate, a higher inflow turbulence intensity enhances flow mixing and this helps the wake reaching a self-similar state in a shorter distance. The length of the near wake reaches long distances, higher than six diameters in some cases, for particularly low turbulence flow. It can be observed that in this range of low turbulence intensities, the variability of the data becomes also greater.

A semi-analytical expression for the length of the near wake of a wind turbine under yaw conditions is presented in Bastankhah and Porté-Agel [19]. The corresponding relationship for zero or negligible yaw angles is:

$$\frac{\ell_{nw}}{D} = \frac{1 + \sqrt{1 - C_T}}{\sqrt{2}(\alpha\, TI_x + \beta\,(1 - \sqrt{1 - C_T}))}, \tag{18}$$

where β is a parameter obtained from analogy with jet flows and has a value of 0.154 and α is obtained from experimental data. The value of $\alpha = 2.32$ prescribed in [19] was obtained from wind-tunnel experiments. The expression for the length of the near wake using this value is represented in Figure 12 as a dashed black line. A value of $\alpha = 3.6$ provides a better fit to the full scale experimental data, as shown by the red dashed line in Figure 12. It should be noted that the experimental data used in order to obtain the value of 3.6 is more exhaustive in terms of range of turbulence intensities covered and number of independent data points obtained.

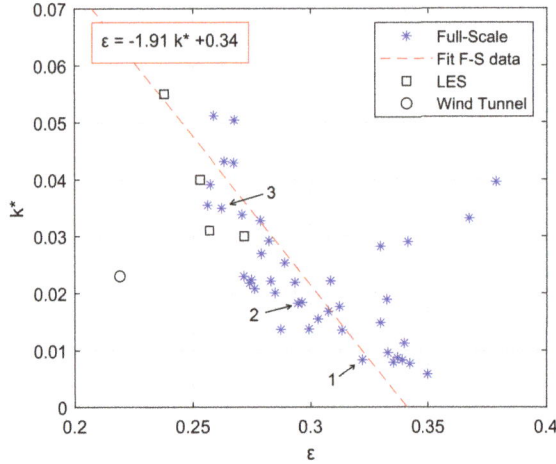

Figure 11. Relationship between wake growth k^* and wake width at the rotor ε. In blue all the data collected during the experiment, in dashed red the linear fit to the full-scale field data presented in Equation (17). In black the data obtained from [11]. Labels numbered 1 to 3 indicate the cases presented in Figure 9.

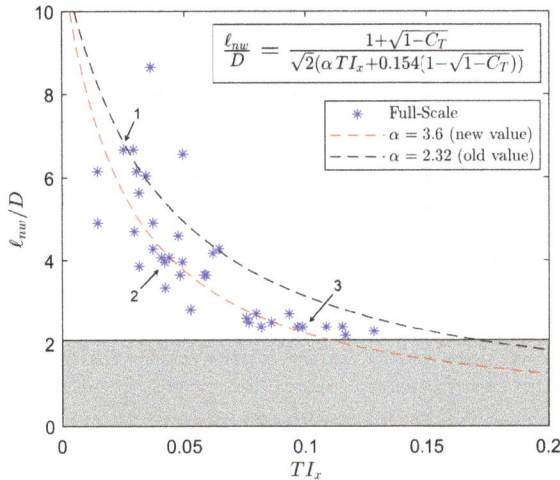

Figure 12. Relationship between near wake length ℓ_{nw} and longitudinal turbulence intensity TI_x. In blue all the data collected during the experiment, in dashed black Equation (18) for the estimation of the near wake length with the value of $\alpha = 2.32$ prescribed in [19] and in dashed red the same equation with the value of $\alpha = 3.6$ that provides a better fit to the full-scale field data. The grey rectangle at the bottom indicates the lowest bound for a possible calculation of the near wake length with the presented setup and analysis. Numbers 1 to 3 indicate the cases presented in Figure 9.

4. Conclusions

The study demonstrates that a measurement setup based on two nacelle-mounted lidars can be used to measure different characteristics of the incoming flow via RHI, PPI and staring-mode scans, while at the same time perform planar scans of the wake. The upstream oriented lidar data were processed with simple algorithms to calculate the vertical profile of horizontal velocity, the yaw angle of the incoming flow, and the longitudinal and transversal turbulence intensities. The downstream oriented lidar data were aggregated statistically and used to reconstruct the longitudinal velocity field in terms of its average and standard deviation. The velocity deficit profiles were fitted to Gaussian functions and provide information about the wake width and the velocity deficit along the center line at different longitudinal distances.

With the calculated inflow and wake parameters described above, it is possible to study the relationships between wake growth rate, wake width at the rotor plane, length of the near wake, and longitudinal turbulence intensity. A higher incoming turbulence increases mixing and the transfer of momentum from the regions outside the wake into it, reduces the length of the near wake, and increases the velocity recovery. A higher wake growth rate, in turn, implies a smaller wake width at the rotor plane. Different relationships have been established for these relationships.

The presented data have been compared to the predictions of the analytical wake model based on Gaussian velocity profiles and mass and momentum conservation developed by Bastankhah and Porté-Agel [11,19]. It has been found that the model predicts the wake expansion and velocity deficit well, as well as the length of the near wake region. The empirical parameters prescribed in the original model have been updated with the field experiment data.

This study can be extended to use volumetric measurements of the wake and using a similar methodology, in order to understand the implications in the development of the far wake under certain atmospheric conditions when the velocity deficit profile is not self-similar in the radial direction (i.e., not circular).

Author Contributions: This study was done as a part of Fernando Carbajo Fuertes doctoral studies supervised by Fernando Porté-Agel, and with the full support and collaboration from Corey D. Markfort.

Acknowledgments: The authors are thankful for the valuable help of Kirkwood Community College for providing access to their wind turbine and the SCADA data. Also to Clipper Windpower for assistance setting up the lidars on the nacelle and for technical specifications of the Liberty wind turbine. This research was supported by the Swiss National Science Foundation [grant 200021-172538], the Swiss Federal Office of Energy [grant SI/501337-01], the Swiss Innovation and Technology Committee (CTI) within the context of the Swiss Competence Center for Energy Research FURIES: Future Swiss Electrical Infrastructure, and the Center for Global and Regional Environmental Research at The University of Iowa.

Conflicts of Interest: The authors declare no conflict of interest. The founding sponsors had no role in the design of the study; in the collection, analysis, or interpretation of data; in the writing of the manuscript, and in the decision to publish the results.

Abbreviations

The following abbreviations are used in this manuscript:

ABL	Atmospheric Boundary Layer
LES	Large Eddy Simulation
LiDAR	Light Detection And Ranging
LoS	Line-of-Sight
RHI	Range Height Indicator
PPI	Plan Position Indicator
SCADA	Supervisory Control And Data Acquisition

References

1. Vermeer, L.; Sørensen, J.; Crespo, A. Wind turbine wake aerodynamics. *Prog. Aerosp. Sci.* **2003**, *39*, 467–510. [CrossRef]
2. Barthelmie, R.J.; Pryor, S.C.; Frandsen, S.T.; Hansen, K.S.; Schepers, J.G.; Rados, K.; Schlez, W.; Neubert, A.; Jensen, L.E.; Neckelmann, S. Quantifying the impact of wind turbine wakes on power output at offshore wind farms. *J. Atmos. Ocean. Technol.* **2010**, *27*, 1302–1317. [CrossRef]
3. Hansen, K.S.; Barthelmie, R.J.; Jensen, L.E.; Sommer, A. The impact of turbulence intensity and atmospheric stability on power deficits due to wind turbine wakes at Horns Rev wind farm. *Wind Energy* **2012**, *15*, 183–196. [CrossRef]
4. Thomsen, K.; Sørensen, P. Fatigue loads for wind turbines operating in wakes. *J. Wind Eng. Ind. Aerodyn.* **1999**, *80*, 121–136. [CrossRef]
5. Markfort, C.D.; Zhang, W.; Porté-Agel, F. Turbulent flow and scalar transport through and over aligned and staggered wind farms. *J. Turbul.* **2012**, *13*, N33. [CrossRef]
6. St. Martin, C.M.; Lundquist, J.K.; Clifton, A.; Poulos, G.S.; Schreck, S.J. Wind turbine power production and annual energy production depend on atmospheric stability and turbulence. *Wind Energy Sci. Discuss.* **2016**, 1–37.
7. Herbert-Acero, J.; Probst, O.; Réthoré, P.E.; Larsen, G.; Castillo-Villar, K. A Review of Methodological Approaches for the Design and Optimization of Wind Farms. *Energies* **2014**, *7*, 6930–7016. [CrossRef]
8. Gebraad, P.; Thomas, J.J.; Ning, A.; Fleming, P.; Dykes, K. Maximization of the annual energy production of wind power plants by optimization of layout and yaw-based wake control. *Wind Energy* **2017**, *20*, 97–107. [CrossRef]
9. Torben Knudsen, T.B.; Automation, M.S. Survey of wind farm control—Power and fatigue optimization. *Wind Energy* **2015**, *18*, 1333–1351. [CrossRef]
10. Chehouri, A.; Younes, R.; Ilinca, A.; Perron, J. Review of performance optimization techniques applied to wind turbines. *Appl. Energy* **2015**, *142*, 361–388. [CrossRef]
11. Bastankhah, M.; Porté-Agel, F. A new analytical model for wind-turbine wakes. *Renew. Energy* **2014**, *70*, 116–123. [CrossRef]
12. Sørensen, J.N.; Shen, W.Z. Numerical Modeling of Wind Turbine Wakes. *J. Fluid. Eng.* **2002**, *124*, 393. [CrossRef]
13. Crespo, A.; Hernández, J.; Frandsen, S. Survey of modelling methods for wind turbine wakes and wind farms. *Wind Energy* **1999**, *2*, 1–24. [CrossRef]
14. Chamorro, L.P.; Porté-Agel, F. A wind-tunnel investigation of wind-turbine wakes: Boundary-Layer turbulence effects. *Bound. Layer Meteorol.* **2009**, *132*, 129–149. [CrossRef]
15. Chamorro, L.P.; Porté-Agel, F. Turbulent flow inside and above a wind farm: A wind-tunnel study. *Energies* **2011**, *4*, 1916–1936. [CrossRef]
16. Zhang, W.; Markfort, C.D.; Porté-Agel, F. Wind-Turbine Wakes in a Convective Boundary Layer: A Wind-Tunnel Study. *Bound. Layer Meteorol.* **2013**, *146*, 161–179. [CrossRef]
17. Lignarolo, L.E.; Ragni, D.; Krishnaswami, C.; Chen, Q.; Simão Ferreira, C.J.; van Bussel, G.J. Experimental analysis of the wake of a horizontal-axis wind-turbine model. *Renew. Energy* **2014**, *70*, 31–46. [CrossRef]
18. Iungo, G.V. Experimental characterization of wind turbine wakes: Wind tunnel tests and wind LiDAR measurements. *J. Wind Eng. Ind. Aerodyn.* **2016**, *149*, 35–39. [CrossRef]
19. Bastankhah, M.; Porté-Agel, F. Experimental and theoretical study of wind turbine wakes in yawed conditions. *J. Fluid Mech.* **2016**, *806*, 506–541. [CrossRef]
20. Bastankhah, M.; Porte-Agel, F. Wind tunnel study of the wind turbine interaction with a boundary-layer flow: Upwind region, turbine performance, and wake region. *Phys. Fluids* **2017**, *29*, 065105. [CrossRef]
21. Miller, M.A.; Kiefer, J.; Westergaard, C.; Hultmark, M. Model Wind Turbines Tested at Full-Scale Similarity. In *Journal of Physics: Conference Series*; IOP Publishing: Bristol, UK, 2016; Volume 753.
22. Barthelmie, R.J.; Hansen, K.; Frandsen, S.T.; Rathmann, O.; Schepers, J.G.; Schlez, W.; Phillips, J.; Rados, K.; Zervos, A.; Politis, E.S.; et al. Modelling and measuring flow and wind turbine wakes in large wind farms offshore. *Wind Energy* **2009**, *12*, 431–444. [CrossRef]

23. Käsler, Y.; Rahm, S.; Simmet, R.; Kühn, M. Wake measurements of a multi-MW wind turbine with coherent long-range pulsed doppler wind lidar. *J. Atmos. Ocean. Technol.* **2010**, *27*, 1529–1532. [CrossRef]
24. Iungo, G.V.; Wu, Y.T.; Porté-Agel, F. Field measurements of wind turbine wakes with lidars. *J. Atmos. Ocean. Technol.* **2013**, *30*, 274–287. [CrossRef]
25. Smalikho, I.N.; Banakh, V.A.; Pichugina, Y.L.; Brewer, W.A.; Banta, R.M.; Lundquist, J.K.; Kelley, N.D. Lidar investigation of atmosphere effect on a wind turbine wake. *J. Atmos. Ocean. Technol.* **2013**, *30*, 2554–2570. [CrossRef]
26. Iungo, G.V.; Porté-Agel, F. Volumetric lidar scanning of wind turbine wakes under convective and neutral atmospheric stability regimes. *J. Atmos. Ocean. Technol.* **2014**, *31*, 2035–2048. [CrossRef]
27. Banta, R.M.; Pichugina, Y.L.; Brewer, W.A.; Lundquist, J.K.; Kelley, N.D.; Sandberg, S.P.; Alvarez, R.J.; Hardesty, R.M.; Weickmann, A.M. 3D volumetric analysis of wind turbine wake properties in the atmosphere using high-resolution Doppler lidar. *J. Atmos. Ocean. Technol.* **2015**, *32*, 904–914. [CrossRef]
28. Aitken, M.L.; Banta, R.M.; Pichugina, Y.L.; Lundquist, J.K. Quantifying Wind Turbine Wake Characteristics from Scanning Remote Sensor Data. *J. Atmos. Ocean. Technol.* **2014**, *31*, 765–787. [CrossRef]
29. Doubrawa, P.; Barthelmie, R.; Wang, H.; Pryor, S.; Churchfield, M. Wind Turbine Wake Characterization from Temporally Disjunct 3-D Measurements. *Remote Sens.* **2016**, *8*, 939. [CrossRef]
30. El-Asha, S.; Zhan, L.; Iungo, G.V. Quantification of power losses due to wind turbine wake interactions through SCADA, meteorological and wind LiDAR data. *Wind Energy* **2017**, *20*, 1823–1839. [CrossRef]
31. Bodini, N.; Zardi, D.; Lundquist, J.K. Three-dimensional structure of wind turbine wakes as measured by scanning lidar. *Atmos. Meas. Tech.* **2017**, *10*, 2881–2896. [CrossRef]
32. Trujillo, J.; Bingöl, F.; Larsen, G.; Mann, J.; Kühn, M. Light detection and ranging measurements of wake dynamics. Part II: two-dimensional scanning. *Wind Energy* **2011**, *14*, 61–75. [CrossRef]
33. Aitken, M.L.; Lundquist, J.K. Utility-scale wind turbine wake characterization using nacelle-based long-range scanning lidar. *J. Atmos. Ocean. Technol.* **2014**, *31*, 1529–1539. [CrossRef]
34. Machefaux, E.; Larsen, G.C.; Troldborg, N.; Gaunaa, M.; Rettenmeier, A. Empirical modeling of single-wake advection and expansion using full-scale pulsed lidar-based measurements. *Wind Energy* **2015**, *18*, 2085–2103. [CrossRef]
35. Machefaux, E.; Larsen, G.C.; Koblitz, T.; Troldborg, N.; Kelly, M.C.; Chougule, A.; Hansen, K.S.; Rodrigo, J.S. An experimental and numerical study of the atmospheric stability impact on wind turbine wakes. *Wind Energy* **2016**, *19*, 1785–1805. [CrossRef]
36. Herges, T.G.; Maniaci, D.C.; Naughton, B.T.; Mikkelsen, T.; Sjöholm, M. High resolution wind turbine wake measurements with a scanning lidar. *J. Phys. Conf. Ser.* **2017**, *854*, 012021. [CrossRef]
37. Sorensen, J.N.; Mikkelsen, R.F.; Henningson, D.S.; Ivanell, S.; Sarmast, S.; Andersen, S.J. Simulation of wind turbine wakes using the actuator line technique. *Philos. Trans. R. Soc. A Math. Phys. Eng. Sci.* **2015**, *373*, 20140071. [CrossRef] [PubMed]
38. Niayifar, A.; Porté-Agel, F. Analytical modeling of wind farms: A new approach for power prediction. *Energies* **2016**, *9*, 741. [CrossRef]

remote sensing

MDPI

Article

Using a Virtual Lidar Approach to Assess the Accuracy of the Volumetric Reconstruction of a Wind Turbine Wake

Fernando Carbajo Fuertes and Fernando Porté-Agel *

Wind Engineering and Renewable Energy Laboratory (WiRE)-École Polytechnique Fedérale de Lausanne (EPFL), 1015 Lausanne, Switzerland; fernando.carbajo@epfl.ch
* Correspondence: fernando.porte-agel@epfl.ch; Tel.: +41-(0)21-693-61-38

Received: 23 March 2018; Accepted: 4 May 2018; Published: 7 May 2018

check for
updates

Abstract: Scanning Doppler lidars are the best tools for acquiring 3D velocity fields of full scale wind turbine wakes, whether the objective is a better understanding of some features of the wake or the validation of wake models. Since these lidars are based on the Doppler effect, a single scanning lidar normally relies on certain assumptions when estimating some components of the wind velocity vector. Furthermore, in order to reconstruct volumetric information, one needs to aggregate data, perform statistics on it and, most likely, interpolate to a convenient coordinate system, all of which introduce uncertainty in the measurements. This study simulates the performance of a virtual lidar performing stacked step-and-stare plan position indicator (PPI) scans on large-eddy simulation (LES) data, reconstructs the wake in terms of the average and the standard deviation of the longitudinal velocity component, and quantifies the errors. The variables included in the study are as follows: the location of the lidar (ground-based and nacelle-mounted), different atmospheric conditions, and varying scan speeds, which in turn determine the angular resolution of the measurements. Testing different angular resolutions allows one to find an optimum that balances the different error sources and minimizes the total error. An optimum angular resolution of 3° has been found to provide the best results. The errors found when reconstructing the average velocity are low (less than 2% of the freestream velocity at hub height), which indicates the possibility of high quality field measurements with an optimal angular resolution. The errors made when calculating the standard deviation are similar in magnitude, although higher in relative terms than for the mean, thus leading to a poorer quality estimation of the standard deviation. This holds true for the different inflow cases studied and for both ground-based and nacelle-mounted lidars.

Keywords: wind energy; atmospheric boundary layer; wind turbine wake; wind lidar; virtual lidar; turbulence; wake modeling; large-eddy simulations

1. Introduction

Wind power growth worldwide is a result of an ever increasing demand for renewable energy. With upper limits of rated power for single turbines reaching the order of 10 MW, the common solution to keep increasing the generated power is to install a larger number of wind turbines. Due to space limitations, in most cases, wind turbines are clustered together in wind farms and effectively this means a higher number of turbines in each wind farm. In large wind farms, most of the wind turbines are affected by the wake flow of others, resulting in a reduction of incoming wind speed and an increase of turbulence [1]. Therefore, a careful evaluation of wind resources is needed for an accurate estimation of not only produced power from a future wind farm [2,3], but also the power losses and increased fatigue loads associated with the wind turbine wakes [4]. Accurate wake models are needed to that end, and the location of each of the wind turbines inside the farm needs to be optimized in order to minimize the losses [5,6].

There are a number of strategies to model the wake of a wind turbine: numerical models [7] discretize the Navier–Stokes (N.S.) equations, either in the physical space or the Fourier space, and solve them with the help of various turbulence models; analytical models [8] use certain assumptions (e.g., the shape of the wind velocity deficit in the wake) and combine them with the N.S. equations either in 1D or 3D; lastly, empirical models can be based on either measurements alone or a mixture between important simplifications of the Navier–Stokes equations and empirical parameters [9]. All of them must, inevitably, be contrasted with measurements of wind turbine wakes in order to estimate their accuracy and, if applicable, be validated under certain conditions.

The interaction between the atmospheric boundary layer (ABL) and a full scale wind turbine is a three-dimensional, dynamic flow phenomenon that extends from approximately two diameters in front of the rotor to several hundred meters, possibly a few kilometers, downwind [1]. The ideal measurement set for a characterization of a single wind turbine wake would include the three components of the wind velocity vector, at all times and at all positions in space within the region of influence. Unfortunately, no measurement technique is able to provide such data. Different instruments provide different sets of measurements, with varying levels of suitability for the characterization of a wind turbine wake.

The standard instrument for turbulence measurements in the atmosphere, the sonic anemometer, is not well suited for the measurement of turbine wake flows. Sonic anemometers only offer point-wise information and it is not practical to cover an extensive volume by mounting them on meteorological masts. Alternatives to the sonic anemometer and the meteorological mast are, in some cases, unmanned aerial vehicle platforms, since they are able to fly and measure in any point in space [10–14]. Nevertheless, they have important limitations: most of them are s till mostly in prototype phase, they are not suitable for long-term statistics, and covering a volume with point-wise measurements would imply thousands of hours of flight.

As an alternative, remote sensing techniques are increasingly popular in atmospheric flows because of their ability to measure where other sensors cannot. Among these, Doppler light detection and ranging (lidar) is the preferred remote sensing technique for atmospheric turbulence measurements due to its accuracy and relatively high spatial resolution and long range. As of today, it is the most suitable measurement technique to study the different characteristics of wind turbine wakes [15–23].

One limitation of the lidar technique arises from the fact that it is based on the Doppler effect and uses backscattered light from aerosol. Therefore, it can only measure the velocity component parallel to the laser beam (radial velocity). A scanning Doppler lidar can orient its laser beam in any direction; this implies that, unless the laser beam is completely vertical or aligned with the direction of the flow or transversal to it, the measured velocity is normally a mixture of the three components of the wind velocity vector. As a consequence, most lidar measurements rely on some assumptions when calculating the relevant variables such as horizontal wind speed, vertical wind speed, or the different turbulence quantities [24–26]. Multiple-lidar techniques exist and can overcome this limitation [27–29] although they multiply the cost, are more cumbersome to use, and require a significantly higher degree of expertise to be properly operated. The use of a single lidar and the necessary assumptions is a source of uncertainty that needs to be estimated.

Another limitation of the lidar technique is the speed of the measurements. A scanning pulsed lidar emits several thousand laser pulses in a particular direction or line-of-sight (LoS), evaluates the backscattered signal, calculates the Doppler shift at different distances from the lidar, and then proceeds to the next laser beam orientation. Since a wind turbine wake is inherently dynamic, in order to cover a volume with lidar measurements, a scanning strategy is needed in order to balance angular resolution and number of measurements at each orientation. This translates into finding a compromise between a spatial interpolation error and a statistical uncertainty. These two need to be estimated as well.

As discussed above, no technique is able to provide the velocity field in order to estimate the errors and uncertainties associated with a volumetric lidar scan of a wind turbine wake. Therefore, a good strategy is to use a virtual lidar technique to perform a virtual experiment. Specifically, a turbulence-resolving large-eddy simulation (LES) is performed and then the characteristics and scanning pattern of a virtual instrument can be programmed to extract information or virtual measurements from LES simulated velocity field. These virtual measurements can be post-processed with the same algorithms used to treat real measurements in order to reconstruct the desired flow feature (in this case, the far wake of the wind turbine) and then it can be compared to the original velocity field from the LES simulation. A virtual lidar technique allows the estimation of the different sources of uncertainties or errors separately. In this manuscript, the sources of error studied are three: the assumptions used to convert the radial velocity to longitudinal velocity, the statistical error, and the interpolation error. Finally, testing different scanning patterns allows one to find an optimum that minimizes the errors. The error magnitude, in turn, will determine the quality of future real (time and resource-consuming) field experiments.

Most of the literature regarding virtual lidar studies is very recent. Stawiarski et al. [30] created a virtual lidar measurement simulator based on LES results in which lidar characteristics such as range gate length, pulse length, total range, and measurement frequency are adjustable. In their simulator, they include the effect of the convolution of the laser pulse as a cylindrical volume centered around the range gate center, and they use a weighted averaging function over the LES data points inside that volume. They discuss extensively the different kinds of errors connected with single- and double-lidar measurements and provide the methodology to study the sources of errors and the optimization of dual-Doppler scan patterns. Stawiarski et al. [31] further performed virtual planar dual-lidar experiments to study the reconstruction and the detection of planar turbulent structures.

Lundquist et al. [32] studied the uncertainty of the Doppler beam swinging (DBS) technique, used by many commercial profiling lidars, when calculating horizontal and vertical velocities while violating the assumption of horizontally homogeneous flow that the technique requires. The calculation of the vertical profiles of horizontal velocity is done using virtual measurements obtained from LES simulations of a wind turbine wake. Similarly, Mirocha et al. [33] have simulated the effect of the inhomogeneity on profiling lidar measurements using a virtual lidar technique approach under different atmospheric stability conditions.

Van Dooren et al. [34] used virtual lidar experiments to explore the possibilities and uncertainty of wind turbine wake reconstruction with ground-based dual-lidar plan position indicator (PPI) scans. For that, they obtained non-synchronous dual-Doppler lidar measurements from LES simulations, and they improved the accuracy of the reconstructed wind field by including a correction based on the mass continuity equation.

Meyer Forsting et al. [35] developed a novel validation methodology for computational fluid dynamics (CFD) models over the wind turbine induction zone using measurements from three synchronous lidars. The validation procedure relied on making the CFD simulation results comparable with the triple lidar data. To that end, they discretized in space the probability density function of the measured free-stream wind speed. Then they reproduced those distributions numerically by weighting the steady-state Reynolds averaged Navier–Stokes simulations. As a last step, the spatial and temporal uncertainty of the triple lidar measurements were quantified and propagated through the data processing.

Lastly, and although the virtual lidar technique is not used, it is worth mentioning that van Dooren et al. [36] studied the uncertainty of synchronous short-range continuous dual-lidar measurements for the measurement of scaled wind turbine wakes placed inside a wind tunnel. They were able to compare the estimated uncertainty of some of their configurations by comparing the lidar measurements to those taken by a triple hot-wire probe.

In this study, a virtual lidar technique based on LES simulations, similar to what has been used in some of the above-mentioned references, is used to calculate different error sources for a single pulsed lidar performing volumetric scans of a full scale wind turbine wake under different atmospheric conditions and for two different lidar locations: nacelle-mounted and ground-based. The technique allows one to optimize the scanning of the wake and evaluate the quality of the measurement strategy and the assumptions used for the treatment of the data.

There are a number of practical advantages to using nacelle-mounted lidars over ground-based ones. Arguably, the most important one is the fact that a lidar fixed to the structure of the nacelle of a wind turbine always maintains the same angular orientation with respect to the rotor. This means that it does not need any reorientation with a changing wind direction and it never gets blocked by the tower of the turbine. Nevertheless, practical considerations of this kind are beyond the scope of this study.

2. Methodology

A virtual lidar approach is based on the idea of performing *virtual experiments* on a flow field in which the three components of the wind velocity vector $\mathbf{v}(x, y, z, t)$ are calculated at each point in space and at each point in time. A convenient way of creating such a flow field is through LES simulations. The results of these simulations can be interpreted as a *virtual reality* from which lidar *virtual measurements* can be extracted, knowing the characteristics of a particular lidar (spatial resolution, repetition rate, range, etc.), its scanning pattern, and that it measures only the projection of the wind velocity vector onto the laser beam direction or LoS. The lidar *virtual measurements* obtained can then be processed by the same algorithm used to treat real lidar measurements in order to reconstruct a particular flow feature. Examples include vertical wind profiles in the ABL, 2D horizontal velocity fields of atmospheric surface layer flow, and 3D velocity fields of wind turbine wakes. The reconstructed flow features can be compared to those obtained from the complete three-dimensional, unsteady LES flow fields as a way to estimate the errors introduced by the assumptions used in the algorithm, by the spatial interpolation, and by the limited number of samples in time (statistical error). This also allows for the optimization of scanning patterns in order to minimize the uncertainty for a particular type of experiment.

Next, details are provided about the LES simulations used in this study, the characteristics of the virtual lidar and its scanning strategy, the algorithm for the reconstruction of the 3D flow field, and the optimization process.

2.1. LES Simulations

The *virtual measurements* are extracted from the results of LES simulations of the interaction of a single turbine with atmospheric boundary layer flow on flat terrain. The WiRE-LES code, described in detail in [37–39], was used for the simulations.

2.1.1. Turbulence Model, Boundary Conditions, and Numerical Methods

LES solves the spatially filtered Navier–Stokes equations and, therefore, solves explicitly all the scales of turbulence greater than the filter scale (same as the grid scale for implicit filters), while the subgrid-scale stresses (SGSs) are parameterized using a subgrid-scale model. In the case of the WiRE-LES code, the spatial derivatives are discretized using a pseudospectral representation for the horizontal directions (hence, periodic lateral boundary conditions) and second-order finite differences for the vertical direction, with a wall modeling based on the log law for the bottom boundary. The top boundary condition is a fixed stress-free lid. The code is fully dealiased using the 3/2 rule and the temporal advancement of the simulation uses a second-order accurate Adams–Bashforth scheme. The SGS turbulence model is the Lagrangian scale-dependent dynamic model detailed in [40,41].

2.1.2. Domain Size and Resolution

The domain size is 3200 m in the longitudinal direction (*x*), 800 m in the transversal direction (*y*), and 500 m in the vertical direction (*z*). The domain is divided uniformly into $160 \times 60 \times 64$ grid points, respectively, which yields a spatial resolution of 20 m in *x*, 13.3 m in *y*, and 7.8 m in *z*.

Since in the horizontal direction, there are periodic boundary conditions, a buffer zone upstream of the wind turbine is required in order to create an undisturbed incoming flow. The inflow condition is obtained via a separate precursor simulation.

2.1.3. Inflow

The boundary-layer flow in the simulations is driven by a constant streamwise pressure gradient over flat homogeneous surfaces and is neutrally stratified. Five inflow conditions are used in order to study the influence of different wind speeds and different turbulence intensities. The different turbulence intensities are recreated by using different surface roughness lengths. They are chosen to cover a very wide range between the lowest value corresponding to water or sand (0.0002 m) to grass-covered land (0.005 m) and finally the highest roughness corresponding to suburban or forestal land (0.5 m) [42]. Different wind speeds at hub height are achieved by modifying the forcing longitudinal pressure gradient dP/dx. The wind speeds selected (6, 7.5, and 9 m/s) cover the range in which the turbine operates at maximum efficiency $C_P \simeq 0.42$ and maximum thrust $C_T \simeq 0.80$.

The nomenclature of the different inflow cases is (X)V-(X)T, where V denotes wind velocity, T denotes turbulence intensity, and finally (X) can be L-low, M-medium, or H-high (e.g., MV-HT indicates medium wind speed and high turbulence intensity). Figure 1 shows the different inflow conditions for the five simulations.

The coordinate system used in this manuscript has its origin at the center of the rotor, and the longitudinal, transversal, and vertical directions are represented by X, Y, and Z, respectively. Distances appear normalized by the rotor diameter D, which is 80 m, as described in the section below.

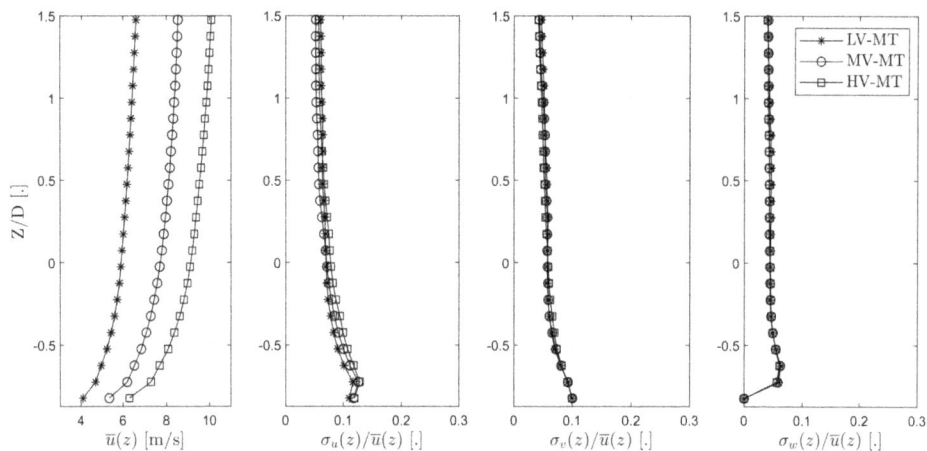

(**a**) Different wind speeds.

Figure 1. *Cont.*

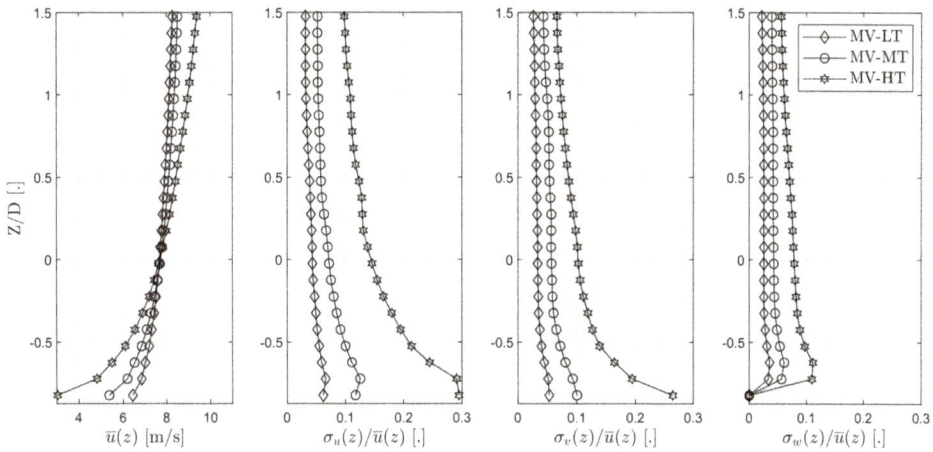

(b) Different turbulence intensities.

Figure 1. Inflow conditions in terms of average horizontal wind speed and longitudinal, transversal and vertical turbulence intensity. The panels on top (**a**) show three cases with the same turbulence intensity (medium) but different wind speeds, while the panels at the bottom (**b**) show three cases with the same wind speed (medium) at hub height but different turbulence intensities. The resulting average transversal and vertical velocities from the LES simulations are negligible.

2.1.4. Turbine Modeling

The simulation of the aerodynamic forces of the wind turbine and their interaction with the ABL flow is performed through an actuator disk model that includes rotation as described in [37,38], where the lift and drag forces of the wind turbine blades are calculated using Blade-Element Momentum theory (BEM) and distributed in a Gaussian manner by convolving the local load and a regularization kernel [43] and integrated over the spatial and temporal resolution of the simulation.

The wind turbine simulated is a V80-2.0 MW Vestas wind turbine with a hub height of 70 m and a rotor diameter of 80 m. All the details of the modeling of the wind turbine can be found in [39]. The results of the simulations are shown in Figure 2 as vertical planes at Y = 0 of the temporal average of the longitudinal component of the wind speed $\bar{u}$ and its standard deviation σ_u.

The simulations were first run for a period of time long enough to achieve stationary flow conditions, and a period of 30 min was then used to sample the simulation results at a resolution of 2 Hz.

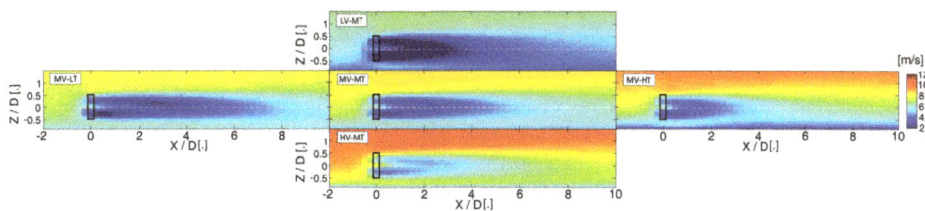

(a) $\bar{u}(x,z)$ for the different cases studied.

Figure 2. *Cont.*

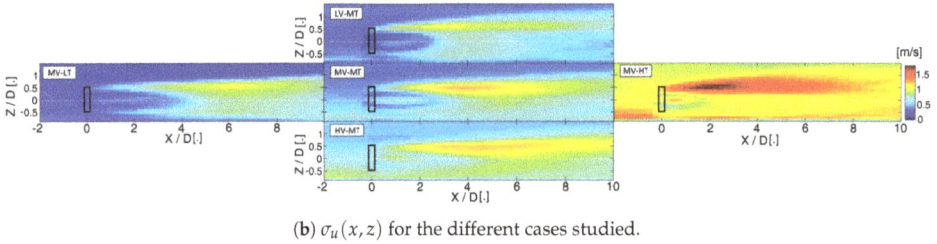

(**b**) $\sigma_u(x,z)$ for the different cases studied.

Figure 2. Flow statistics of the simulated wind turbine wakes for the five different inflow cases studied. The figures present vertical planes at Y = 0 of the longitudinal component of the wind velocity in terms of its average (**a**) and its standard deviation (**b**). The black boxes correspond to the location of the wind turbine rotor.

2.2. Lidar Virtual Measurements

The lidar simulated in this study is a Halo-Photonics Streamline lidar. It is a pulsed Doppler scanning lidar, which means that it provides measurements of the radial velocity Vr (the projection of the wind velocity into the LoS) at regular distance intervals along the laser beam direction. In this case, the Streamline lidar is able to provide a spatial resolution of 18 m along its LoS, at an acquisition frequency of 2 Hz. The range has been set to infinite since, on our own experience, it exceeds 1.5 km under most atmospheric conditions.

The virtual radial velocity measurements taken by the lidar are then calculated as

$$Vr(t) = u(x,y,z,t)cos(\theta(t))cos(\varphi(t)) + v(x,y,z,t)cos(\theta(t))sin(\varphi(t)) + w(x,y,z,t)sin(\theta(t)) \quad (1)$$

where u, v, and w are the components of the wind velocity vector field from the LES simulations, and are defined in space (x,y,z) and time (t), φ is the azimuth angle (angle between the laser beam and the the vertical plane at Y = 0 of the wake), and θ is the elevation angle (angle between the laser beam and a horizontal plane) that define the orientation of the laser beam in time.

The formally correct calculation of the virtual radial velocity for each laser gate of 18 m would be a convolution of the envelope of the laser pulse on the simulated and projected velocity field [44]. In this study, the convolution is not calculated and the result is instead a point-wise calculation in the linearly interpolated projected field every 18 m. The reason for this is that the resolution of the LES simulation (20 m in the longitudinal direction) is almost the same as the lidar one. This means that the lidar convolution is similar to the spatial filtering already performed by the LES.

The effect of the spatial convolution of high-resolution (20 m) lidar measurements in the attenuation of the longitudinal turbulence intensity has been studied by comparing lidar measurements to sonic anemometry [27]. These experiments, under arguably less favorable conditions than the virtual setup presented here, have shown that the attenuation is in the order of a few tenths of a percentage point, since most of the energy-containing scales are resolved at that resolution. These results may not be extended to lower resolution measurements. No significant effect is expected when reconstructing average velocity fields.

It must be noted that in this study the accuracy of the lidar (instrument error) is neglected. The virtual measurement of the radial velocity (Equation (1)) could include a random error that varies for each instrument and that it depends on the signal-to-noise ratio (SNR) of each measurement. This, in turn, depends on the quality of the different optical parts of the instrument, the power of the laser pulse, the aerosol content, the humidity, and other parameters related to the processing of the Doppler signal such as the length of the laser gate or the number of pulses averaged. It has been shown that a similar pulsed Doppler lidar to the one discussed in this study can achieve, with favorable SNR, accuracies in the order of ±0.1 m/s or lower [45]. In theory, the effect of simulating this error for any particular instrument will have the effect of slightly increasing the statistical error (in a random manner, therefore no bias introduced) and slightly increasing the standard deviation measurements (therefore, inducing a small bias). Nevertheless, the instrument error of an accurate lidar should be significantly smaller than the turbulent velocity fluctuations found in the wake of a wind turbine and can, in most of the cases, be neglected.

Finally, if a continuous scanning strategy is chosen, the radial velocity calculation has to include the convolution along the arc that the lidar gate covers between two consecutive measurements. This effect can be particularly important when calculating the standard deviation of the radial velocity for locations far downstream and for low angular resolutions of the measurements. On the other hand, a step-and-stare scan avoids this convolution and arguably provides a better estimation of the turbulent fluctuations. This is why only step-and-stare scans are considered in this study.

2.3. Scanning Strategy and Reconstruction of 3D Fields

Depending on the position of the lidar, the ranges of the azimuth and elevation angles are calculated in order to cover the whole volume of interest and are shown in Table 1. A pulsed scanning lidar can only measure along a particular direction each time, which implies that there will have to be a compromise between the number of measurements at each point and the angular resolution. Due to the wake being quasi-axial symmetric, the angular resolution is kept the same for the elevation and the azimuth angles. Once a particular angular resolution is set, the virtual lidar scans the volume of interest in consecutive step-and-stare swipes at constant elevation angles (equivalent to PPI scans) at a frequency of 2 Hz between measurements until the end of the 30 min period of each simulation.

Table 1. Ranges of the azimuth and elevation angles needed to cover the whole volume of interest for both lidar locations.

	Nacelle-Mounted	Ground-Based
Azimuth angle range	$-16.7°$ to $+16.7°$	$-16.7°$ to $+16.7°$
Elevation angle range	$-16.7°$ to $+16.7°$	$+1.3°$ to $+34.0°$

The reconstruction of the longitudinal velocity field at the volume of interest is similar to the procedure described in [16,18] and starts by calculating the average of the radial velocity measurements $\overline{Vr}(\varphi, \theta, r)$ for each laser beam orientation determined by the angles φ and θ and distance r, creating a regular spherical grid with the origin at the lidar location. It continues with the assumption of a negligible effect of the average transversal and vertical components of the velocity into the projection on the laser beam direction: $\overline{v}(x, y, z)\cos(\theta)\sin(\varphi) = 0$; $\overline{w}(x, y, z)\sin(\theta) = 0$. This allows one to reconstruct the longitudinal component of the velocity from the radial velocity measurement simply by

$$\overline{u}(\varphi, \theta, r) = \frac{\overline{Vr}(\varphi, \theta, r)}{\cos(\theta)\, \cos(\varphi)} \, . \tag{2}$$

The error of this assumption is simply calculated as

$$\varepsilon = \overline{v}(x, y, z)\cos(\theta)\sin(\varphi) + \overline{w}(x, y, z)\sin(\theta) \, . \tag{3}$$

The reconstruction of the standard deviation field of the longitudinal velocity component uses a different assumption (stronger than the previous one), which directly equates the variations of the instantaneous radial velocity to those of the longitudinal velocity component:

$$\sigma_u(\varphi, \theta, r) = \sigma_{Vr}(\varphi, \theta, r) \, . \tag{4}$$

Lastly the values in the regular spherical grid (φ, θ, r) are converted to the original Cartesian grid of the LES simulations (x, y, z) via a linear interpolation. These values constitute the final reconstructed velocity fields from virtual lidar measurements that can be compared directly to the original LES fields.

2.4. Optimization

The approach detailed in this section allows one to study three different error sources:

- the error of the assumption of unidirectional flow, which depends on the average spanwise and vertical velocity components $\bar{v}(x, y, z)$ and $\bar{w}(x, y, z)$ of the wake flow field and the position of the lidar, which determines the angles θ and φ at which, in turn, the laser beam operates;
- the statistical error when calculating the average radial velocity at each point in the spherical grid $\overline{Vr}(\varphi, \theta, r)$, which depends on the number of independent lidar measurements for each orientation of the laser beam;
- the interpolation error when converting $\bar{u}(\varphi, \theta, r)$ in the spherical grid to $\bar{u}(x, y, z)$ in the original Cartesian grid, which depends on the angular resolution between consecutive laser beam orientations.

The error associated with the unidimensional average flow assumption is independent from the other two and the only way to minimize it is by locating the lidar in a different position, effectively changing the orientation angles θ and φ of the laser beam, as expressed in Equation (3). Two lidar locations are tested in this study: ground-based at the tower base and nacelle-mounted.

The statistical error and the interpolation error are linked by the fact that the lidar can only measure at one laser beam orientation at a time. A coarser angular resolution will mean more measurements along each orientation (thus, a higher interpolation error but lower statistical one) and vice versa. An optimum compromise can be found in which the sum of both errors is minimum. Seven different angular resolutions are tested, and they imply a number of measurements along each orientation during a 30 min period for a 2 Hz sampling rate, as shown in Table 2. For both lidar locations (ground and nacelle) the range of angles that the lidar has to cover in order to scan the whole region of interest is very similar (see Section 2.3) and therefore the number of repetitions is the same in both cases. For each of these cases, the total error is studied and an optimum compromise is found.

Table 2. Angular resolutions used for the 3D scan optimization and corresponding number of repetitions for each laser beam orientation. Note that the elevation and azimuthal angular resolutions are kept the same.

Angular resolution	1°	1.5°	2°	2.5°	3°	4°	5°
Number of repetitions	3	7	12	21	29	44	73

As detailed above, this study does not include other sources of error, such as the effect of the laser pulse convolution, other errors associated with possible non-stationarity and non-uniformity of the atmospheric flow or the error on the radial velocity measurement (instrument error).

3. Results

This section presents the results of all the calculations of the errors for the reconstruction of the longitudinal velocity field in terms of its average value $\bar{u}(x, y, z)$ and standard deviation $\sigma_u(x, y, z)$. All errors calculated in this study are presented as absolute values (no difference between positive and negative values), whether they are expressed dimensionally (in m/s) or as a percentage value. The average of the real values of the errors within the volume of interest is close to zero, excluding significant biases in the reconstruction of the different cases.

3.1. Error of the Average Longitudinal Velocity Component

The error associated with the reconstruction of the average longitudinal velocity component has, as discussed in Section 2.4, three sources. The first one, which is the assumption of unidirectional average flow, is independent from the other two and it can be treated separately. The remaining two are the statistical and interpolation errors, which are linked by the lidar measurement frequency, as described previously. Together, the three error sources conform the total error.

3.1.1. Error Associated with the Assumption of Unidirectional Average Flow

The calculation of the error associated with the assumption of unidirectional average flow has three steps: The first step is fixing the location of the lidar as a point in the virtual flow field and calculating the angles φ and θ that correspond to the laser orientation from the lidar to every point in the flow field. The second one is calculating the radial velocity $\overline{Vr}$ from Equation (1) using the average values $\overline{u}, \overline{v}, \overline{w}$ instead of the instantaneous ones. The third one is applying the assumption of unidirectional average flow and reconstructing the longitudinal velocity field by using Equation (2). Since the error is evaluated at exactly the same grid points as the LES simulation, there is no interpolation error, and since the average velocity components are used, there is no statistical error.

One example of error fields for the inflow case MV-MT is shown in Figure 3 for a nacelle-mounted lidar and for a ground-based lidar situated at the base of the tower. The origin of the error within the volume of interest can be divided into two: first, the deviation from the assumption of unidirectional average flow (i.e., non-zero spanwise and vertical average velocity components), which is greater in the near wake due mostly to the tangential induction of the rotor, and, second, the angle between the laser beam orientation and the x axis (given by φ and θ), whose magnitude is larger for those points at the most upstream outer edges of the region of interest, and it is smaller for the nacelle-mounted lidar case. Therefore, it is easy to notice that the errors are greater when measuring with a ground-based lidar, and it is particularly well illustrated in the transversal planes at a downstream distance of 3D, as shown in Figure 3a,b . While the $\overline{v}$ and $\overline{w}$ components are the same in both cases, a ground-based lidar requires higher elevation angles (θ). This is responsible for errors reaching up to 1.5%, while for the nacelle-based lidar they never exceed 0.25%.

Table 3 shows the errors associated with the assumption of unidirectional average flow for the five different inflow cases studied and for both nacelle-mounted and ground based lidars. The table shows average and maximum errors for the volume of interest already defined in Section 2.3. It is possible to extract three main conclusions from it:

- The assumption of unidirectional average flow when reconstructing the average longitudinal component of the wind velocity $\overline{u}$ is a good approximation, because of the low average and maximum errors, for the study of the far wake for all inflow cases and both lidar locations.
- The nacelle-mounted lidar yields lower average and maximum errors (between two and five times lower) than the ground-based lidar situated at the bottom of the tower.
- Different inflow conditions do not affect significantly the relative magnitude of the error associated with the assumption of unidirectional average flow.

From the results presented in this section, it can be concluded that the best option to minimize only the error associated with the assumption of unidimensional average flow is to use a nacelle-mounted lidar, although it does not give a very significant advantage since both lidar positions provide acceptably low errors for most analysis purposes.

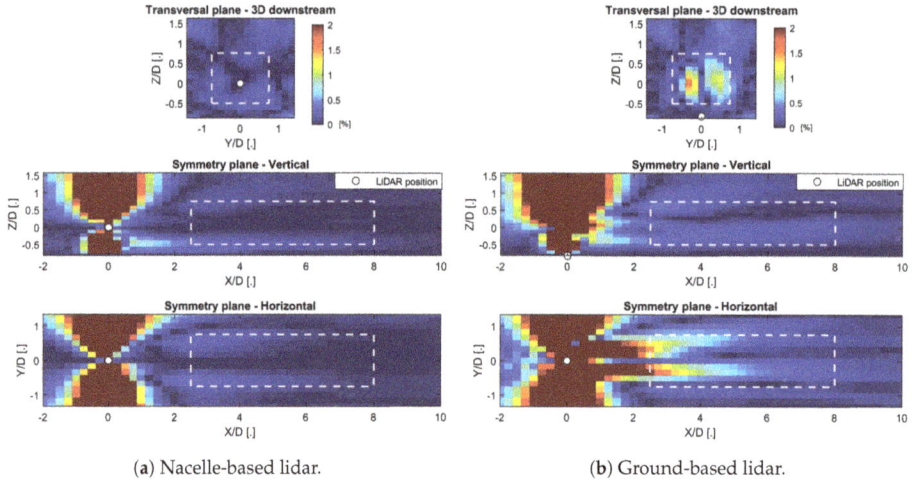

(a) Nacelle-based lidar.   (b) Ground-based lidar.

Figure 3. Example of errors in the reconstruction of the longitudinal average velocity component associated only with the assumption of unidimensional flow. The errors correspond to the MV-MT inflow case and for the lidar installed on the nacelle (**a**) and at the base of the tower (**b**). The volume of interest is delimited by the white dashed line. The figures show the error as a percentage of the undisturbed wind speed at hub height (see Section 2.1.3).

Table 3. Errors in the reconstruction of the average longitudinal velocity component associated only with the assumption of unidimensional flow. The errors are presented for the different inflow cases studied in terms of average and maximum values found within the volume of interest (see Section 2.3 and white dashed lines in Figure 3) for both nacelle-mounted and ground-based lidars. Percentages are based on the undisturbed wind speed at hub height (see Section 2.1.3).

	Average Error [%]		Maximum Error [%]	
	Nacelle	Ground	Nacelle	Ground
MV-MT	0.09	0.29	0.29	1.73
MV-LT	0.10	0.29	0.28	1.87
MV-HT	0.12	0.24	0.59	1.23
LV-MT	0.16	0.25	0.72	1.47
HV-LT	0.06	0.22	0.30	1.44

3.1.2. Total Error

The total error includes, on top of the error discussed in the previous section, the effects of the limited number of samples for statistics calculations and the error made when interpolating back from a spherical grid to a Cartesian one. The calculation of the total error follows these steps:

1. For a given lidar position (nacelle or ground) the range of maximum and minimum values of the azimuth φ and elevation θ angles is calculated. Then, for a given angular resolution, the laser beam performs consecutive PPI scans until the end of the 30 min periods. This determines the evolution of the elevation and azimuth angles in time $\theta(t)$ and $\varphi(t)$. The number of repetitions or samples for each orientation is shown in Table 2.
2. For each time step, the virtual lidar measurement is simulated by calculating the radial velocity Vr at each point along the laser beam using Equation (1).

3. The average of the radial velocity is calculated at each point in space in which the virtual lidar obtains the measurements, creating a spherical regular grid with its origin at the lidar location $\overline{Vr}(\varphi, \theta, r)$.

4. The assumption of unidirectional average flow is used to calculate the average longitudinal component of the wind speed $\overline{u}(\varphi, \theta, r)$ as shown in Equation (2).

5. The data in the spherical grid are interpolated linearly to the original Cartesian grid to obtain $\overline{u}(x, y, z)$.

6. The difference between the reconstructed $\overline{u}(x, y, z)$ and the same variable obtained from the original LES results conforms the error at each point in space. The average and maximum values of the error inside the volume of interest are computed.

The statistical uncertainty of the average of the radial velocity is inversely proportional to the square root of the number of independent samples $\sqrt{N}$ [46]. Since the measurement frequency of the lidar has an upper limit, a higher angular resolution means fewer measurements at each point in space for a given period and vice versa. Thus, a higher angular resolution yields a lower interpolation error, but a higher statistical error. This holds true until the time between measurements at each point in space approaches the integral time scale, which is considered the time between statistically independent measurements. It must be noted, though, that samples along the same laser beam are correlated, and those corresponding to the same PPI scans are unlikely to be independent of each other since they are consecutive measurements in time.

The linear interpolation error, in turn, is proportional to the gradient of the spatial derivative of the average flow field and proportional to the distance between the measurement points of the spherical grid. The horizontal gradient of the wind speed is greater close to the wind turbine, while the distance between measurement points increases with increasing distance to the lidar location and decreasing angular resolution.

Figure 4 shows an example of the total errors for different angular resolutions for the case MV-MT and a nacelle-based lidar. The first noticeable fact is that the angular resolution of 3° (Figure 4b) shows the lowest errors and therefore is a good compromise between angular resolution and the number of samples or repetitions at each point. On the other hand, a resolution of 1° (Figure 4a) shows a high statistical error, evident by the fact that the errors are randomly distributed (except in the horizontal plane, since it is based on the same PPI scan as explained above), while a resolution of 5° (Figure 4c) shows a high interpolation error, visible between the white dots that represent the points at which the virtual lidar takes measurements and from which interpolation then takes place (this is most noticeable at approximately Z/D= 0.1 and Y/D = −0.1).

Table 4 presents an example of an optimization of the angular resolution for the inflow case MV-MT for both lidar locations. As discussed in the previous paragraphs, when scanning the wake with either a high angular resolution (1°) or a low one (5°), the errors are greater than when using a compromise resolution. The optimum value in the two cases presented is 3°, which balances the statistical errors and the interpolation errors. An extension of this table for all inflow cases studied is presented in Appendix A, Table A1, where it is possible to identify the optimum angular resolution for each case. For most cases, the optimum value is still 3°, while for a few cases it is 2.5° or 4°, although the error is not significantly sensitive in this range of angular resolutions. It can therefore be concluded that 3° is the overall optimum angular resolution.

Table 5 shows the error associated with each inflow case when using an optimum angular resolution of 3°. It can be seen that the errors are reasonably low for most analysis purposes and similar for all inflow cases. Contrary to what is discussed in Section 3.1.1, the nacelle-mounted and ground-based lidars show similar results when considering the total error.

Four main conclusions can be derived from all the information presented in this section:

*   An optimum angular resolution which balances the statistical error and the interpolation error can be found.

- The optimum angular resolution is nearly the same for all inflow cases and both lidar locations, and the overall optimum value is 3°.
- Different inflow cases or lidar locations do not affect significantly the optimized total error.
- The total errors found inside the volume of interest are low (average error smaller than 2% and maximum error lower than 8%), which are deemed acceptable for most applications.

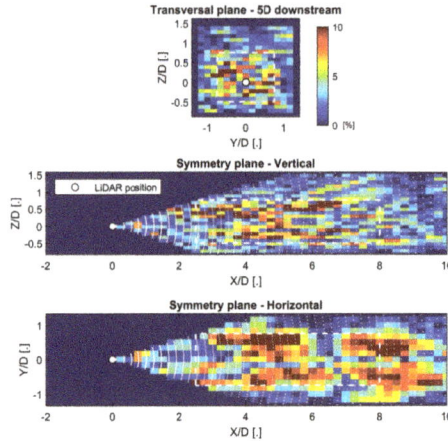

(**a**) Angular resolution 1°.

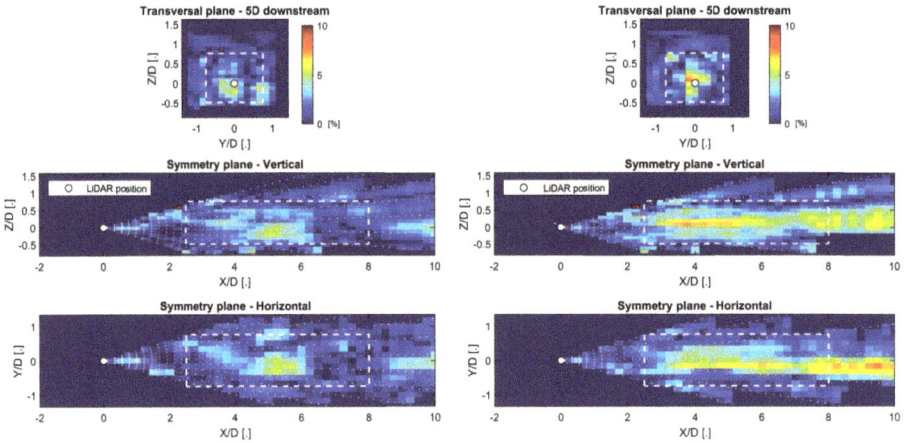

(**b**) Angular resolution 3°.

(**c**) Angular resolution 5°.

Figure 4. Example of total errors in the reconstruction of $\bar{u}(x, y, z)$. The errors correspond to the MV-MT inflow case with the nacelle-mounted lidar and for three different angular resolutions: 1° (**a**), 3° (**b**), and 5° (**c**). The volume of interest is delimited by the white dashed line, and spherical grid points are marked as white dots. The figures show the error as a percentage of the undisturbed wind speed at hub height (see Section 2.1.3).

Table 4. Example of optimization of the angular resolution for one inflow case and both lidar locations. The table presents the average and maximum total errors for the reconstruction of the average longitudinal velocity component inside the volume of interest. The errors are shown as a percentage of the undisturbed wind speed at hub height (see Section 2.1.3).

Angular Resolution	1°	1.5°	2°	2.5°	3°	4°	5°
	MV-MT nacelle-mounted						
Average error [%]	5.2	2.6	1.6	1.8	1.5	1.7	2.5
Maximum error [%]	18.6	12.2	5.8	6.9	5.4	5.7	8.0
	MV-MT ground-based						
Average error [%]	4.7	2.5	1.7	1.7	1.6	1.8	2.5
Maximum error [%]	18.9	15.3	7.7	7.8	6.3	6.9	8.5

Table 5. Total average and maximum errors for the reconstruction of $\bar{u}(x, y, z)$ inside the volume of interest for all inflow cases studied and both lidar locations with the overall optimum angular resolution of 3°. The errors are shown as a percentage of the undisturbed wind speed at hub height (see Section 2.1.3). The optimum values found in Table 4 correspond to the first column of this table.

	Nacelle-Mounted				
	MV-MT	MV-LT	MV-HT	LV-MT	HV-MT
Average error [%]	1.5	1.9	1.4	1.6	1.7
Maximum error [%]	5.4	7.1	5.7	5.8	5.4
	Ground-Based				
	MV-MT	MV-LT	MV-HT	LV-MT	HV-MT
Average error [%]	1.6	2.0	1.8	1.5	1.4
Maximum error [%]	6.3	7.6	6.5	7.0	6.6

3.2. Error of the Standard Deviation of the Longitudinal Velocity Component

The total error of the reconstruction of the standard deviation of the longitudinal velocity component includes the error of the assumption detailed in Equation (4), the statistical error, and the interpolation error. The calculation follows the same steps presented in Section 3.1.2, except for Point 3, where $\sigma_{Vr}(\varphi, \theta, r)$ is calculated instead of $\bar{u}(\varphi, \theta, r)$, and Point 4, in which Equation (4) is used instead of Equation (2). The same considerations regarding the statistical uncertainty and the interpolation error explained previously are also valid for the reconstruction of the standard deviation of the longitudinal velocity component.

Figure 5 illustrates the impact of the total error on the reconstruction of the standard deviation of the longitudinal velocity component field $\sigma_u(x, y, z)$ for different angular resolutions for the case MV-MT and the nacelle-based lidar. It must be noted that, in this figure, for the sake of clarity, the magnitude of $\sigma_u(x, y, z)$ is used instead of the error. It can be seen that a high angular resolution (Figure 5b) results in a poor performance, while coarser resolutions (Figure 5c,d) yield visibly lower errors.

All the errors associated with each inflow case and both lidar locations are shown in Appendix A, Table A2. The minimization of the average and maximum errors found inside the volume of interest indicates that the optimum values for the angular resolution are always between 2.5 and 4°, so 3° is chosen again as an overall optimum angular resolution.

(**a**) LES simulation.

(**b**) Angular resolution 1°.

(**c**) Angular resolution 3°.

(**d**) Angular resolution 5°.

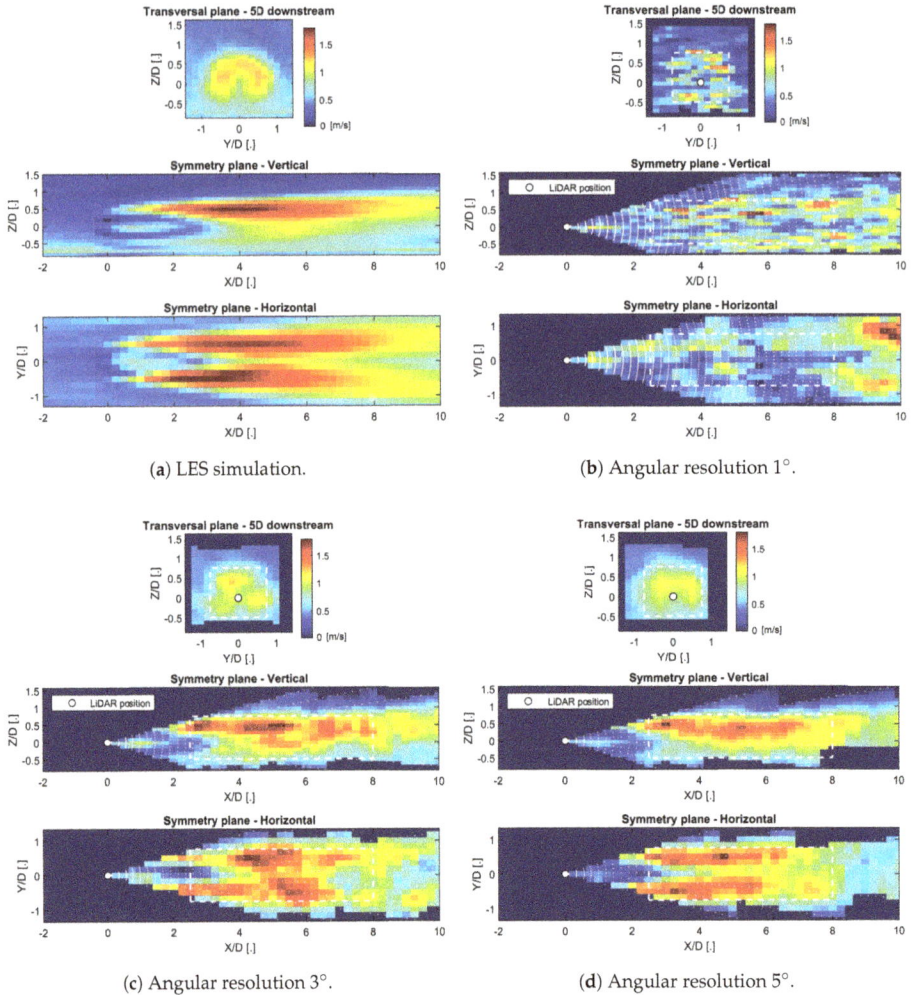

Figure 5. $\sigma_u(x,y,z)$ from (**a**) LES simulations and an example of its reconstruction using three different angular resolutions: (**b**) 1°, (**c**) 3°, and (**d**) 5°. The example corresponds to the MV-MT inflow case and to the nacelle-mounted lidar. The volume of interest is delimited by the white dashed line, and spherical grid points are marked as white dots. The figures show the magnitude of $\sigma_u(x,y,z)$ in m/s.

Table 6 shows only the errors associated with an angular resolution of 3°. The errors are close in magnitude, although slightly higher, to the errors of $\bar{u}(x,y,z)$ (see Table 5). This fact might induce some confusion when comparing them. However, it should be noted that the values of σ_u are significantly smaller than the values of $\bar{u}$, so the accuracy of the calculation of the standard deviations is considerably poorer than that of the mean velocity. It can also be observed that a nacelle-mounted lidar offers a slightly better performance than a ground-based one.

Table 6. Total average and maximum errors for the reconstruction of $\sigma_u(x, y, z)$ inside the volume of interest for all inflow cases studied and both lidar locations with the overall optimum angular resolution of 3°. The errors are shown as a percentage of the undisturbed wind speed at hub height (see Section 2.1.3).

	Nacelle-Mounted				
	MV-MT	**MV-LT**	**MV-HT**	**LV-MT**	**HV-MT**
Average error [%]	1.9	1.7	1.9	2.1	1.6
Maximum error [%]	6.0	6.4	6.7	7.2	5.7
	Ground-Based				
	MV-MT	**MV-LT**	**MV-HT**	**LV-MT**	**HV-MT**
Average error [%]	2.4	2.0	2.9	2.5	2.0
Maximum error [%]	7.1	6.5	8.9	8.1	5.8

The findings in this section can be summarized in three main points:

- The overall optimum angular resolution for the reconstruction of σ_u is also 3°.
- The reconstruction of the $\sigma_u(x, y, z)$ field is of worse quality than that of $\bar{u}(x, y, z)$.
- A nacelle-mounted lidar offers a slight advantage over a ground-based one.

4. Conclusions

This study exploits the potential of the virtual lidar technique to explore the uncertainty associated with a volumetric scan of the wake of a wind turbine. When performing a succession of PPI scans, an optimum angular resolution can be found that minimizes the errors when calculating the longitudinal velocity field in terms of its average and its standard deviation. This kind of analysis is important during the experiment design part, prior to a lidar measurement campaign, in order to optimize the scanning pattern. First, the optimization dictates the best way to perform the scan, and the quantification of the error estimates the expected quality of the measurements, which will determine if they are acceptable (accurate enough) or not for a particular purpose.

Our analysis has found that, when performing a volumetric scan of the far wake of an 80 m diameter wind turbine, an angular resolution close to 3° provides the best overall results. This holds true for a lidar with a measurement frequency of 2 Hz and a spatial resolution of 18 m. The accuracy of experiments with lidars with different characteristics can be studied using the same methodology. The study has also shown that different turbulence intensity conditions and different wind speeds do not seem to affect the quality of the measurements significantly. The location of the lidar does not seem to play a significant role on the magnitude of the errors, although a nacelle-mounted lidar has a slight advantage over a ground-based one. Other practical advantages associated with a nacelle-mounted lidar, such as a constant angular orientation with the turbine rotor, are not discussed in this study.

Finally, the errors found when reconstructing the average longitudinal velocity component are low, regardless of the configuration or inflow case studied. This fact suggests that full scale measurements using the setup and data processing detailed previously should be of high quality and potentially acceptable for most applications. On the other hand, the error associated with the reconstruction of the standard deviation of the longitudinal velocity component is almost identical in magnitude to that of the average velocity, which means that, in relative terms, it is higher. This should be considered when assessing the acceptability of the lidar measurement of this variable, depending on its final purpose.

Future studies could address the effect of the thermal stability of the ABL, the wind veer, non-stationary conditions, or the horizontal inhomogeneity of the undisturbed boundary layer within the wind turbine wake region.

Author Contributions: This study was done as a part of Fernando Carbajo Fuertes doctoral studies supervised by Fernando Porté-Age.

Funding: This research was supported by the Swiss National Science Foundation [grant 200021-172538], the Swiss Federal Office of Energy [grant SI/501337-01], and the Swiss Innovation and Technology Committee (CTI) within the context of the Swiss Competence Center for Energy Research 'FURIES: Future Swiss Electrical Infrastructure.

Acknowledgments: The authors are thankful for the valuable help of Ka Ling Wu, who performed the LES simulations used in this study.

Conflicts of Interest: The authors declare no conflict of interest. The founding sponsors had no role in the design of the study; in the collection, analyses, or interpretation of data; in the writing of the manuscript; or in the decision to publish the results.

Abbreviations

The following abbreviations are used in this manuscript:

ABL	Atmospheric Boundary Layer
BEM	Blade Element Momentum
DBS	Doppler Beam Swinging
CFD	Computational Fluid Dynamics
LES	Large Eddy Simulation
Lidar	LIght Detection And Ranging
LoS	Line-of-Sight
SGS	Subgrid Scale
SNR	Signal-to-Noise Ratio
PPI	Plan Position Indicator

Appendix A. Total Errors for All Cases Studied

Table A1. Total average and maximum errors for the reconstruction of $\bar{u}(x,y,z)$ inside the volume of interest for all inflow cases studied and both lidar locations. The errors are shown as a percentage of the undisturbed wind speed at hub height (see Section 2.1.3). The errors corresponding to an optimum angular resolution for each case are identified with an asterisk.

Angular Resolution	1°	1.5°	2°	2.5°	3°	4°	5°	1°	1.5°	2°	2.5°	3°	4°	5°
	MV-MT nacelle-mounted							MV-MT ground-based						
Average error [%]	5.2	2.6	1.6	1.8	1.5*	1.7	2.5	4.7	2.5	1.7	1.7	1.6*	1.8	2.5
Maximum error [%]	18.6	12.2	5.8	6.9	5.4*	5.7	8	18.9	15.3	7.7	7.8	6.3*	6.9	8.5
	MV-LT nacelle-mounted							MV-LT ground-based						
Average error [%]	3.9	2.4	2.5	1.7*	1.9	2.5	3.7	3.4	2.4	2.4	1.9*	2	2.5	3.6
Maximum error [%]	17.2	10.7	12.1	5.2*	7.1	8.5	11.7	21.9	10.4	11.2	7.3*	7.6	8.9	11.5
	MV-HT nacelle-mounted							MV-HT ground-based						
Average error [%]	5.8	3.3	2.5	2.1	1.4	1.3*	1.5	6.2	3.6	3	1.8	1.8	1.3*	1.5
Maximum error [%]	21.7	20.7	10.6	7.1	5.7	5*	12.6	26.9	15.6	11.2	6.3	6.5	5.9*	6.6
	LV-MT nacelle-mounted							LV-MT ground-based						
Average error [%]	4.6	3.5	2.8	2	1.6	1.4*	2.2	4.4	3.8	2.7	2.5	1.5*	1.6	2.3
Maximum error [%]	24.8	21.2	11	7.5	5.8*	6.3	9	20.7	18	11.7	11.2	7	6.2*	8.8
	HV-MT nacelle-mounted							HV-MT ground-based						
Average error [%]	4.1	3.3	2	1.8	1.7	1.6*	2.3	4.7	2.6	2.2	1.8	1.4*	1.7	2.2
Maximum error [%]	28.7	17.6	8.3	8.6	5.4*	6.3	9.2	18	13.2	9.6	7.4	6.6	6.1*	8.7

Table A2. Total average and maximum errors for the reconstruction of $\sigma_u(x,y,z)$ inside the volume of interest for all inflow cases studied and both lidar locations. The errors are shown as a percentage of the undisturbed wind speed at hub height (see Section 2.1.3). The errors corresponding to an optimum angular resolution for each case are identified with an asterisk.

Angular Resolution	1°	1.5°	2°	2.5°	3°	4°	5°	1°	1.5°	2°	2.5°	3°	4°	5°
	MV-MT nacelle-mounted							MV-MT ground-based						
Average error [%]	5.5	2.9	2.2	2.0	1.9 *	2.0	2.1	5.2	3.2	2.5	2.4	2.4	2.2 *	2.5
Maximum error [%]	15.0	11.9	8.5	6.6	6.0	5.0 *	5.3	14.6	11.3	9.8	8.8	7.1	6.4 *	6.5
	MV-LT nacelle-mounted							MV-LT ground-based						
Average error [%]	4.2	3.1	2.5	1.7 *	1.7 *	1.9	2.1	4.4	2.7	2.8	2.0 *	2.0 *	2.0 *	2.3
Maximum error [%]	12.7	9.2	8.8	6.4 *	6.4 *	6.7	7.7	13.5	9.3	8.3	7.2	6.5	6.4 *	7.2
	MV-HT nacelle-mounted							MV-HT ground-based						
Average error [%]	6.3	3.6	2.9	2.0	1.9	1.7 *	2.0	6.9	3.9	3.6	2.8	2.9	2.6 *	2.7
Maximum error [%]	19.8	12.8	11.5	8.1	6.7	6.1	5.4 *	19.6	13.3	13.4	9.4	8.9	7.2 *	8.2
	LV-MT nacelle-mounted							LV-MT ground-based						
Average error [%]	5.3	3.5	2.8	2.3	2.0 *	2.1	2.3	5.9	3.8	2.8	2.7	2.5	2.4 *	2.7
Maximum error [%]	16.0	12.2	9.2	7.2	8.0	7.2	6.8 *	16.8	13.4	11.0	10.0	8.1	7.2 *	8.0
	HV-MT nacelle-mounted							HV-MT ground-based						
Average error [%]	4.6	2.7	1.8	1.6 *	1.6 *	1.6 *	1.7	4.7	2.7	2.3	2.0	2.0	1.9 *	2.1
Maximum error [%]	14.2	10.8	6.7	5.6	5.7	5.7	5.1 *	13.9	9.7	8.1	5.9	5.8 *	5.9	6.0

References

1. Vermeer, L.; Sørensen, J.; Crespo, A. Wind turbine wake aerodynamics. *Prog. Aerosp. Sci.* **2003**, *39*, 467–510. [CrossRef]
2. Barthelmie, R.J.; Pryor, S.C.; Frandsen, S.T.; Hansen, K.S.; Schepers, J.G.; Rados, K.; Schlez, W.; Neubert, A.; Jensen, L.E.; Neckelmann, S. Quantifying the impact of wind turbine wakes on power output at offshore wind farms. *J. Atmos. Ocean. Technol.* **2010**, *27*, 1302–1317. [CrossRef]
3. Hansen, K.S.; Barthelmie, R.J.; Jensen, L.E.; Sommer, A. The impact of turbulence intensity and atmospheric stability on power deficits due to wind turbine wakes at Horns Rev wind farm. *Wind Energy* **2012**, *15*, 183–196. [CrossRef]
4. Thomsen, K.; Sørensen, P. Fatigue loads for wind turbines operating in wakes. *J. Wind Eng. Ind. Aerodyn.* **1999**, *80*, 121–136. [CrossRef]
5. Herbert Acero, J.; Probst, O.; Réthoré, P.E.; Larsen, G.; Castillo-Villar, K. A Review of Methodological Approaches for the Design and Optimization of Wind Farms. *Energies* **2014**, *7*, 6930–7016. [CrossRef]
6. Gebraad, P.; Thomas, J.J.; Ning, A.; Fleming, P.; Dykes, K. Maximization of the annual energy production of wind power plants by optimization of layout and yaw-based wake control. *Wind Energy* **2017**, *20*, 97–107. [CrossRef]
7. Sørensen, J.N.; Shen, W.Z. Numerical Modeling of Wind Turbine Wakes. *J. Fluids Eng.* **2002**, *124*, 393. [CrossRef]
8. Bastankhah, M.; Porté-Agel, F. A new analytical model for wind-turbine wakes. *Renew. Energy* **2014**, *70*, 116–123. [CrossRef]
9. Crespo, A.; Hernández, J.; Frandsen, S. Survey of modelling methods for wind turbine wakes and wind farms. *Wind Energy* **1999**, *2*, 1–24. [CrossRef]
10. Kocer, G.; Mansour, M.; Chokani, N.; Abhari, R.; Müller, M. Full-Scale Wind Turbine Near-Wake Measurements Using an Instrumented Uninhabited Aerial Vehicle. *J. Sol. Energy Eng.* **2011**, *133*, 041011. [CrossRef]
11. Reuder, J.; Jonassen, M.O. First Results of Turbulence Measurements in a Wind Park with the Small Unmanned Meteorological Observer SUMO. *Energy Procedia* **2012**, *24*, 176–185. [CrossRef]
12. Wildmann, N.; Hofsäß, M.; Weimer, F.; Joos, A.; Bange, J. MASC—A small Remotely Piloted Aircraft (RPA) for wind energy research. *Adv. Sci. Res.* **2014**, *11*, 55–61. [CrossRef]
13. Subramanian, B.; Chokani, N.; Abhari, R. Experimental analysis of wakes in a utility scale wind farm. *J. Wind Eng. Ind. Aerodyn.* **2015**, *138*, 61–68. [CrossRef]

14. Subramanian, B.; Chokani, N.; Abhari, R.S. Drone-Based Experimental Investigation of Three-Dimensional Flow Structure of a Multi-Megawatt Wind Turbine in Complex Terrain. *J. Sol. Energy Eng.* **2015**, *137*, 051007. [CrossRef]

15. Käsler, Y.; Rahm, S.; Simmet, R.; Kühn, M. Wake measurements of a multi-MW wind turbine with coherent long-range pulsed doppler wind lidar. *J. Atmos. Ocean. Technol.* **2010**, *27*, 1529–1532. [CrossRef]

16. Iungo, G.V.; Wu, Y.T.; Porté-Agel, F. Field measurements of wind turbine wakes with lidars. *J. Atmos. Ocean. Technol.* **2013**, *30*, 274–287. [CrossRef]

17. Smalikho, I.N.; Banakh, V.A.; Pichugina, Y.L.; Brewer, W.A.; Banta, R.M.; Lundquist, J.K.; Kelley, N.D. Lidar investigation of atmosphere effect on a wind turbine wake. *J. Atmos. Ocean. Technol.* **2013**, *30*, 2554–2570. [CrossRef]

18. Iungo, G.V.; Porté-Agel, F. Volumetric lidar scanning of wind turbine wakes under convective and neutral atmospheric stability regimes. *J. Atmos. Ocean. Technol.* **2014**, *31*, 2035–2048. [CrossRef]

19. Banta, R.M.; Pichugina, Y.L.; Brewer, W.A.; Lundquist, J.K.; Kelley, N.D.; Sandberg, S.P.; Alvarez, R.J.; Hardesty, R.M.; Weickmann, A.M. 3D volumetric analysis of wind turbine wake properties in the atmosphere using high-resolution Doppler lidar. *J. Atmos. Ocean. Technol.* **2015**, *32*, 904–914. [CrossRef]

20. Aitken, M.L.; Banta, R.M.; Pichugina, Y.L.; Lundquist, J.K. Quantifying Wind Turbine Wake Characteristics from Scanning Remote Sensor Data. *J. Atmos. Ocean. Technol.* **2014**, *31*, 765–787. [CrossRef]

21. Doubrawa, P.; Barthelmie, R.; Wang, H.; Pryor, S.; Churchfield, M. Wind Turbine Wake Characterization from Temporally Disjunct 3-D Measurements. *Remote Sens.* **2016**, *8*, 939. [CrossRef]

22. El-Asha, S.; Zhan, L.; Iungo, G.V. Quantification of power losses due to wind turbine wake interactions through SCADA, meteorological and wind LiDAR data. *Wind Energy* **2017**, *20*, 1823–1839. [CrossRef]

23. Bodini, N.; Zardi, D.; Lundquist, J.K. Three-dimensional structure of wind turbine wakes as measured by scanning lidar. *Atmos. Meas. Tech.* **2017**, *10*, 2881–2896. [CrossRef]

24. Sathe, A.; Mann, J.; Gottschall, J.; Courtney, M.S. Can Wind Lidars Measure Turbulence? *J. Atmos. Ocean. Technol.* **2011**, *28*, 853–868. [CrossRef]

25. Sathe, A.; Mann, J. Measurement of turbulence spectra using scanning pulsed wind lidars. *J. Geophys. Res. Atmos.* **2012**, *117*. [CrossRef]

26. Newman, J.F.; Clifton, A. An error reduction algorithm to improve lidar turbulence estimates for wind energy. *Wind Energy Sci.* **2017**, *2*, 77–95. [CrossRef]

27. Carbajo Fuertes, F.; Iungo, G.V.; Porté-Agel, F. 3D Turbulence Measurements Using Three Synchronous Wind Lidars: Validation against Sonic Anemometry. *J. Atmos. Ocean. Technol.* **2014**, *31*, 1549–1556. [CrossRef]

28. Newman, J.F.; Bonin, T.A.; Klein, P.M.; Wharton, S.; Newsom, R.K. Testing and validation of multi-lidar scanning strategies for wind energy applications. *Wind Energy* **2016**, *19*, 2239–2254. [CrossRef]

29. Vasiljević, N.; Lea, G.; Courtney, M.; Cariou, J.P.; Mann, J.; Mikkelsen, T. Long-Range WindScanner System. *Remote Sens.* **2016**, *8*, 896. [CrossRef]

30. Stawiarski, C.; Traumner, K.; Knigge, C.; Calhoun, R. Scopes and challenges of dual-doppler lidar wind measurements-an error analysis. *J. Atmos. Ocean. Technol.* **2013**, *30*, 2044–2062. [CrossRef]

31. Stawiarski, C.; Träumner, K.; Kottmeier, C.; Knigge, C.; Raasch, S. Assessment of Surface-Layer Coherent Structure Detection in Dual-Doppler Lidar Data Based on Virtual Measurements. *Bound.-Layer Meteorol.* **2015**, *156*, 371–393. [CrossRef]

32. Lundquist, J.K.; Churchfield, M.J.; Lee, S.; Clifton, A. Quantifying error of lidar and sodar doppler beam swinging measurements of wind turbine wakes using computational fluid dynamics. *Atmos. Meas. Tech.* **2015**, *8*, 907–920. [CrossRef]

33. Mirocha, J.D.; Rajewski, D.A.; Marjanovic, N.; Lundquist, J.K.; Kosović, B.; Draxl, C.; Churchfield, M.J. Investigating wind turbine impacts on near-wake flow using profiling lidar data and large-eddy simulations with an actuator disk model. *J. Renew. Sustain. Energy* **2015**, *7*, 043143. [CrossRef]

34. Van Dooren, M.F.; Trabucchi, D.; Kühn, M. A methodology for the reconstruction of 2D horizontal wind fields of wind turbinewakes based on dual-Doppler lidar measurements. *Remote Sens.* **2016**, *8*, 809. [CrossRef]

35. Meyer Forsting, A.R.; Troldborg, N.; Murcia Leon, J.P.; Sathe, A.; Angelou, N.; Vignaroli, A. Validation of a CFD model with a synchronized triple-lidar system in the wind turbine induction zone. *Wind Energy* **2017**, *20*, 1481–1498. [CrossRef]

36. Van Dooren, M.F.; Campagnolo, F.; Sjöholm, M.; Angelou, N.; Mikkelsen, T.; Floris, M. Demonstration and uncertainty analysis of synchronised scanning lidar measurements of 2-D velocity fields in a boundary-layer wind tunnel. *Wind Energy Sci.* **2017**, *2*, 329–341. [CrossRef]
37. Porté-Agel, F.; Wu, Y.T.; Lu, H.; Conzemius, R.J. Large-eddy simulation of atmospheric boundary layer flow through wind turbines and wind farms. *J. Wind Eng. Ind. Aerodyn.* **2011**, *99*, 154–168. [CrossRef]
38. Wu, Y.T.; Porté-Agel, F. Large-Eddy Simulation of Wind-Turbine Wakes: Evaluation of Turbine Parametrisations. *Bound.-Layer Meteorol.* **2011**, *138*, 345–366. [CrossRef]
39. Wu, Y.T.; Porté-Agel, F. Modeling turbine wakes and power losses within a wind farm using LES: An application to the Horns Rev offshore wind farm. *Renew. Energy* **2015**, *75*, 945–955. [CrossRef]
40. Porté-Agel, F.; Meneveau, C.; Parlange, M.B. A scale-dependent dynamic model for large-eddy simulation: application to a neutral atmospheric boundary layer. *J. Fluid Mech.* **2000**, *415*, 261–284. [CrossRef]
41. Stoll, R.; Porté-Agel, F. Dynamic subgrid-scale models for momentum and scalar fluxes in large-eddy simulations of neutrally stratified atmospheric boundary layers over heterogeneous terrain. *Water Resour. Res.* **2006**, *42*, 1–18. [CrossRef]
42. Wieringa, J. Updating the Davenport roughness classification. *J. Wind Eng. Ind. Aerodyn.* **1992**, *41*, 357–368. [CrossRef]
43. Mikkelsen, R. Actuator Disc Methods Applied to Wind Turbines. Ph.D. Thesis, Technical University of Denmark, Lyngby, Denmark, 2003.
44. Sathe, A.; Mann, J. A review of turbulence measurements using ground-based wind lidars. *Atmos. Meas. Tech.* **2013**, *6*, 3147–3167. [CrossRef]
45. Pearson, G.; Davies, F.; Collier, C. An analysis of the performance of the UFAM pulsed Doppler lidar for observing the boundary layer. *J. Atmos. Ocean. Technol.* **2009**, *26*, 240–250. [CrossRef]
46. Benedict, L.H.; Gould, R.D. Towards better uncertainty estimates for turbulence statistics. *Exp. Fluids* **1996**, *22*, 129–136. [CrossRef]

remote sensing

MDPI

Article

Wind Gust Detection and Impact Prediction for Wind Turbines

Kai Zhou [1,*], Nihanth Cherukuru [2], Xiaoyu Sun [3] and Ronald Calhoun [1]

[1]   Mechanical Engineering, Arizona State University, Tempe, AZ 85287, USA; ronjcalhoun@gmail.com
[2]   National Center for Atmospheric Research, Boulder, CO 80305, USA; c.n.wagmi@gmail.com
[3]   Data Science Group, SAP, 1101 W Washington St #401, Tempe, AZ 85281, USA; sun.xiaoyu25@gmail.com
*   Correspondence: kzhou6@asu.edu; Tel.: +1-480-335-3211

Received: 25 January 2018; Accepted: 23 March 2018; Published: 25 March 2018

check for
updates

Abstract: Wind gusts on a scale from 100 m to 1000 m are studied due to their significant influence on wind turbine performance. A detecting and tracking algorithm is proposed to extract gusts from a wind field and track their movement. The algorithm utilizes the "peak over threshold method," Moore-Neighbor tracing algorithm, and Taylor's frozen turbulence hypothesis. The algorithm was implemented for a three-hour, two-dimensional wind field retrieved from the measurements of a coherent Doppler lidar. The Gaussian shape distribution of the gust spanwise deviation from the streamline was demonstrated. Size dependency of gust deviations is discussed, and an empirical power function is derived. A prediction model estimating the impact of gusts with respect to arrival time and the probability of arrival locations is introduced, in which the Gaussian plume model and random walk theory including size dependency are applied. The prediction model was tested and the results reveal that the prediction model can represent the spanwise deviation of the gusts and capture the effect of gust size. The prediction model was applied to a virtual wind turbine array, and estimates are given for which wind turbines would be impacted.

Keywords: wind gusts; Doppler lidar; detecting and tracking; impact prediction

1. Introduction

Rapid changes of wind speed in the atmosphere, also called wind gusts, cause large fatigue loads on wind turbines. These loads reduce the lifetime of wind turbine components. Oscillations or ramping of the generated power can result in fast fluctuations of grid voltage and may pose additional burdens to the electric grid. Researchers have proposed adaptive and feed-forward control systems, which can adjust wind turbine settings for approaching winds [1–4]. A feed-forward control system requires accurate and fast gust detection system. Our purpose is to provide the information of wind gusts to the control systems.

There are a variety of gust detecting and tracking algorithms in the literature. The international standard IEC 61400-1 specifies standardization of several temporal gust models for wind turbine design. Branlard defined gusts as a short-term wind speed variation within a turbulent wind field [5,6]. Although different definitions exist, all suggest gusts invoke rapid wind speed changes. As many atmospheric sensors measure winds in relatively small volumes, changes in wind speed have been considered temporally. Temporal variations of wind speed can be converted to spatial variations using Taylor's frozen turbulence hypothesis. Note that spatial variations in pressure on wind speed also damage buildings [7]. However, fewer studies directly measure spatial variations. The lack of such studies could be due to the limitation of the available instruments. Anemometers on met masts are the most common instruments for gust studies, but they are limited to measuring wind speed at fixed points. On the other hand, Doppler lidar can address this limitation. A long-range Doppler lidar can provide wind velocity field in a 3D domain up to 10 km with high temporal and spatial resolution.

The scale of a spatial gust is an important factor. Kelley et al. [8] and Chamorro et al. [9] studied the flow-structure interaction between wind turbines and atmospheric coherent structures with a scale ranging from the size of a wind turbine rotor to the thickness of the atmospheric boundary layer. They found that structures primarily in that range have a high correlation with the generated power and can induce strong structural responses. For structures with scales smaller than the size of a wind turbine rotor, the effects of their high-frequency components will be averaged out along the turbine blades and will not propagate to the drive train of a wind turbine and affect the power generation. For atmospheric gusts larger than 1000 m, varying winds in these large-scale gusts can be classified and captured as "meandering of the wind." Effects of the wind meandering on wind turbines are complicated and begin to become relevant for yaw control. Therefore, we believe the gusts with scales between 100 m and 1000 m have the most significant effects on wind turbine performance.

In addition to the definition of wind gusts, the literature contains a variety of ideas for gust detecting and tracking. Mayor adapted two computer-vision methods for flow motion estimation: the cross-correlation method and the wavelet-based optical flow method [10,11]. However, the cross-correlation method has limitations for non-uniform velocity fields, and the optical flow method requires relatively small (few pixels) movement and is computationally demanding. These requirements make them impractical. On the other hand, Branlart [6] proposed several detection methods for different gust models, but gusts are defined in the time domain. He also estimated the arrival time of the gusts and presented an exponential probability distribution of the gust's spanwise propagation. However, the gust distribution may not be representative, as the signal collected behind the turbine rotor would be heavily contaminated by the turbulence induced by the blades. The number of measurement points from anemometers is also limited. Therefore, a fast detecting and tracking algorithm and a reliable prediction model that can be used in real-time prediction of spatial gusts need to be developed.

In this work, we focus on spatial wind gusts within a limited range of scales. We propose a practical gust detecting and tracking algorithm with low computational cost. The novelty of the algorithm is to utilize dispersion and transport theory to create a practical tool which can provide short-range predictions of probable impact zones downstream for puffs or gusts detected upstream with a long-range Doppler lidar. The tool can provide real-time short-term predictions of impact time and location for gusts approaching a wind farm. The propagation of the wind gusts through the wind farm is not considered in this paper. By taking the gust size into consideration, the accuracy of the prediction can be increased, and valuable wind forecasting information can be provided to the control system.

2. Materials and Methods

2.1. Definition of Spatial Gusts

The spatial gust studied in the present work is defined mathematically in this section. The scale of the gust of interest ranges from 1D to 10D (D is the diameter of a wind turbine. D = 100 m is used, hereinafter). The magnitude of the wind speed fluctuation of a gust region, $|v'|$, should be 1.5 times larger than the standard deviation σ of the wind speed over the wind field, i.e., $|v'| > 1.5\sigma$ and $v' = v - \bar{v}$, where v is the local velocity vector and $\bar{v}$ is the mean wind velocity. The gust regions should also have good temporal coherency and local spatial connectivity around their centers. Gusts are assumed to advect roughly along the mean streamline, and the major structure should preserve during traveling.

2.2. Data Information and Wind Field Retrieval

The wind field utilized in this work was retrieved by a new proposed two-dimensional variational analysis method (2D-VAR) from the measurements collected by a Lockheed Martin Coherent

Technologies (LMCT) WindTracer® Doppler lidar (Louisville, CO, USA) during 25–27 June 2014, at Tehachapi, California [12]. The specification of the Doppler lidar is listed in Table 1. The lidar was located on a hill (1450 m above sea level (ASL)) at a wind farm near Tehachapi City, which is at an altitude of 1220 m (ASL). The data was collected in a horizontal plane at 1453 m ASL which included the height of the lidar system (3 m).

Table 1. Specifications of the Doppler lidar.

Parameters	Settings
Wavelength	1.6 μm
Pulse energy	2 mJ
Pulse repetition frequency	750 Hz
Range resolution	100 m
Blind zone	436 m
Max range	10 km

The 2D-VAR method used in the present work is based on a variational parameter identification formulation [12]. The method involves finding the best fit 2D wind velocity vector (X) which minimizes a cost function: $J(X) = \frac{1}{\Omega} \int \sum W_i C_i^2 d\Omega$, where W is a pre-defined weight matrix which determines the relative importance of the terms in the cost function. The constrains, C, are functions of the wind vector X, and are comprised of radial velocity equation, tangential velocity equation for low elevation angles and the advection equation. And Ω represents the analysis domain. A quasi-Newton method is implemented for the minimization. This retrieval algorithm has the advantage of preserving local structures in complex flows while being computationally efficient with possible real-time applications.

The retrieval algorithm was used to convert the measured radial velocities to two-dimensional (2D) wind field in Cartesian coordinate in a domain of size 6 km × 4 km. A sample contour plot of the retrieved wind field is shown in Figure 1. The temporal resolution of the retrieved results is 30 s as determined by the lidar scanning pattern, and the size of the spatial grid is 80 m × 80 m as specified by the retrieval algorithm. Note that even though the 2D-VAR method was used in the present work, the proposed detecting and tracking algorithm can be applied to any 2D wind field retrieved from any algorithm or obtained from any experimental instrument, given sufficient spatial and temporal resolution.

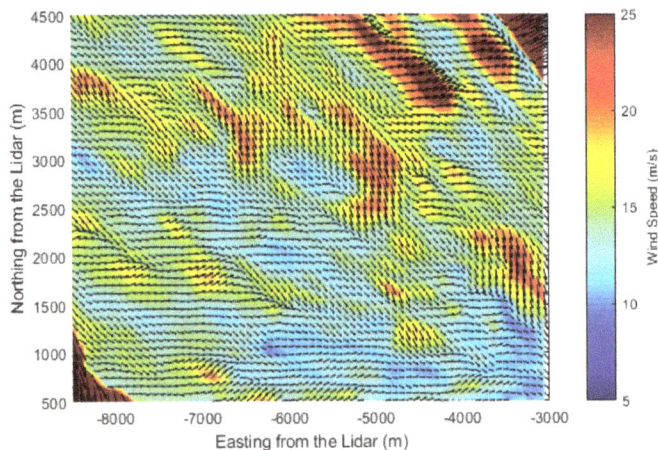

Figure 1. Wind field retrieved by the two-dimensional variational analysis method (2D-VAR) algorithm. Background colors indicate the magnitude of the local wind speed, and the arrows show the local wind direction.

2.3. Data Preprocessing

Data quality control was first performed in the wind field. The dark red regions at the northeast and southwest corners in Figure 1 were removed. The reason the data at the two corners were removed is two-fold: First, the hills were treated as hard targets, and therefore, due to possible contamination in the range-gates immediately in front of the hard targets, the data at the hills and 1–2 range gates before the hills were removed before running the retrieval algorithm. Second, the retrieval algorithm requires the data to be treated with a Gaussian filter to minimize the effect of noise on the gradients for the advection term in the cost function. Therefore, the Gaussian filter could not be applied at the boundaries of the lidar scan due to the missing data, which caused the artifacts at the corners in Figure 1, so the results 1–2 more range gates before the hills were removed.

Additionally, data with very high magnitudes above 30 m/s were rejected since they were judged to be spurious for this dataset. While a small amount of still valid data might be filtered out, we found empirically 30 m/s could be a reasonable trade-off value between removing most of the noises and keeping sufficient valid data points.

Moreover, the spatial resolution of the dataset needs to be considered. The current grid size is 80 m, implying any structures smaller than the grid size will be smoothed. Only very few data points were left for atmospheric structures with scales of a few hundred meters. Because the gust extraction process in the next step (Section 2.4) is based on the wind speed at the data points and the size and shape of the gust patch, the truncation process may cause unnatural-looking contours of the extracted gusts. Because the shape of the small-scale gusts is reduced to few pixels connected only at the vertices, some small patches could be easily neglected during the boundary tracing process. Therefore, to avoid losing many valid small-scale patches, linear interpolation using Delaunay triangulation was used to keep the shape of the gust contour to some extent [13]. Delaunay triangulation is a method that triangulates the discrete points in a plane such that no discrete point is inside the circumcircle of any triangle in the triangulation of the points. In the Delaunay triangulation, same weights were assigned to the vertices of the triangular. The wind field after the interpolation is shown in Figure 2. However, the interpolated boundaries should not be interpreted as better approximations to true boundaries than the coarse grid. Ideally, one may avoid the interpolation step, if the lidar range-gate size were significantly smaller.

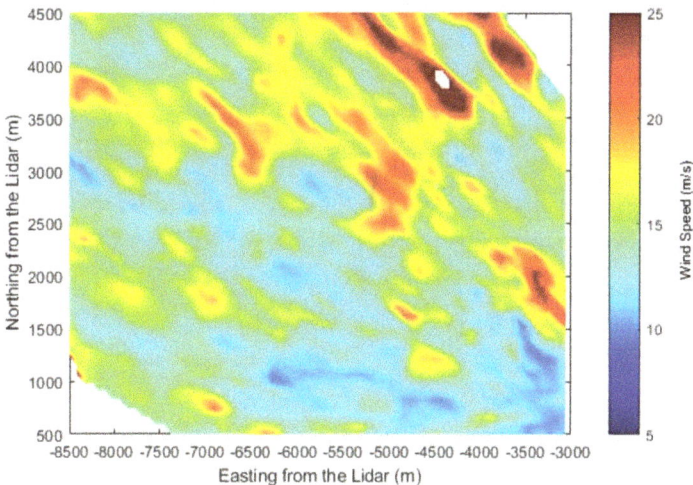

Figure 2. Wind field in Figure 1 after data preprocessing. Colors indicate the magnitude of the local wind speed.

2.4. Detection of Gust Patches

Peak over threshold (POT) method was used to extract gust regions from the wind field [6]. The POT method is to detect any wind event that has an amplitude over a predefined threshold. Here, we used the POT method for both up-crossing and down-crossing the threshold. The POT method removes the data points not satisfying the velocity threshold mentioned in Section 2.1 and only keeps the gust regions. The resultant data was converted to binaries to facilitate the boundary tracing algorithm. The extracted and converted gust regions are shown in Figure 3, with gust regions in white. Note that no scale filters were applied to the patches presented in Figure 3.

Next, the Moore-Neighbor tracing algorithm with Jacob's termination condition was applied to the binary data to trace the boundaries of the patches [14]. In the Moore-Neighbor tracing algorithm, an important concept is the Moore neighborhood which is a set of 8 pixels around a target pixel that share a vertex or edge with the target pixel. To track the boundary of a patch, as in Figure 3, the tracking algorithm uses any one of the white pixels on the boundary of the patch as the target pixel, and visits (moving clockwise for example) its Moore-neighbor pixels (black pixels) before entering another white pixel. Then, it uses the next white pixel (moving clockwise for example) as the target pixel and repeats the procedure. When it revisits the first white pixel it entered originally, the algorithm stops and all the visited black pixels comprise the boundary of the patch. There are two widely used termination conditions for the algorithm. The original one is to stop the algorithm after reentering the first white pixel for the second time. The other one is called Jacob's stopping condition, which also stops the algorithm after reentering the first white pixel for the second time, but in the same direction one originally enters it. Since concave shapes are common for gust regions as seen in Figure 3, Jacob's stopping condition was applied because it is more powerful to trace such shapes than the original stopping condition.

Moreover, since the patches with holes inside due to the above filtering still need to be considered as a whole, only exterior boundaries of the patches were traced. Additionally, patches with scale out of the range from 1D to 10D were filtered out and the centroids of the patches were calculated during the tracing process. The scale of a patch is defined as the square root of its area calculated by multiplying the grid area (10 m × 10 m) by the number of the grid points inside or on the boundary. The retrieved wind field along with detected gust boundaries is shown in Figure 4.

Figure 3. Extracted gust regions in black-white scale from the wind field in Figure 2. The gust regions are in white.

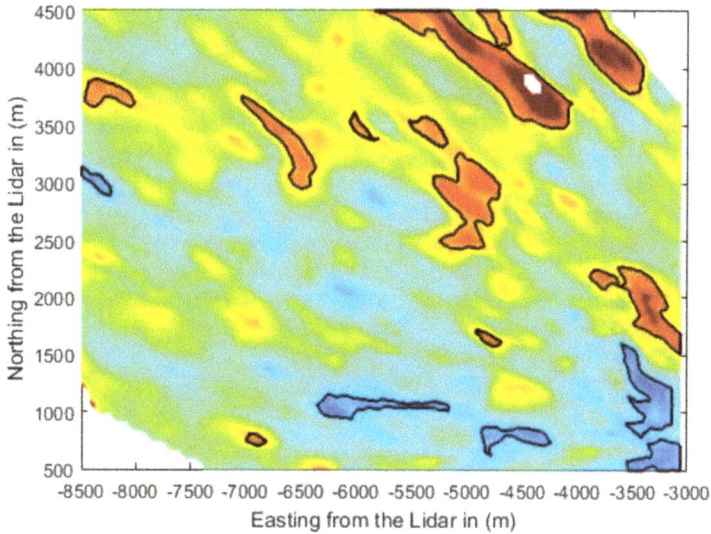

Figure 4. Retrieved wind field with detected boundaries of gust regions. Black contours are the boundaries of the detected patches.

2.5. Tracking of Gust Patches

To predict the future location of the gust patches, their advective characteristics need to be confirmed at first. Therefore, the next question is how to associate the patches between time frames given the detected regions. Given the formation of the spatial gust is contributed by the atmospheric turbulence, we assume that the gust regions propagate along the mean wind direction and their turbulent properties remain unchanged according to Taylor's frozen turbulence hypothesis. Nevertheless, since the time intervals between two wind fields are integers, multiples of 30 s, they are relatively long compared to the eddies in the atmosphere, so Taylor's hypothesis is relaxed to some extent. It means slight changes in wind speed and wind direction, variations of scales, and deformation of the shapes are allowed in the proposed algorithm.

After extracting the gust regions from the wind field at two time frames, the tracking algorithm takes the detected gusts as inputs. The patches at an earlier frame are called "original patches," and patches at later frame are called "candidate patches." The time interval between the two frames should not be larger than 90 s. Otherwise, the tracking algorithm will fail due to large changes in wind speed or shape caused by the evolution of the wind field.

Next, a searching zone is assigned to each original patch. In the searching zone, the algorithm searches for the corresponding target patch on the second frame over all the candidate patches. The candidate patches close to the boundaries of the studied wind field are neglected because of partial observation of the patches. The searching zone is placed downstream of the original patches and oriented along the mean wind direction of the measurement domain. The distance between the original patch and the center of the searching zone is set to the product of the mean wind speed of the measurement domain and the time interval between the frames. The size of the searching zone changes adaptively because a patch could deviate further from the streamline as time elapses. In that case, a large searching zone is needed to cover the possible area that the target patch could reach. The spanwise searching range is set proportional to the product of time and the standard deviation of wind velocity in the spanwise direction, and the same method is applied in the streamwise direction. The spanwise and streamwise velocity components can be calculated by projecting the retrieved x and y velocity components to the streamwise and spanwise directions. The searching zone constrains the tracking algorithm to a small window instead of the whole domain, which avoids unnecessary calculation. Examples of the original-target patch pairs along with their searching zones are presented in Figure 5c. The wind fields are shown in Figure 5 a,b.

For an original patch, the algorithm searches for the target patch over all the candidate patches. Then, it loops the same procedure over all the other original patches. The target patch of each original patch is chosen from the candidate patches according to predefined criteria, and scores are assigned to the candidate patches for each criterion. The criteria include the changes of wind speed, the directional difference between the mean wind direction and the real moving direction, size difference, and shape deformation. For the changes of wind speed and direction, the root-mean-square deviations (RMSD) of the wind speed and direction between the points in the original and candidate patch are calculated using Equation (1), and the scores are assigned according to the RMSDs in Equation (2).

$$RMSD_p = \sqrt{\frac{\sum_{i=1}^{N}\left(x_{i,\,original} - x_{i,\,target}\right)^2}{N}} \tag{1}$$

$$Score_p = \begin{cases} 1, & RMSD_p < 0.5 \\ 0, & RMSD_p \geq 0.5 \end{cases} \tag{2}$$

Here, the subscript "p" represents the calculated variable name, wind speed or wind direction, and x_i stands for the value of the corresponding variable of the point i inside or on the original or candidate patch. In the calculation, we assume that the original and candidate patches share the same centroid. N represents the number of the points that are in both original and candidate patches. If a point is only in either the original or candidate patch, the point is ignored during the calculation.

For size changes and shape deformation, the intersection of union (IoU) value for each original-candidate pair is obtained by calculating the ratio of the number of intersection points to the number of union points. Here, we assume the original and candidate patch share the same centroid as well. The score for the shape change is calculated by Equation (3).

$$Score_{shape} = \begin{cases} 0, & IoU < 0.5 \\ IoU, & IoU \geq 0.5 \end{cases} \tag{3}$$

The threshold value (0.5) in Equations (2) and (3) is determined empirically. After the individual score for each criterion is obtained, the total score for each candidate patch m, $TotalScore_m$, is calculated by summarizing over all the individual scores.

$$TotalScore_m = Score_{speed} + Score_{direction} + Score_{shape} \tag{4}$$

Patches with small difference earn high total scores, and patches with zero total scores are rejected. If multiple candidates have non-zero total scores, the one with the highest total score is chosen as the target. If such a target is identified, we claim that the original patch is advecting, and the target patch is the counterpart of the original at the second frame. Conversely, if no target is detected, the original patch loses its temporal coherency during traveling.

The tracking algorithm was tested on the wind fields shown in Figure 5a,b. Three examples of tracked original-target pairs are shown in Figure 5c. The time interval between the original and target patches is 60 s. The other patch pairs are not shown. In Figure 5, we can see many small-scale patches shown in Figure 5a are not shown inside the corresponding searching zones in Figure 5b where they should arrive after 60 s by following Taylors' hypothesis. It provides the evidence that smaller patches are more difficult to track than larger ones. The reason would be that larger-scale patches have strong connectivity inside their internal structures, while the spatial correlation of small patches is weak and unstable. The coherency can be easily destroyed by the stretching and compression of larger-scale turbulent eddies. The destructive effect can reduce the velocity of the patches to a value beyond the threshold, deform or separate the patches, or merge small patches with other structures. All these situations can make the detecting and tracking algorithm ineffective.

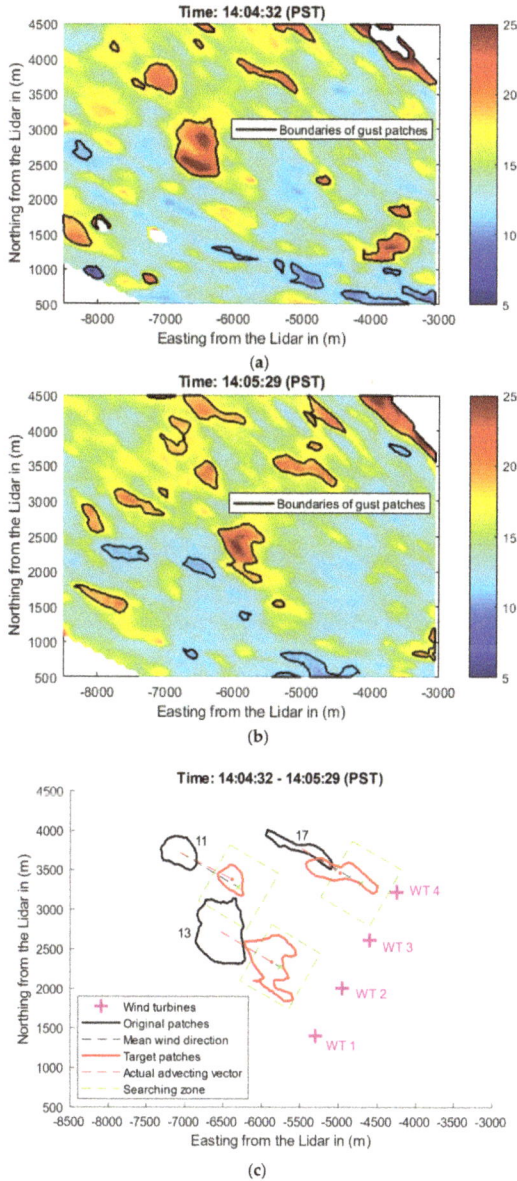

Figure 5. (a) Wind field at the first time frame. Colors indicate the wind speed. Black contours are the boundaries of detected patches; (b) Wind field at the second time frame; (c) Example original- target patch pairs. Black patches are the original gust regions extracted from (a), and red patches are the corresponding target gust regions extracted from (b). Green dashed boxes are the searching zones. Black and red dashed lines indicate the moving direction of the streamline and the real movement of the patches, respectively. Pink crosses indicate the locations of a virtual wind turbine array.

2.6. Prediction of Impact

After identifying advecting patches, the arrival time and probability of arrival locations of the patches can be predicted.

2.6.1. Prediction of Impact Time

The impact time of gust patches at a certain location can be predicted using Taylor's hypothesis, that is, dividing the distance between the gust and the location by the mean velocity. The equation is shown in Equation (5):

$$t = \frac{d_{stream}}{v_{stream}} . \tag{5}$$

Both the distance d_{stream} and velocity v_{stream} are the values projected on the mean wind direction. The digression of wind speed along the streamline during traveling to the location of interest is ignored due to the negligible difference for high wind speed (>15 m/s) [6]. Equation (5) provides us with fair estimation of the arrival time, since the gust generally follows Taylor's hypothesis in a short time period without much deviation, and this simple estimation makes the real-time prediction possible. However, spatial gusts do not always travel with the same velocity in reality. Alternatively, one can track the same patch for several time steps, calculate the actual traveling speed, and replace the mean wind speed with the obtained wind speed. Using the actual speed could improve the accuracy of the prediction, but it might also increase the complexity of the algorithm. Since the present work is to provide a fast prediction within a few seconds for wind turbine control, we focus on using the mean wind speed to calculate the arrival time.

2.6.2. Prediction of the Probability of Impact Location

If Taylor's hypothesis holds, the gust region should arrive at a point downstream from the original patch, called the "predicted point." However, due to the turbulent nature of the atmosphere, the real arrival point can only be estimated by a statistical probability centered at the predicted point. Here, we treat the gust patches as massless passive particles that are released from a point source, and use the concentration distribution model of plumes to model the spanwise distribution of gust patches dispersing in a turbulent wind field.

Several plume models and puff models have been developed by researchers to investigate the particle distribution in the air, such as Gaussian plume model [15,16], K-model, statistical model, and similarity model [17] etc., and more advanced but computationally more expensive models like SCIPUFF [18], RIMPUFF [19], and LODI [20]. Since this work is aimed at estimating the impact of gusts in a few seconds or minutes in advance of wind turbine control, the fast Gaussian plume model is used.

The reason for using the Gaussian plume model for wind gust studies is explained in the following. First, since the gusts travel in a relatively short distance in this study, we assume the interaction between the large-scale structures and the background turbulence is small. Therefore, we can treat the gusts as passive quantities. Second, we assume the eddy viscosity and eddy diffusivity are of comparable value, i.e., the Prandtl number is close to 1. The reason is that, according to the Reynold analysis, under stable or neutral boundary conditions, the turbulent transport of scalar quantities is similar to the transport of momentum, given the transport agents are the same [21]. In our case, the mean wind speed during the studied time period is relatively high (around 15 m/s), so the neutral boundary condition dominated the studied time period (The strong mechanical mixing due to the high wind speed balanced the buoyance-driven turbulence in the atmosphere.). Therefore, we can apply the plume model that was used for passive particles to the transportation of gust structures studied in the present work.

The original Gaussian model (Equation (6)) is modified to suit the current case: vertical distribution is neglected since only the deviation of gusts in the horizontal plane is of interest.

The velocity term is removed as the probabilities will be normalized by the maximum value, meaning only relative probabilities matter. The modified equation is shown in Equation (7).

$$P(y,z) = \frac{1}{2\pi\sigma_y\sigma_z u}e^{-y^2/2\sigma_y^2} \times [e^{-(z-h)^2/2\sigma_z^2} + e^{-(z+h)^2/2\sigma_z^2}] \tag{6}$$

$$P(y) = \frac{1}{\sqrt{2\pi}\sigma_y}e^{-y^2/2\sigma_y^2} \tag{7}$$

where y refers to the spanwise deviation of a gust patch from the streamline with y equal to zero on the streamline, and z indicates the height above the ground. In Equation (6) h is the distance from the source to the ground. P is the probability at the deviation y, and σ_y is the standard deviation of y. We assume the turbulence field driving the gust patches is stationary and homogeneous. Taylors' dispersion theorem provides an expression of σ_y^2 as a function of the releasing time t for stationary turbulence [17].

$$\sigma_y^2 = 2\sigma_{v_L'}^2 \int_0^t \int_0^\tau r(\tau)d\tau dt \tag{8}$$

where $\sigma_{v_L'}$ is the standard deviation of the spanwise Lagrangian velocity fluctuation in which the subscript "L" stands for Lagrangian scale, and r is the autocorrelation of the spanwise velocity fluctuation. Using the random walk theory [22], the standard deviation of the deviation y, σ_y in the Lagrangian system can be approximated by Equation (8) when $t \rightarrow \infty$

$$\sigma_y = \sqrt{2\sigma_{v_L'}^2 T_L t} \tag{9}$$

and T_L is the Lagrangian integral time scale.

Here, we assume the gust patches are 'released' for sufficient time ($t \rightarrow \infty$) in the atmosphere, so the gust structures only advect but not disperse (grow) during traveling. It indicates that the diffusivity is constant, i.e., $K \sim \sigma_{v'}^2 T_L$. As the diffusivity is defined as $K = \frac{1}{2}\frac{d\sigma_y^2}{dt}$, Equation (9) can be deduced from the diffusivity. On the other hand, because of the stationary and homogenous turbulence assumption, the temporal correlation of the spanwise velocity fluctuation is approximated by instantaneous spatial correlation. The standard deviation of the Lagrangian velocity fluctuation $\sigma_{v_L'}$ was replaced by the standard deviation of Eulerian velocity fluctuation $\sigma_{v'}$, due to instantaneous velocities being studied in this work [23]. The spanwise velocity fluctuation and its standard deviation were obtained from the given wind field. Equation (9) is transformed to

$$\sigma_y = \sqrt{2\sigma_{v'}^2 T_L t} \tag{10}$$

where the Lagrangian time scale $T_L = 200\ s$, adapted from the work of Dosio et al. [24]. The reason of the adoption is that, we did not have sufficient statistical measurements for the time scale calculation in the present work, since the time interval of our lidar measurements is relatively long (30 s) and we did not have any other supplementary device installed locally. Therefore, instead of calculating the time scale directly, we adapted the result from Dosio's simulation of a convective boundary layer. The data used in this paper is between 3 pm and 6 pm (local time), so the atmospheric flow may be expected to contain some convective activity. Therefore, we believe the selected value could be considered as the best available approximation of the time scale during the studied period. However, we suggest to use meteorological measurements (like a met mast) if available to calculate the time scale, which would improve the estimations.

Additionally, we assume the atmospheric characteristics, with respect to the meandering of gusts, are preserved during the studied period so that the Lagrangian time scale remains constant. Furthermore, the random walk theory used above is only for particles. The effect of the scales of the gusts needs to be considered in Equation (10) since we are modeling gusts. The details will be illustrated in Section 3, and the modified random walk theory is given in Equation (11). The modified random walk theory (Equation (11)) and the Gaussian model (Equation (7)) comprise the impact prediction model for advecting gust patches with different sizes.

3. Results and Discussion

3.1. Preliminary Results

Three hours of retrieved wind fields (during 3–6 pm PST on 26 June) were processed. A time interval of 60 s between original patches and target patches was used. The original patches and corresponding target patches were extracted and associated using the detecting and tracking algorithm. Original random walk theory, Equation (10) was used for prediction. The spanwise deviation (L_{real}) of the target patches was measured. The spanwise deviation is defined as the deviation of the centroid of the gust from the streamline of the wind field, in the direction perpendicular to the mean wind direction of the measurement domain. Then, the spanwise deviation was normalized by the rotor diameter, i.e., $l_{real} = L_{real} / D$, and l_{real} is the dimensionless spanwise deviation. The histogram of the distribution probability of the dimensionless deviation is shown in Figure 6. The probability is normalized by the value at the 'predicted point'. The predicted point is assumed to have the maximum probability. Figure 6 presents a Gaussian shape distribution of the gust spanwise deviation. It demonstrates our assumption that the spanwise distribution of advecting patches can be represented by a Gaussian model, rather than the exponential shape found by Branlard [6].

However, a large discrepancy was observed between the modeled distribution of the gust spanwise deviation (dashed line) and the real deviation using the detecting and tracking algorithm (histogram). The reason is that the original random walk theory does not consider the size of the particles, as the random walk theory assumes that the size difference between particles is negligible. However, the assumption does not hold for gust patches with scales in the order larger than 100 m. The size of the gusts significantly affects their movement. Small gusts are easy to move because many large-scale eddies in the atmosphere move them around, which makes their movement relatively random. On the other hand, large-scale gusts that need even larger-scale eddies to move them are more likely to flow along the streamline of the wind field. Therefore, size dependency of the gust patches needs to be considered.

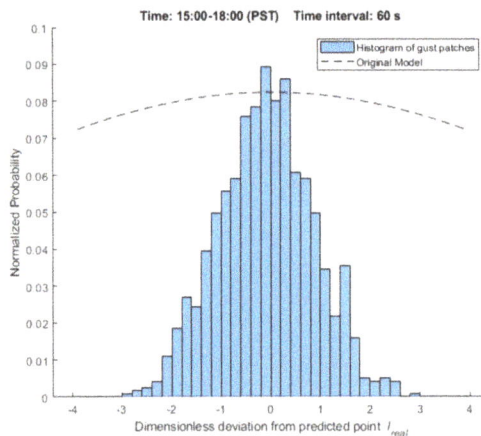

Figure 6. Normalized distribution probability of the spanwise deviation of patches from the predicted points. The dashed line represents the normalized probability modeled by Equation (10), and the histogram shows the normalized probability of the real spanwise deviations.

3.2. Size Dependency

To study the effect of the size of the gusts on the spanwise deviation, the dimensionless deviation l_{real} was further normalized by the square root of the time interval, i.e., $l_{real, \Delta t} = \frac{l_{real}}{\sqrt{\Delta t}}$, to remove the dependency of the deviation on time. The usage of the square root is according to the relationship between the deviation and time presented in Equation (10). The patch scale (L_{pat}) was normalized by the rotor diameter, i.e., $l_{pat} = L_{pat}/D$, and l_{pat} is the dimensionless scale. A scatter plot of the normalized deviation $l_{real, \Delta t}$ of all the detected patches with the dimensionless scale l_{pat} is shown in Figure 7. Then, the dimensionless deviations were grouped to bins according to their sizes, and inside each bin, the deviations and the dimensionless scales were averaged, respectively. The averaged values were fitted by a power function. The results and the fitting curve are plotted in Figure 8. Figures 7 and 8 present a negative dependency of the deviation on the patch scale. The relationship can be well expressed by a function $l_{real, \Delta t} = \alpha l_{pat}^{-\beta}$, where α and β are fitting parameters, and $\alpha = 1.0$ and $\beta = -0.36$ is found for the current dataset.

After the size dependency was empirically derived, Equation (10) was modified to include the derived relationship by multiplying it by the derived power function.

$$\tilde{\sigma}_{y,mod} = \tilde{\sigma}_y \cdot \alpha \left(l_{pat} \right)^{\beta}$$
$$\tilde{\sigma}_y = \sqrt{2\sigma_{v'}^2 T_L t / D} \tag{11}$$
$$l_{pat} = L_{pat}/D$$

where $\tilde{\sigma}_y$ is the dimensionless standard deviation of the deviation y, modeled by the original random walk theory.

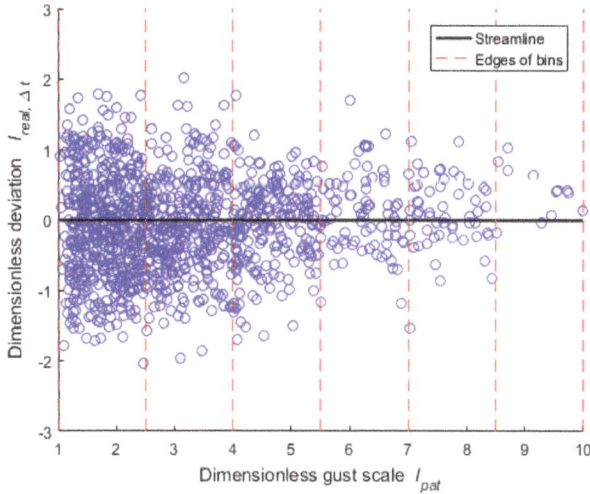

Figure 7. Scatter plot of normalized deviations and patch scales. The streamline overlays the line with the deviation $l_{real, \Delta t}$ equal to zero. The red dashed lines imply the edges of the bins. The average dimensionless scales at the bins are 1.75, 3.25, 4.75, 6.25, 7.75, and 9.25.

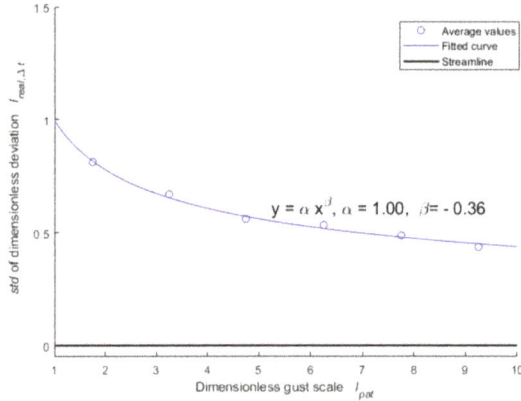

Figure 8. The relationship between dimensionless deviations and dimensionless gust scales. Circles are the average dimensionless deviations at different average gust scales, and the solid line is the fitting curve.

3.3. Improved Results Using the Modified Prediction Model

After the size dependency correction was added to Equation (10), the modified model was retested on the data presented in Figure 6. The patches were separated into three groups according to their sizes: 1D–2.5D, 2.5D–4D, and 4D–5.5D. Patches larger than 5.5D are not included due to insufficient sample points. The results are shown in Figure 9. Gaussian functions (red lines) were fitted to the histograms. The modeled distribution (black lines) were computed using Equations (7) and (11). A good match for different sizes of patches are shown in the comparison between the real measured distribution and the modeled distribution. It illustrates that the distribution probability of gust spanwise deviation can be well represented by the proposed prediction model at least for the current dataset.

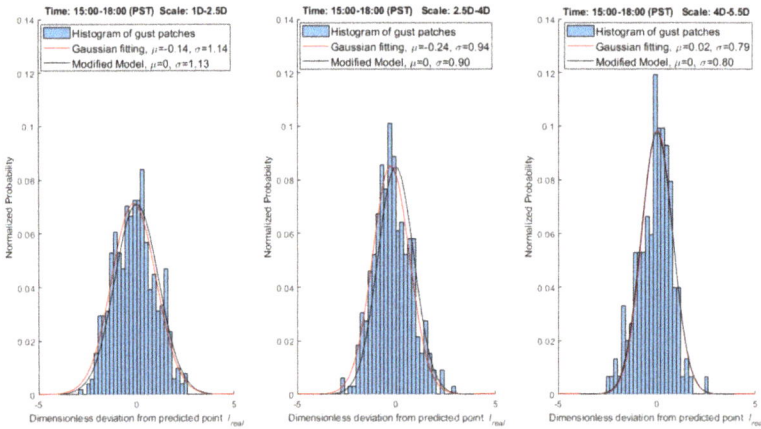

Figure 9. Normalized distribution probability of the modeled deviation (black lines) and real deviation (histogram) of the patches. The red lines are the fitting curves to the histograms. The same dataset was used as in Figure 6. The patches were separated into three groups according to their sizes.

3.4. Application of the Modified Prediction Model

By applying the detecting and tracking algorithm to the retrieved wind field, we monitored the movement of the gusts during the three-hour period. Next, we positioned a virtual wind turbine array in the wind field shown in Figure 5a, and used the three original gusts in Figure 5c as examples, to illustrate what the time of impact and probable arrival locations of the gusts are for the hypothetical turbines. Here, the impact time and locations were calculated by the prediction model. The wind turbine array consists of four turbines that align perpendicular to the mean wind direction. The distance between two adjacent wind turbines is 700 m. The distance between the gusts and the wind turbine locations were measured, and the time taken by the gusts to reach the turbines was estimated using Equation (5). The real arrival time was also obtained by applying the detecting and tracking algorithm to the wind field at different time frames. Additionally, the spanwise distribution probabilities of the gusts at different wind turbines were predicted and are shown in Figure 10. Both predicted and real arrival time and positions are listed in Table 2. Patch 11 disappeared before reaching any wind turbine, so the real arrival time and position were left blank in Table 2 and the real arrival position is not shown in Figure 10. This fact also verifies that small patches are more difficult to track. Table 2 shows a good match between the predicted and real arrival time and locations within reasonable tolerances, which validates the proposed prediction model. However, a further quantitative validation is needed in the future work.

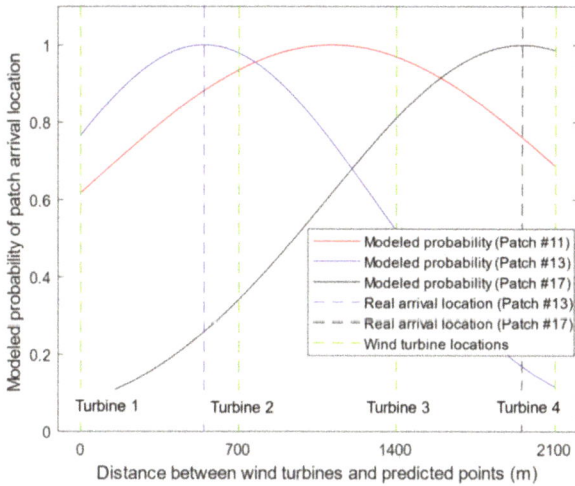

Figure 10. Real and predicted arrival positions, and predicted normalized spanwise distribution probability of the gusts. The patch numbers correspond to the gust numbers in Figure 5c. Green dash lines indicate the locations of the wind turbines relative to Turbine 1, and the distance is measured along the wind turbine array. The blue and black dash lines represent the real arrival positions of Patch 13 and 17, respectively. Since Patch 11 disappeared before arriving any wind turbine, no line indicates its real arrival location in the figure.

Table 2. Predicted and real arrival time and position of patches.

Patch Number	#11	#13	#17
Impact wind turbine	Turbine 3	Turbine 2	Turbine 4
Predicted arrival time	183.03 s	116.37 s	92.9 s
Real arrival time	-	~119.54 s	~119.54 s
Spanwise distance between predicted arrival position and wind turbine of impact	285.2 m (to its left)	153.7 m (to its left)	144.9 m (to its left)
Spanwise distance between real arrival position and wind turbine of impact	-	106.16 m (to its left)	87.48 m (to its left)

4. Conclusions

Due to the importance of wind gusts for wind turbine performance, spatial wind gusts with a scale from 100 m to 1000 m are studied using a Doppler lidar. A detecting and tracking algorithm for the spatial gusts is introduced. The proposed algorithm is computationally inexpensive and can provide real-time tracking and prediction. However, it is sensitive to the temporal and spatial resolution of the dataset. The algorithm was performed on a three-hour wind field data obtained from the measurements of a WindTracer Doppler lidar. The results reveal a Gaussian shape distribution of gust spanwise deviation from the streamline and a negative power-law dependency of the deviation on gust scales. Furthermore, a prediction model using the Gaussian plume model and random walk theory was introduced and corrected by the size dependency. The prediction model was tested on the same dataset, and the results further demonstrate the power law relationship. The prediction model was applied to a virtual wind turbine array to estimate the impact time and locations of the gusts on the turbines. The model was validated in a limited set of conditions, due to the lack of the dataset with a longer time period. Therefore, more quantitative validation of the model is suggested for future work. However, the major value of this work is to provide a novel approach for the study of large-scale wind gusts using a long-range Doppler lidar. In real-life applications, the proposed algorithm could be efficiently integrated to feed-forward wind turbine control systems given 2D wind fields in time series, and the wind field could be provided by a long-range Doppler lidar. Given the knowledge of the movement of potential hazardous gusts, the control system would have sufficient leading time to adjust wind turbines to optimal settings. In this case, the power generation could remain stable at the desired value, and the loads on the turbine components could be reduced. Consequently, the impact on the electric grid due to the power fluctuation could be mitigated.

Future research can be focused on the validation of the detecting and tracking algorithms on other datasets with a shorter time interval between frames and a higher spatial resolution. Using Kalman filters or other multiple object tracking algorithms, the detecting and tracking algorithm might be improved. Deep learning approaches might be expected to recognize patches quickly with acceptable accuracy. For the prediction model since the size dependency was derived empirically from the current dataset, the parameters need fine calibration for other datasets and the dependency of gusts with scales larger than 5.5D is not presently clear. In addition, the quality of the prediction model needs to be validated, which could be addressed in a dataset with longer time and in more varied atmospheric conditions. Finally, using real-time measurements, like lidar or meteorological tower, the Lagrangian time scale estimation can be improved.

Acknowledgments: We recognize the support of the Electric Power Research Institute (EPRI), sponsor award 10001389 (Program Officer: Dr. Aidan Tuohy), and a Navy Neptune Grant, Office of Naval Research (Program Officer: Dr. Rich Carlin).

Author Contributions: Kai Zhou is responsible for the study. He participated in the data collection, developed the algorithm, produced the results, and was the main writer of the manuscript. Nihanth W. Cherukuru converted the raw lidar measurements to 2D wind field. Xiaoyu Sun provided suggestions on the detecting and tracking algorithm and revisited the manuscript. Ronald Calhoun provided advisement and guidance.

Conflicts of Interest: The authors declare no conflict of interest.

References

1. Schlipf, D.; Pao, L.Y.; Cheng, P.W. Comparison of feedforward and model predictive control of wind turbines using LIDAR. In Proceedings of the Conference on Decision and Control, Maui, HI, USA, 10–13 December 2012; pp. 3050–3055.

2. Scholbrock, A.; Fleming, P.; Fingersh, L.; Wright, A.; Schlipf, D.; Haizmann, F.; Belen, F. Field testing LIDAR based feed-forward controls on the NREL controls advanced research turbine. In Proceedings of the 51th AIAA Aerospace Sciences Meeting Including the New Horizons Forum and Aerospace Exposition, Grapevine, TX, USA, 7–10 January 2013.

3. Kumar, A.A.; Bossanyi, E.A.; Scholbrock, A.K.; Fleming, P.; Boquet, M.; Krishnamurthy, R. Field Testing of LIDAR-Assisted Feedforward Control Algorithms for Improved Speed Control and Fatigue Load Reduction on a 600-kW Wind Turbine: Preprint. Presented at the EWEA 2015 Annual Event, Golden, CO, USA, 17–20 November 2015.

4. Kristalny, M.; Madjidian, D. Decentralized feedforward control of wind farms: Prospects and open problems. In Proceedings of the 2011 50th IEEE Conference on Decision and Control and European Control Conference, Orlando, FL, USA, 12–15 December 2011; pp. 3464–3469.

5. International Electrotechnical Commission. *International Standard, IEC 61400-1, Wind Turbines—Part 1: Design Requirements*, 3rd ed.; International Electrotechnical Commission: Geneva, Switzerland, 2005.

6. Branlard, E. Wind energy: On the statistics of gusts and their propagation through a wind farm. ECN-Wind-Memo-09-005Master's Thesis, Energy research Centre of the Netherlands, Petten, The Netherlands, February 2009.

7. Beljaars, A.C. *The Measurement of Gustiness at Routine Wind Stations: A Review*; World Meteorological Organization: Geneva, Switzerland, 1987.

8. Kelley, N.; Shirazi, M.; Jager, D.; Wilde, S.; Adams, J.; Buhl, M.; Patton, E. *Lamar Low-Level Jet Project Interim Report*; Technical Paper No. NREL/TP-500-34593; National Renewable Energy Laboratory, National Wind Technology Center: Golden, CO, USA, 2004.

9. Chamorro, L.P.; Lee, S.J.; Olsen, D.; Milliren, C.; Marr, J.; Arndt, R.E.A.; Sotiropoulos, F. Turbulence effects on a full-scale 2.5 MW horizontal-axis wind turbine under neutrally stratified conditions. *Wind Energy* **2015**, *18*, 339–349. [CrossRef]

10. Mayor, S.D.; Lowe, J.P.; Mauzey, C.F. Two-component horizontal aerosol motion vectors in the atmospheric surface layer from a cross-correlation algorithm applied to scanning elastic backscatter lidar data. *J. Atmos. Ocean. Technol.* **2012**, *29*, 1585–1602. [CrossRef]

11. Mayor, S.D.; Dérian, P.; Mauzey, C.F.; Hamada, M. Two-component wind fields from scanning aerosol lidar and motion estimation algorithms. In Proceedings of the SPIE Optical Engineering+ Applications, San Diego, CA, USA, 17 September 2013; p. 887208.

12. Cherukuru, N.W.; Calhoun, R.; Krishnamurthy, R.; Benny, S.; Reuder, J.; Flügge, M. 2D VAR single Doppler lidar vector retrieval and its application in offshore wind energy. *Energy Proced.* **2017**, *137*, 497–504. [CrossRef]

13. Amidror, I. Scattered data interpolation methods for electronic imaging systems: A survey. *J. Electron. Imaging* **2002**, *11*, 157–177. [CrossRef]

14. Moore-Neighbor Tracing. Available online: http://www.imageprocessingplace.com/downloads_V3/root_downloads/tutorials/contour_tracing_Abeer_George_Ghuneim/moore.html (accessed on 9 October 2017).

15. Batchelor, G.K. Diffusion in a field of homogeneous turbulence: II. The relative motion of particles. In *Mathematical Proceedings of the Cambridge Philosophical Society*; Cambridge University Press: Cambridge, UK, 1952; pp. 345–362.

16. Petersen, W.B.; Catalano, J.A.; Chico, T.; Yuen, T.S. *INPUFF-A Single Source Gaussian Puff Dispersion Algorithm: User's Guide*; US EPA: Research Triangle Park, NC, USA, 1984.

17. Hanna, S.R.; Briggs, G.A.; Hosker, R.P., Jr. *Handbook on Atmospheric Dispersion*; Technical Information Center, U.S. Department of Energy: Washington, DC, USA, 1982.

18. Sykes, R.I.; Parker, S.F.; Henn, D.S.; Cerasoli, C.P.; Santos, L.P. *PC-SCIPUFF Version 1.2 PD Technical Documentation*; ARAP Report No. 718; Titan Corporation, Titan Research & Technology Division, ARAP Group: Princeton, NJ, USA, 1998.

19. Thykier-Nielsen, S.; Deme, S.; Mikkelsen, T. Description of the atmospheric dispersion module RIMPUFF. 1999. Available online: https://s3.amazonaws.com/academia.edu.documents/39632120/Description_of_ the_atmospheric_dispersio20151103-1858-w1thfx.pdf?AWSAccessKeyId=AKIAIWOWYYGZ2Y53UL3A& Expires=1521966958&Signature=vnkIK16SU2hvkTa3L7PZWujwIB8%3D&response-content-disposition= inline%3B%20filename%3DDescription_of_the_atmospheric_dispersio.pdf (accessed on 25 March 2018).

20. Leone, J.M., Jr.; Nasstrom, J.S.; Maddix, D.M.; Larson, D.J.; Sugiyama, G.; Ermak, D.L. *Lagrangian Operational Dispersion Integrator (LODI) User's Guide*; Report UCRL-AM-212798; Lawrence Livermore National Laboratory: Livermore, CA, USA, 2005.

21. Li, D.; Katul, G.G.; Zilitinkevich, S.S. Revisiting the turbulent Prandtl number in an idealized atmospheric surface layer. *J. Atmos. Sci.* **2015**, *72*, 2394–2410. [CrossRef]

22. Kundu, P.K.; Cohen, I.M.; Dowling, D.W. *Fluid Mechanics*, 4th ed.; Academic Press: Burlington, MA, USA, 2008.

23. Price, J.F. *Lagrangian and Eulerian representations of fluid flow: Part I, kinematics and the equations of Motion*; Clark Laboratory Woods Hole Oceanographic Institution: Woods Hole, MA, USA, 2005.

24. Dosio, A.; Guerau de Arellano, J.V.; Holtslag, A.A.; Builtjes, P.J. Relating Eulerian and Lagrangian statistics for the turbulent dispersion in the atmospheric convective boundary layer. *J. Atmos. Sci.* **2005**, *62*, 1175–1191. [CrossRef]

remote sensing

MDPI

Comment

Comments on "Wind Gust Detection and Impact Prediction for Wind Turbines"

Shane D. Mayor [1,*] and Pierre Dérian [2]

1   California State University Chico, 400 West First Street, Chico, CA 95929, USA
2   Independent researcher, 44000 Nantes, France; contact@pierrederian.net
*   Correspondence: sdmayor@csuchico.edu

Received: 12 July 2018; Accepted: 10 October 2018; Published: 12 October 2018

check for
updates

Abstract: We refute statements in "Zhou, K., et al. Wind gust detection and impact prediction for wind turbines. *Remote Sens.* **2018**, *10*, 514." about the impracticality of motion estimation methods to derive two-component vector wind fields from single scanning aerosol lidar data. Our assertion is supported by recently published results on the performance of two image-based motion estimation methods: cross-correlation (CC) and wavelet-based optical flow (WOF). The characteristics and performances of CC and WOF are compared with those of a two-dimensional variational (2D-VAR) method that was applied to radial velocity fields from a single scanning Doppler lidar. The algorithmic aspects of WOF and 2D-VAR are reviewed and we conclude that these two approaches are in fact similar and practical.

Keywords: lidar; wind; Doppler; aerosol; motion estimation; optical flow; cross-correlation; wind energy; gust prediction; variational analysis

1. Introduction

Zhou et al. [1] provided an excellent example of the value of remote observations of two-component wind fields in the lower atmospheric boundary layer. They obtained two-component wind fields by applying a 2D variational (2D-VAR) method [2] to single component velocity data from a single Doppler lidar. While their paper mentions cross-correlation (CC) and wavelet-based optical flow (WOF) algorithms as alternative ways to obtain such wind fields, statements about these two methods are inaccurate and do not consider the latest published research on this subject. In this note, we correct statements in [1] based on our recent work [3,4] and put the CC, OF, and 2D-VAR methods into perspective. We contend that the WOF and 2D-VAR algorithms are in fact similar and that motion estimation methods should not be discounted for remote wind sensing applications such as the one they developed.

2. Statements in Zhou et al. [1] About Motion Estimation Methods

Zhou et al. [1] described an algorithm to detect and track wind gusts. Their algorithm operates on two-dimensional (2D) two-component (2C) wind fields which must first be made available. They chose to compute the wind fields by applying a 2D variational analysis method (2D-VAR) to radial component velocities measured by a single Doppler lidar ([1], Section 2.2). Alternative approaches to remotely observe similar 2D-2C wind fields are also mentioned in the introduction of the paper:

> "*Mayor adapted two computer-vision methods for flow motion estimation: the cross-correlation method and the wavelet-based optical flow method [5,6]."*—(Zhou et al. [1], Section 1).

However, after the above sentence, the paper continues with:

"[. . .] the cross-correlation method has limitations for non-uniform velocity fields and the optical flow method requires relatively small (few pixels) movement and is computationally demanding. These requirements make them impractical."—(Zhou et al. [1], Section 1).

We thank Zhou et al. [1] for citing these two papers [5,6]. Their statements regarding our research results are however inaccurate and do not include the latest peer-reviewed research in the field. The purpose of this note is: (1) to correct their statements; and (2) to demonstrate that the method they chose and the motion estimation approach that we have been focused on are, in fact, comparable in terms of temporal and spatial discretization, accuracy, and computational load, and thus are similarly practical given the presently available information.

3. The Practicality of Motion Estimation Methods

The quoted statements above ignore the abundance of published evidence of the strong skill of motion estimation algorithms and dismiss the approach for use in meteorological applications [3–5,7–9]. The literature shows that two numerical techniques (cross-correlation (CC) and wavelet-based optical flow (WOF)) are capable of extracting horizontal 2D-2C vector wind fields from near-horizontal aerosol lidar scans. We believe that these flow fields could be used in applications such as gust detection and prediction for wind turbines. In fact, the WOF motion estimation technique is currently in use for other short-term wind prediction applications [10].

First, we begin with the statement: *"[. . .] the cross-correlation method has limitations for non-uniform velocity fields"*. While this statement is correct in that the CC approach has limitations, as does any measurement technique, it fails to acknowledge the vastly more important point that the algorithm works well over a wide range of wind conditions. This includes moderate and strong winds, when non-uniformity in wind direction is small compared to light wind conditions that may occur in boundary layers driven exclusively by convection [4]. Zhou et al. [1] focused on detecting and tracking gusts in the context of wind power production. Given that windy conditions are favorable for generating electricity, our opinion is that the CC method would not be limited by non-uniformity of the wind field within the interrogation window in such applications. Furthermore, Hamada et al. [4] (Section 4.a.1 "Light wind case") showed good results for cross-correlation even when winds are light and variable.

Second, we have shown that *"small (few pixels) movement"* is not a problem for WOF applied to non-Doppler scanning elastic lidars [3]. It is true that the application of optical flow algorithms are often limited to image sequences in which features move with small displacements from frame to frame. However, the WOF algorithm is designed to handle large displacements, first by taking advantage of the multiscale nature of the wavelet framework [11], and second by incorporating the algorithm into a more classical pyramidal approach ([3], (Appendix B)). Velocities as high as 15 m·s^{-1} have been successfully measured [3] (Section 5.a.3, "Strong wind case"). With a typical scan time of 17 s and a 8 m·pixel^{-1} grid spacing, this amounts to displacements above 30 pixel which cannot be deemed "small".

Third, all of the calculations for CC and WOF are performed in real-time during the time between scans (10 s to 20 s) using general purpose graphical processing units (GPGPUs). In fact, the appendix of the paper by Mayor et al. [5] describes the methodology used to accelerate the process using common computer architectures. Further, two conference presentations detail the steps to achieve this performance [12,13]. GPGPUs are available for almost any computer and the models with the required performance cost about 1000 USD.

4. Comparing CC, WOF, and 2D-VAR Methods

We now put the three CC, WOF, and 2D-VAR methods into perspective. Let us first clarify that the objective here is not to state whether any method is better or worse than another.

We first consider the spatial and temporal discretization of the 2D-2C wind fields produced by the three approaches: parameters used in Hamada et al. [4], Dérian et al. [3], and Zhou et al. [1] are listed in Table 1. The time steps are comparable (17 s for CC and WOF, and 30 s for 2D-VAR). (By *time steps*, we mean the elapsed time between scans, which is dependent upon the angular width of the sector scanned and the performance of the lidar used.). The biggest differences lie in the spatial domain: the 2D-VAR method covers a much larger domain than CC and WOF (24 km^2 for 2D-VAR versus 4.6–13 km^2 for the other two). (The area observed is dependent mostly upon the angular width of the sector scan and the maximum range of the useful data. These, in turn, are strongly dependent upon aerosol signal-to-noise ratio for aerosol lidars or the coherent signal-to-noise ratio for Doppler lidars, both of which vary with geographic location and weather conditions.) However, the vector spacing is much finer for WOF than for 2D-VAR: 8 m versus 80 m. WOF provides 10–50 times more vectors than 2D-VAR while operating in a shorter time window. Therefore, it can be safely assumed that WOF could deliver wind fields comparable to those given by 2D-VAR in terms of domain size, spatial, and temporal discretizations.

Table 1. Characteristics of 2D-2C wind field retrieval algorithms.

Method	Hamada et al. [4] (Cross-Correlations)	Dérian et al. [3] (Wavelet-Based Optical Flow)	Zhou et al. [1] (2D-VAR)
Domain shape	60° sector, 0.5 km to 3–5 km range	60° sector, 0.5 km to 3–5 km range	6 km × 4 km
Domain area	4.6 km^2 to 13 km^2	4.6 km^2 to 13 km^2	24 km^2
Vector spacing	sparse	dense, 8 m	dense, 80 m
Number of vectors	variable	$\approx 5 \times 10^4$ to 2×10^5	$\approx 4 \times 10^3$
Time step	$\approx$17 s	$\approx$17 s	30 s
Real-time computations	yes	yes	"possible"

We now examine the accuracy of the estimated wind vectors. Zhou et al. [1] did not comment on the accuracy of the wind fields retrieved with the 2D-VAR algorithm. They cited Cherukuru et al. [2] for details, so we turn to it instead. Cherukuru et al. [2] validated their approach based on 10-min averages, the standard averaging period for the wind energy industry, and compared the 2D-VAR results with measures from a cup and vane anemometer. Hamada et al. [4] and Dérian et al. [3] also considered 10-min averages to validate the CC and WOF algorithms, respectively, by comparing measurements from a Doppler lidar operating in vertical profiling mode. Statistics for the CC, WOF, and 2D-VAR methods are gathered in Table 2. The methods are not directly comparable as Hamada et al. [4] and Dérian et al. [3] gave per-component values, whereas Cherukuru et al. [2] reported on wind speed and direction. Nevertheless, they appear to be similarly accurate.

Finally, we briefly mention algorithmic aspects for WOF and 2D-VAR. They are more comparable as they both deliver dense fields, whereas CC is intrinsically a sparse, local approach. Moreover, optical flow is actually a form of 2D variational analysis. For both algorithms, the wind field is obtained by minimizing a cost function defined as an integral over the spatial domain, and the minimization is achieved by a quasi-Newton method. The differences lie in the choice of the data model—which connects the partial observations (aerosol backscatter for WOF and radial velocity component for 2D-VAR) to the unknown (the 2D-2C horizontal wind field)—and the regularizer (which closes the problem and facilitates a successful minimization). In WOF, the data model is based on the advection of a passive scalar field ([3], Equation (4)): the aerosol backscatter, itself similar to a concentration of particles [14]. In 2D-VAR, the data-model involves the advection of an active scalar field and the radial velocity component ([2], Section 3, Equations (5)–(8)). The regularizer used in WOF is a first-order smoothing term that penalizes strong gradients in the estimated velocity fields ([3], Equation (5)). Authors of the 2D-VAR used the departure of the estimated wind field from a prior solution obtained using another method ([2], the "background constraint" in Section 3). We note that the latter requires solving another wind reconstruction problem before running 2D-VAR, which adds to the computational load. To summarize, both approaches solve inverse problems that are different in terms of models but comparable in terms of structure.

Table 2. Accuracy of wind field retrieval algorithms measured on 10-min averages. Values for Hamada et al. [4] are from Tables 4 and 5. Values for Dérian et al. [3] are from Tables 2 and 3 and Section 5.a.4. Zhou et al. [1] did not report on the accuracy of wind retrieval; values listed here for the 2D-VAR are obtained from Cherukuru et al. [2] (Figure 2 and Table 1). "n.r." stands for "not reported". Note that horizontal wind vectors derived from the Streamline Doppler lidar resulted from a Doppler beam swinging technique that placed off-zenith sample volumes for radial velocities approximately 210 m apart, whereas the cup and vane anemometer is sampled at essentially one point in the atmosphere.

Method	Hamada et al. [4] (Cross-Correlations)	Dérian et al. [3] (Wavelet-Based Optical Flow)	Cherukuru et al. [2] (2D-Var)
Reference measure	Streamline Doppler lidar	Streamline Doppler lidar	cup and vane anemometer
Number of points (duration)	891 ($\approx$150 h)	892 ($\approx$150 h)	120 (20 h)
Wind speed RMS error	n.r.	n.r.	0.383 m·s^{-1}
Wind speed correl. coeff.	n.r.	$\sqrt{0.991} \approx 0.995$	0.96
Wind direction error	n.r.	1.1°	−1.4°
Wind direction correl. coeff.	n.r.	$\sqrt{0.944} \approx 0.976$	0.98
West–east component u RMS error	0.36 m·s^{-1}	0.29 m·s^{-1}	n.r.
West–east component u correl. coeff.	$\sqrt{0.993} \approx 0.996$	$\sqrt{0.995} \approx 0.997$	n.r.
South–north component v RMS error	0.37 m·s^{-1}	0.29 m·s^{-1}	n.r.
South–north component v correl. coeff.	$\sqrt{0.995} \approx 0.997$	$\sqrt{0.997} \approx 0.998$	n.r.

In our opinion, the largest weaknesses of the motion estimation approaches (CC and WOF) to wind field measurement using current aerosol lidars are: (1) the failure of the technique during periods when small-scale aerosol inhomogeneities are not present; and (2) the rarity and expense of commercially available scanning aerosol lidars with the required high-performance. Stable stratification can suppress the production of turbulent eddies and the formation of aerosol inhomogeneities that are required. Based on only two field experiments that we have conducted over land in the Central Valley of California, this condition often occurs at night. We also note that at least one compact and portable aerosol lidar with the required performance is now commercially available and that wind fields derived from it are shown in [7]. As laser, detector, and optical technologies mature, we expect to see improvements in the performance and accessibility to this approach of wind field measurement.

5. Conclusions

Motion estimation approaches for computing 2D-2C wind fields are not impractical for the reasons stated in [1]. Two recent peer-reviewed publications document the performance of CC and WOF techniques which are comparable to those published for their 2D-VAR. We have also taken this opportunity to provide a comparison of the WOC and 2D-VAR techniques to show that the two methods actually have much in common and that both are practical.

Author Contributions: S.D.M. and P.D. shared all aspects of producing this note.

Conflicts of Interest: The authors declare no conflict of interest.

References

1. Zhou, K.; Cherukuru, N.; Sun, X.; Calhoun, R. Wind gust detection and impact prediction for wind turbines. *Remote Sens.* **2018**, *10*, 514. [CrossRef]
2. Cherukuru, N.W.; Calhoun, R.; Krishnamurthy, R.; Benny, S.; Reuder, J.; Flügge, M. 2D VAR single Doppler lidar vector retrieval and its application in offshore wind energy. *Energy Procedia* **2017**, *137*, 497–504. [CrossRef]
3. Dérian, P.; Mauzey, C.F.; Mayor, S.D. Wavelet-based optical flow for two-component wind field estimation from single aerosol lidar data. *J. Atmos. Ocean. Technol.* **2015**, *32*, 1759–1778. [CrossRef]
4. Hamada, M.; Dérian, P.; Mauzey, C.F.; Mayor, S.D. Optimization of the cross-correlation algorithm for two-component wind field estimation from single aerosol lidar data and comparison with Doppler lidar. *J. Atmos. Ocean. Technol.* **2016**, *33*, 81–101. [CrossRef]
5. Mayor, S.D.; Lowe, J.P.; Mauzey, C.F. Two-component horizontal aerosol motion vectors in the atmospheric surface layer from a cross-correlation algorithm applied to scanning elastic backscatter lidar data. *J. Atmos. Ocean. Technol.* **2012**, *29*, 1585–1602. [CrossRef]
6. Mayor, S.D.; Dérian, P.; Mauzey, C.F.; Hamada, M. Two-component wind fields from scanning aerosol lidar and motion estimation algorithms. *SPIE Lidar Remote Sens. Environ. Monit. XIV* **2013**. [CrossRef]
7. Mayor, S.D.; Dérian, P.; Mauzey, C.F.; Spuler, S.M.; Ponsardin, P.; Pruitt, J.; Ramsey, D.; Higdon, N.S. Comparison of an analog direct detection and a micropulse aerosol lidar at 1.5-micron wavelength for wind field observations—with first results over the ocean. *J. Appl. Remote Sens.* **2016**, *10*. 031. [CrossRef]
8. Mayor, S.D.; Eloranta, E.W. Two-dimensional vector wind fields from volume imaging lidar data. *J. Appl. Meteorol.* **2001**, *40*, 1331–1346. [CrossRef]
9. Schols, J.L.; Eloranta, E.W. The calculation of area-averaged vertical profiles of the horizontal wind velocity from volume imaging lidar data. *J. Geophys. Res.* **1992**, *97*, 18395–18407. [CrossRef]
10. Bieringer, P.E.; Higdon, S.; Bieberbach, G.; Hurst, J.; Mayor, S.D. Assimilation of lidar backscatter and wind data into atmospheric transport and dispersion model. In Proceedings of the Eighth Symposium on Lidar Atmospheric Applications, American Meteorological Society, Seattle, WA, USA, 2017. Available online: https://ams.confex.com/ams/97Annual/webprogram/Paper308839.html (accessed on 11 October 2018).
11. Dérian, P.; Héas, P.; Herzet, C.; Mémin, E. Wavelets and optical flow motion estimation. *Numer. Math. Theory Methods Appl.* **2012**, *6*, 116–137.
12. Mauzey, C.F.; Dérian, P.; Mayor, S.D. Wavelet-based optical flow for real-time wind estimation using CUDA. In Proceedings of the GPU Technology Conference, San Jose, CA, USA, 24–27 March 2014.
13. Mauzey, C.F.; Lowe, J.P.; Mayor, S.D. Real-time wind velocity estimation from aerosol lidar data using graphics hardware. In Proceedings of the GPU Technology Conference, San Jose, CA, USA, 14–17 May 2012.
14. Held, A.; Seith, T.; Brooks, I.M.; Norris, S.J.; Mayor, S.D. Intercomparison of lidar aerosol backscatter and in-situ size distribution measurements. Presentation number B-WG01S2P05. In Proceedings of the European Aerosol Conference, Granada, Spain, 2–7 September 2012.

remote sensing

MDPI

Article

Reducing the Uncertainty of Lidar Measurements in Complex Terrain Using a Linear Model Approach

Martin Hofsäß[1],[*],[†],[‡], Andrew Clifton [2],[‡] and Po Wen Cheng [1],[‡]

[1] Stuttgarter Lehrstuhl für Windenergie, Universität Stuttgart, Allmandring 5b, 70569 Stuttgart, Germany; cheng@ifb.uni-stuttgart.de
[2] WindFors—Universität Stuttgart, Allmandring 5b, 70569 Stuttgart, Germany; clifton@windfors.de
[*] Correspondence: hofsaess@ifb.uni-stuttgart.de; Tel.: +49-(0)711-68568308
[†] Current address: Allmandring 5b, 70569 Stuttgart, Germany.
[‡] These authors contributed equally to this work.

Received: 2 July 2018; Accepted: 10 September 2018; Published: 13 September 2018

check for updates

Abstract: In complex terrain, ground-based lidar wind speed measurements sometimes show noticeable differences compared to measurements made with in-situ sensors mounted on meteorological masts. These differences are mostly caused by the inhomogeneities of the flow field and the applied reconstruction methods. This study investigates three different methods to optimize the reconstruction algorithm in order to improve the agreement between lidar measurements and data from sensors on meteorological masts. The methods include a typical velocity azimuth display (VAD) method, a leave-one-out cross-validation method, and a linear model which takes into account the gradients of the wind velocity components. In addition, further aspects such as the influence of the half opening angle of the scanning cone and the scan duration are considered. The measurements were carried out with two different lidar systems, that measured simultaneously. The reference was a 100 m high meteorological mast. The measurements took place in complex terrain characterized by a 150 m high escarpment. The results from the individual methods are quantitatively compared with the measurements of the cup anemometer mounted on the meteorological mast by means of the three parameters of a linear regression (slope, offset, R^2) and the width of the 5th–95th quantile. The results show that expanding the half angle of the scanning cone from 20° to 55° reduces the offset by a factor of 14.9, but reducing the scan duration does not have an observable benefit. The linear method has the lowest uncertainty and the best agreement with the reference data (i.e., lowest offset and scatter) of all of the methods that were investigated.

Keywords: complex terrain; complex flow; lidar; VAD; remote sensing; wind energy

1. Introduction

Doppler wind lidar (light detection and ranging) systems emit laser pulses that are reflected by aerosols in the atmosphere. This backscattered light has a frequency shift (the Doppler effect) due to the movement of the aerosols. The wind speed in direction of the laser beam can be calculated from the frequency shift using the Doppler equation. The alignment of the laser can be described by the azimuth and elevation angle. Ground-based lidar systems can be operated with different operating modes, such as staring, velocity azimuth display (VAD), plan position indicator (PPI) or range height indicator (RHI). The operating modes differ in how the two angles to each other are moved. In PPI mode the elevation angle is fixed and only the azimuth angle is varied over a small sector range. In RHI mode the azimuth angle is fixed and the elevation angle is varied. Both angles are fixed when a lidar operates in a so-called staring mode. The most common mode for commercial wind profilers is the VAD mode. In VAD mode, the azimuth angle is varied over a circle with a fixed elevation angle. The number of measuring points over the circle are specified by the manufacturer. Also, the backscattered light does not come from a single point but instead along a finite length. This probe length depends on the pulse duration and the measurement technology, for example the use of a pulsed or continuous wave (CW) light source [1]. The CW lidars measure the heights one after the other, while pulsed lidar systems measure several heights simultaneously. It is then possible to estimate a wind vector at one or more points in the lidar scan volume using the wind speed data and azimuth, elevation, and range information and by making assumptions about homogeneity and stationarity. This process is known as wind field reconstruction [1].

Ground-based doppler wind lidar systems have many advantages over wind measurements made with in-situ instruments mounted on meteorological masts (met masts). Commercial ground-based lidar devices can measure at heights up to 300 m, which is above the tip heights of modern wind turbines, while met masts frequently only reach 80–100 m. The lidar can also be reconfigured for different tasks [2], and moved easily so that several locations can be investigated with short measurement campaigns.

Previous comparisons between lidar measurements and measurements made with cup anemometers on met masts show an almost perfect match between the two: [3] estimated a slope of 0.9558 and an offset of 0.1577 m·s^{-1} and a R^2 of 0.9984 in flat terrain and homogeneous flow conditions, while [4] found a slope of 1.004, offset of −0.079 m·s^{-1} and an R^2 of 0.996 based on 10 min averages.

In complex terrain, the correlations between wind speed measurements with single points sensors on met masts and lidar systems show considerable differences. The reason for the large differences are the inhomogeneous flow conditions, which can be caused, for example, by the terrain (momentum-induced turbulence) or by thermal effects (buoyancy-induced turbulence). Further causes can be found in [5]. The magnitude of the differences depends on local conditions:

- Fluctuations in the flow field: Changes in terrain roughness upstream of the measurements cause large variations and changes in wind speed with height lead to uncertainty in the wind speed [6].
- Lidar technology: The weighting function of the lidar signal processing also leads to over- or under estimation of the wind speed.

In this study pulsed ground-based lidar systems were used, which performed modified VAD scans. In a VAD scan, the elevation angle is the half opening angle ϕ of the cone. The lidar devices determine, at selected points, the line-of-sight velocity $v_{los}(\theta, \phi, f_d)$, which is the projection of the wind velocity vector $\vec{v}(x, y, z)$ onto the normal vector $\vec{n}(\theta, \phi)$ of the laser beam (Equation (1), the index indicates the axis direction). The line-of-sight velocity depends on the half opening angle ϕ, the azimuth angle θ and the measuring distance or focal distance f_d. The normal vector is calculated as $\vec{n}(\theta, \phi) = [\cos(\theta)\sin(\phi), \sin(\theta)\sin(\phi), \cos(\phi)] = [n_x, n_y, n_z] = \vec{n}(x, y, z)$ where n_i is the component of the normal vector in $i = x, y, z$ direction. The vector $\vec{n}(\theta, \phi)$ can also be expressed in Cartesian coordinates, which are mainly used in this study. The line-of-sight velocity is given by:

$$v_{los} = u \cdot n_x + v \cdot n_y + w \cdot n_z \tag{1}$$

If v_{los} is measured from multiple azimuthal directions and elevations, the wind velocity components u, v, w can be calculated. If these measurements are made from a single point and encompass a volume of the atmosphere, it must be assumed that u, v, and w apply at all points in the wind field, i.e., that the flow is homogeneous. This leads to errors in the determination of wind speeds in heterogeneous, "complex" flow. Bingöl [7] has shown with a simple mathematical model of a flow over a hill, that the error in the horizontal wind speed can be in the order of magnitude of 10%.

In order to reduce the error during data collection, multi-lidar systems [8–10] can be used. The multi-lidar systems measure simultaneously $v_{los}(x, y, z)$ at the same point. From this system of equations, the wind velocity components can be calculated without assuming the flow is homogeneous. The first simultaneous measurements with several lidar systems were carried out at the *Musketeer Experiment 2007* in flat terrain [8]. Measurements in complex terrain were made during the *Kassel*

Experiment 2014 [9] and during the *NEWA* campaign in Perdigão [11]. However, this measurement configuration is very cost-intensive (greater than 200,000€ per lidar device), which is not affordable for site assessments. Furthermore, a large effort has to be made for a successful operation of the multi-lidar configuration [10].

Another possibility to improve the measurements is to use wind field reconstruction models to better convert v_{los} data into a wind vector or wind field [5,12,13].

Finally, it is possible to use flow modelling to try to reduce differences between lidar measurement and traditional measurements when a campaign has already been completed. These typically focus on the use of commercial Computational Fluid Dynamics (CFD) programs to model flows and "correct" (in reality, "adjust") lidar measurements in complex terrain to better match point measurements from anemometers. Despite a continuing need to validate such tools in complex terrain, a 2015 study [14] concluded that with proper parametrization and flow modelling, flow simulations can be a useful tool in the post-processing of lidar measurements in complex terrain and this is likely to be an important approach in future [2].

In this study, the main focus is on the use of ground-based lidar systems with VAD scan patterns with additional vertical measuring points. This paper introduces several modified measurement strategies and adapted evaluation algorithms to show that it is possible to reduce uncertainties without having to forego the advantages of ground-based lidar systems. The results were also used to explore three questions about the effect of the scan design:

1. Is the length of the trajectory circulation time important in the comparisons with the met mast?
2. Does the half opening-angle ϕ have an influence on the accuracy of the correlations?
3. Is the local resolution of the measurement points important for the result?

Section 2 gives an overview over the measurement campaign, the test site and the methods used for the wind field reconstruction. In Section 3 the results of the comparison between the results of the investigated methods and the reference in-situ measurements are presented. The discussion of the results is in Section 4 and conclusions are presented in Section 5.

2. Materials and Methods

This study uses lidar data obtained during a measurement campaign in complex terrain in Southern Germany. This section provides an overview of the test site, the used measuring systems and the settings and scan patterns. Furthermore, the data preprocessing and data filtering will be discussed. Finally, the wind field reconstruction methods are presented.

2.1. Test Site and Measurement Campaign

The experiment was carried out in south-west Germany on the Swabian Alb. The Swabian Alb extends for around 100 km and can be categorized as a very hilly area consisting of high plateaus surrounded by a pronounced 100–150 m tall wooded escarpment known in the region as the *Albtrauf*. The measurements described here are centred on a 100 m high met mast in relatively flat land less than 1 km from a section of *Albtrauf* (Figure 1).

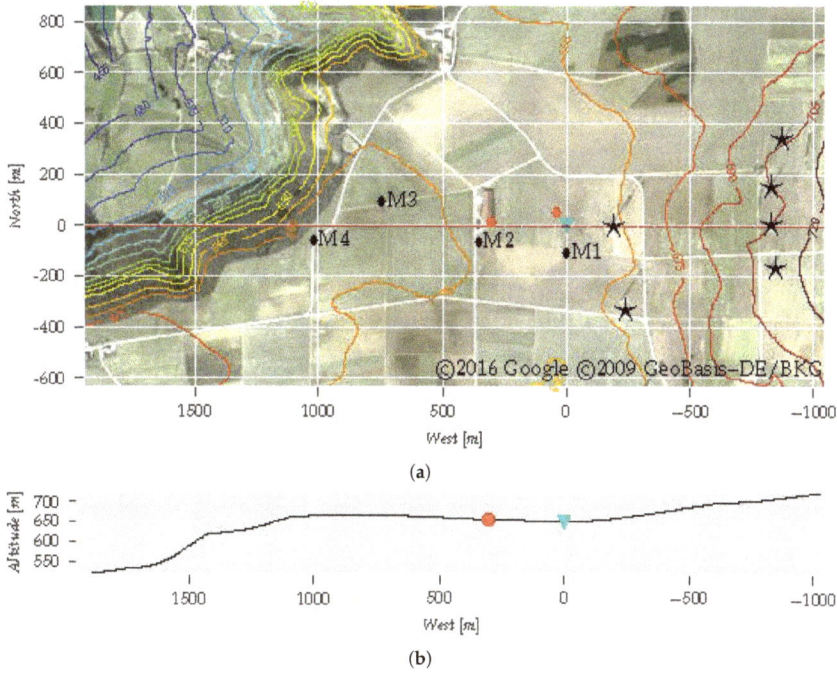

Figure 1. (a) The test site topography. An aerial picture (©2016 Google) is overlain with elevation contours (©2009 GeoBasis–DE/BKG, LGL (Geobasisdaten ©LGL Landesamt für Geoinformation und Landentwicklung Baden-Württemberg, Az.: 2851.9-1/19)). The locations of the met mast (▼), lidars (●), wind turbines (★), and vertical profiles from [15,16] are also shown. The *Albtrauf* is between the two thicker elevation contours. (b) Terrain cross section. The locations of the met mast (▼) and lidars (●) are shown. The experiment site. (a) Topography and structures (b) Terrain section along red dashed line.

Two wind lidars were installed near the mast. These lidar were the SWE Scanner [17] and a Galion all sky long-range scanning wind lidar. The measurement campaign ran from March 2015 until February 2016. Not all systems were available during the whole measurement period. The SWE Scanner measured from March 2015 to February 2016 and the Galion system measured from March 2015 until June 2015.

Ideally the lidar and met mast would have been co-located or the lidar would been positioned a few 10s of meters upwind of the met mast to avoid wakes. However, this was not possible in this study because the lidars required around 2 kW of power, which could not be supplied directly at the mast's location. Instead the lidars were located at a nearby farm building (visible in Figure 1a), approximately ≈300 m from the met mast.

In a second short measurement period from November 2015 to February 2016 the SWE Scanner was installed at the foot of the met mast guy wires, approximately 48 m from the met mast on a bearing of 330°. During this time the same configuration as in the main measurement period was used, but the number of pulses were lowered from 10,000 to 3000 to reduce the the scanning time. The met mast and both lidar systems were time synchronized with a GPS time signal.

2.1.1. Local Conditions

This site has been the subject of many coordinated studies [16,18–20]. A small remotely-piloted aircraft was used to measure the flow conditions [18]. Detailed Detached-Eddy Simulations (DES) of the flow have been validated using different measurement methods [15,16]. Wind tunnel tests and CFD simulations have also been to model the flow over the site [20].

The wind at this site is mostly from the west to north-west (Figure 2). Around 38% of winds are perpendicular to the *Albtrauf*, and so changes in wind direction of 10° can result in significantly different inflow profiles downwind of the *Albtrauf* [21].

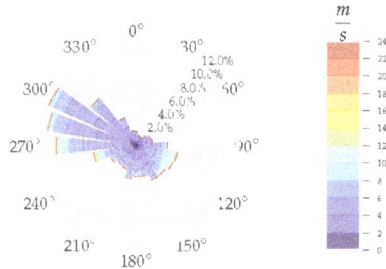

Figure 2. Windrose at the experiment site.

The goal of this study is to compare the data from the met mast and the lidar to assess different measurement and data processing approaches. It is therefore essential that there is minimal external influence on the results, for example as a result of differences in flow between the lidar and tower that might be introduced by the ground cover or the terrain. The impact of these differences can be judged using the International Electrotechnical Committee (IEC) 61400-12-1 Standard (2005, [22]) for power performance testing of wind turbines, which quantifies the potential impact of obstacles as a function of size and distance. At 1400 m south-west of the mast, there is a small forest (not visible in Figure 1), which, according to the Standard, is far away enough that it should not have an influence. East of the met mast, the terrain rises again up to a height of 720 m above sea level. There are also several commercial wind turbines near to the mast which could influence the flow. These obstructions would therefore reduce the free stream sector (according to [22]) from 171° to 341°. However, this large a sector still introduces the chance for flow differences between the devices.

Other studies of this site can be used to understand the podar location and the met mast. The vertical wind profile at the points M1–M4 in Figure 1a was simulated for wind from 295° using CFD that was validated against measurements from an unmanned aerial vehicle [15,16]. Results from the CFD (Figure 3) showed a speed-up of ≈20% at 100 m above ground at M4, closest to the escarpment. At M3 the speed-up factor at 100 m above ground is ≈1.15, while at M1 the speed up effect has disappeared and the vertical profile is fully recovered. From these profiles it can also be seen that the 100-m wind speed at M2 is within 2% of M1, i.e., the mean wind speed difference between M2 (near the lidar) and M1 (near the met mast) is minimal. Based on the results from the CFD study, it was decided to limit the data used to a 70° inflow wind direction sector from 245° to 315°.

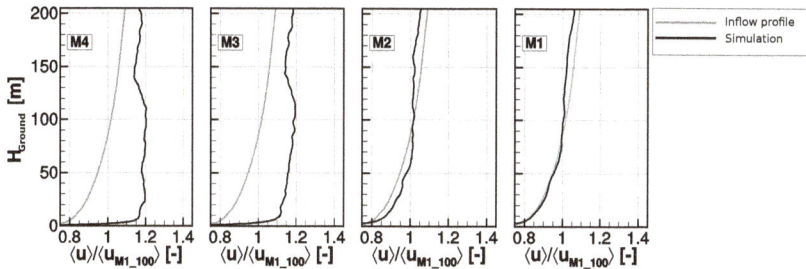

Figure 3. The vertical wind profile near the escarpment and on the plateau. M4 is at the escarpment, M2 is closest to the lidar, and M1 is closest to the met mast (Figure adapted and translated from Figure 5.8 in Ref. [16]).

2.1.2. Met Mast

The 100 m met mast is located 1000 m east of the escarpment and is equipped with numerous measurement instruments, which are mounted at several heights. It is a met tower that meets the guidelines of the IEC 61400-12-1 (2005) standard for power performance measurements [22]. These include various first class cup anemometers (at 10 m, 25 m and 100 m), first class wind vanes (at 25 m, 50 m, 75 m and 92 m), barometers (at 5 m and 98 m), thermometers (at 5 m, 50 m, 75 m and 98 m), hygrometers (at 5 m, 50 m, 75 m and 98 m) and three three-dimensional (3-D) ultra sonic anemometers (*USA*) mounted at 50 m, 75 m and 98 m. In north-south direction the mast is equipped with booms of a length of 5 m to reduce the influence and the flow interaction with the met mast main body. The sampling frequency of the met mast sensors has been set to 20 Hz. In this context, first class means that the sensor has an accuracy class of 0.5 and meets all requirements of the IEC 61400-12-1 (2005) standard [22] for wind turbine power performance testing. The met mast therefore meets or exceeds wind industry best practices for wind measurements.

2.1.3. Lidar

The SWE Scanner system and Galion lidar were installed next to each other ≈300 m west of the met mast. Both lidar systems used pulsed lasers and so can measure several heights simultaneously.

The global coordinate system is a right hand system where the u or x_1-axis points in north direction, the v or x_2-axis points west direction and the w or x_3-axis points vertical to the sky. The wind direction corresponds to the Geographic Coordinates from north clock wise. The device coordinate systems ($\theta = 0°$) correspond to the global coordinate system (North).

The SWE Scanner was configured to use a modified VAD method known as the *six beam* trajectory [23]. The trajectory has five points on a circle with a half-opening angle of $\phi = 15°$ and one vertical point in the centre of the circle (Figure 4a). The azimuth angle θ is equally spaced on the circle with $\triangle\theta = 72°$. The scan duration of the trajectory was 8.8 s. Five focal distances were measured simultaneously (50 m, 75 m, 100 m, 125 m, 150 m) at each azimuthal position.

The Galion lidar has a maximum measurement range of approximately 4000 m. The device is fully configurable, with freely chosen azimuth and elevation angles and a maximum half opening angle of $\phi_{max} = 92°$. In this campaign, the device measured three *six beam* trajectories with three different opening angles $\phi_{1,2,3} = [20°, 39.2°, 55°]$ one after another (Figure 4b). These measurements were also used to test a method for determining the second statistical moments of the measured wind speed in complex terrain [24,25]. The opening angles used were therefore the same as used in [24]. The azimuth angle was equally spaced with $\triangle\theta = 72°$. The order of trajectory points was chosen to give the least time for every measurement sequence. The trajectory looks like a pentagram (Figure 4c) because the elevation angle can be changed faster than the azimuth angle. For a given combination of θ and ϕ the device is able to measure the radial velocity v_{los} at multiple focal distances df. The data includes measurements for different distances ranging from 45 m to 735 m in 30 m intervals. The scan duration for all 18 points within the trajectory was 49 s.

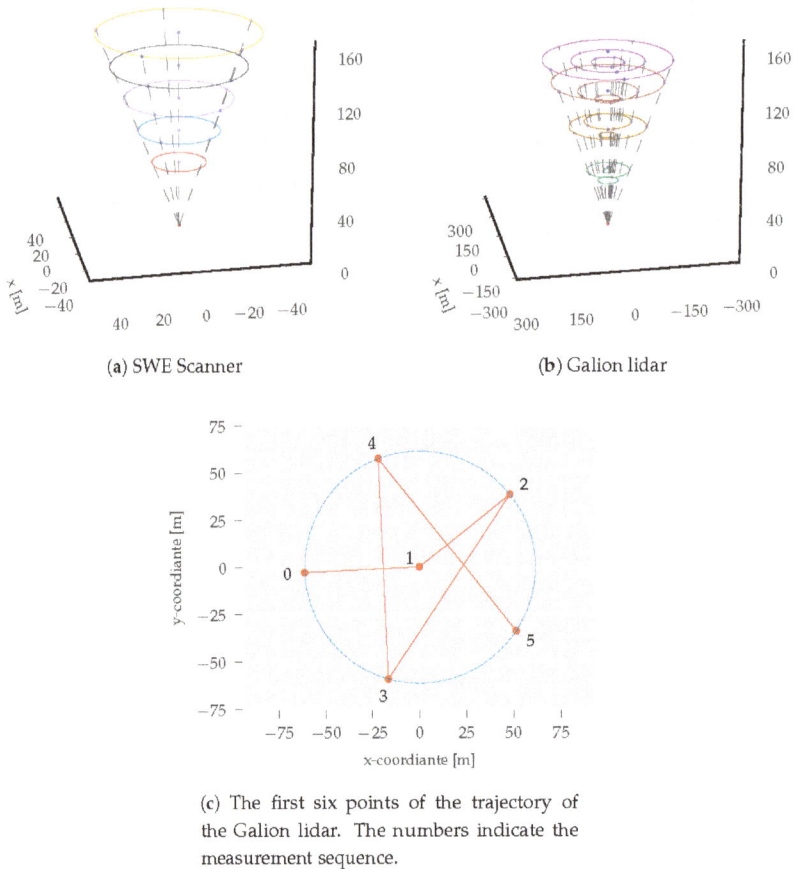

(a) SWE Scanner

(b) Galion lidar

(c) The first six points of the trajectory of the Galion lidar. The numbers indicate the measurement sequence.

Figure 4. Measurement trajectories for (**a**) the SWE Scanner and (**b**,**c**) the Galion lidar.

The two lidar systems differ in trajectory duration and in the configuration of the half opening angle ϕ. This difference in configuration enables the influence of these parameters on the measurement quality to be studied.

2.1.4. Data Filtering and Selection

In order to compare the different wind field reconstruction methods with the reference data from the met mast, these must be prepared and filtered in the pre-process. It is necessary to filter the data in order to accurately compare the different systems. The process of filtering and selection is shown in Figure 5. Data from the lidar and met mast are treated differently:

Lidar: The high-resolution recorded lidar data are processed with the wind field reconstruction methods: The last N data points are used for each time step. These methods determine both the wind velocity components u, v, w and statistical parameters (CNR_{mean}, CNR_{min} and CNR_{max}) for the carrier-to-noise ratio (CNR). The CNR parameters are still required for later data processing. The horizontal wind speed is calculated from the wind speed components u, v (Equation (2)).

These data are now subjected to a CNR filter to exclude samples outside of the CNR range. It is important that both the CNR_{min} and the CNR_{max} are within the CNR limits. Values outside these limits are not taken into account for further consideration. From these data the 10 min statistics are calculated and selected.

$$v_h = \sqrt{(u^2 + v^2)} \tag{2}$$

Met mast: The recorded data is first subjected to a plausibility test where the system checks that values are within a realistic range. If data is available from a second sensor at nearly the same height a comparison of the data from those sensors is also made, e.g., between the cup anemometer at 100 m and the horizontal wind speed from the sonic at 98 m. Then the 10 min statistics are calculated, and the data is selected according to the selection criteria in Table 1 .

The directional dependency of the lidar met mast data is checked in a different way, represented by the green path in Figure 5.

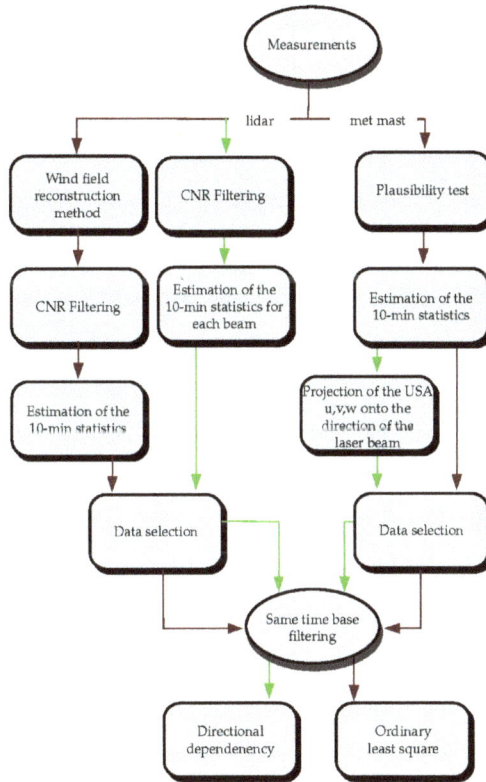

Figure 5. Steps in processing and selecting data from the lidar and met mast.

Table 1. Selection criterion used in Figure 5.

Parameter	Acceptable Range	
	Minimum	Maximum
Lidar measurements		
CNR (min, mean) [dB]	−22	10
Weather conditions		
Wind direction [°]	245	315
Temperature [°]	>2	-
Mast function		
Std. cup anemometer	>0.01	-
Data availability		
lidar [%]	90	100
met mast [%]	100	-

After filtering and selection, both data streams will contain gaps. Only lidar and met mast measurement data that are available at the same time step are used.

2.2. Wind Field Reconstruction Methods

Two different wind field reconstruction methods were used, the *Continuous least-square* method and a *linear* model, which additionally use the velocity gradient. In addition, these two algorithms are combined with predicted residual error sum of squares (*PRESS*) statistics.

$$\underbrace{\begin{bmatrix} v_{los1} \\ \vdots \\ v_{losN} \end{bmatrix}}_{\text{b}} = \underbrace{\begin{bmatrix} n_{x1} & n_{y1} & n_{z1} \\ \vdots & \vdots & \vdots \\ n_{xN} & n_{yN} & n_{zN} \end{bmatrix}}_{\text{A}} \underbrace{\begin{bmatrix} u \\ v \\ w \end{bmatrix}}_{\text{x}} \tag{3}$$

2.2.1. Continuous Least-Square

This procedure, here referred to as continuous least-square (*CLS*), is a modified variant of the standard VAD method as used in commercial systems [26] to determine the wind velocity components. The idea behind this is that the velocity components are calculated continuously and not after just one trajectory cycle. This means that if the trajectory has N data points, the last N data points will be used for the calculation. Considering Equation (1), a system of equations (Equation (3)) can then be established from these N data points to determine the wind velocity components u, v, w from the lidar data. The matrix **A** of the equation system consists of the components of the normal vectors $\vec{n}(x, y, z)_i$ of the used N data points. To solve Equation (3), the vector **b** is multiplied by the inverse Matrix $\mathbf{A}^{-1}$. If the trajectory has more than three data points (which are at least necessary) the Moore-Penroe pseudoinverse $\mathbf{A}^{-1}$ will be used to solve the system of equations at the time step t. In order to solve this system of equations, it is not necessary to assume the homogeneous flow, since a solution is estimated which has the smallest absolute error. As a result, measurement errors (unrealistic data (e.g., due to bad CNR) strongly distort the solution of the equation system (Equation (3)). These unrealistic data must then be filtered in post-processing.

2.2.2. Predicted Residual Error Sum of Squares (PRESS)

The *PRESS* method [27,28] can be used to make the *CLS* procedure more robust. *PRESS* detects outliers and excludes them from the evaluation using leave-one-out cross validation. The *PRESS* method has been used for comparable applications such as cross-validation of samples [29], but this paper is the first known application of the *PRESS* method to lidar wind data.

In *PRESS*, the initial record—containing N data points—is reduced by one value and the reduced record N-1 is used to calculate the underlying model. The calculation scheme is represented in Figure 6 and repeats itself N times for each time step. The new estimated reconstructed line-of-sight velocity $\hat{v}_{los}$ are calculated from this solution and the inverted model. Then the coefficient of determination R^2 (Equation (4)) is calculated to quantify the goodness of fit. The R^2 is calculated as:

$$R^2 = \frac{SS_{reg}}{SS_{tot}} \tag{4}$$

$$SS_{reg} = \sum_i (\hat{v}_{los,i} - \overline{v}_{los})^2 \tag{5}$$

$$SS_{tot} = \sum_i (v_{los,i} - \overline{v}_{los})^2 \tag{6}$$

where SS is the sum of the squares for the regression (reg) and the total (tot); $\overline{v}_{los}$ represents the mean values and $\hat{v}_{los}$ is the reconstructed line-of-sight velocity. In each time step a different point is omitted. From these N iterations, the variant with the highest R^2 is selected for further post-processing.

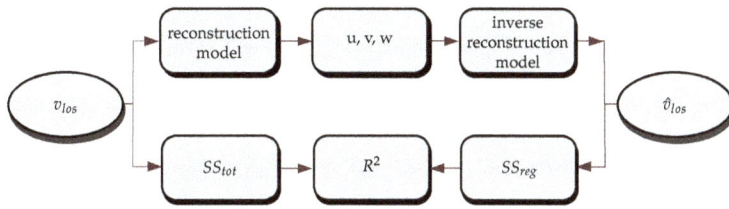

Figure 6. Overview of the workflow to estimate the combination of data points the maximized the coefficient of determination R^2, based on the input variables v_{los}, the estimated wind velocity components u, v, w, the model predicted $\hat{v}_{los}$ and the total sum of squares (SS_{tot}) and the regression sum of squares SS_{reg}.

2.2.3. Linear VAD Model

This procedure is based on the idea of converting the turbulent wind vector $\vec{v}(x, y, z)$ at the point x_0, y_0, z_0 into a linear function through a Taylor polynomial up to the first order to represent the inhomogeneous flow conditions in complex terrain more accurately. The resulting Taylor polynomial for the wind velocity vector $\vec{v}(x, y, z)$ is represented in Equation (7). This is made up of a stationary value $\vec{v}(x_0, y_0, z_0)$ and three directional derivative in space at the point of development (index 0).

$$\vec{v}(x,y,z) = \vec{v}(x_0,y_0,z_0) + \frac{\partial \vec{v}}{\partial x}(x - x_0) + \frac{\partial \vec{v}}{\partial y}(y - y_0) + \frac{\partial \vec{v}}{\partial z}(z - z_0) \tag{7}$$

$$
\begin{aligned}
v_{los} = {} & u_0 n_x + u_x' n_x (x - x_0) + u_y' n_x (y - y_0) + u_z' n_x (z - z_0) \\
& + v_0 n_y + v_x' n_y (x - x_0) + v_y' n_y (y - y_0) + v_z' n_y (z - z_0) \\
& + w_0 n_z + w_x' n_z (x - x_0) + w_y' n_z (y - y_0) + w_z' n_z (z - z_0)
\end{aligned} \tag{8}
$$

The linear approximation (Equation (8)) consists of twelve unknown variables. These are the three wind vector components (u, v, w with index 0) at the point x_0, y_0, z_0 and the other nine variables are the partial derivatives (mark with a *ı* in Equation (8)) of the three wind vector components in the Cartesian coordinate space (index x, y, z). This equation cannot be solved without further assumptions. Some terms have the same geometric identity and are linearly dependent: the components u_y' and v_x' and the terms u_z' and w_x' and the v_z' and w_y' terms had the same prefactors.

Each gradient pair has the same prefactors as shown in the example for u_y and v_x:

$n_x y = xy/f_d = x n_y$ with $n_x = x/f_d$ (f_d is the distance of the lidar measuring point).

To solve the Equation (8), two important assumptions have to be made:

1. The components are only evaluated along the vertical axis of the lidar device (coordinate origin), which means that $x_0 = y_0 = 0$.
2. The gradient of the vertical wind speed components in x- and y-direction w_x and w_y would be zero.

CFD simulations and the results show very small changes in the inclination angle at this test site that support the second assumption [16]. There are no more assumptions possible for the components u_y and v_x to simplify the Equation (8) and so the two terms are not separable. A case to simplify Equation (8) is when all trajectory points are located on a horizontal plane with the height $z_0 = z$, then the partial derivatives in the direction of z would disappear. In this work, a lidar system is used which works with constant focal lengths f_d and thus places the measuring points along a sphere in space.

With these assumptions it was possible to reduce the number of unknown variables from twelve to nine and a system of equations with these nine unknown variables is obtained. These nine variables consists of the three wind velocity components and the six simplified components of the strain tensor. In order to calculate the unknown components, at least nine linearly independent measuring points are required. The reduced system of equations is represented in Equation (9). On the left hand side is the known vector **b** which contains the N measured v_{losi} and on the right hand side are the components of the normal vector in matrix **A** and the nine components of the linear approximation of the velocity vector in vector **x**. Equation (9) can be solved by inverting matrix $\mathbf{A}^{-1}$ and multiplying by the vector **b**.

$$
\underbrace{\begin{bmatrix} v_{los1} \\ \vdots \\ v_{losN} \end{bmatrix}}_{b} = \underbrace{\begin{bmatrix} (n_x)_1 & (n_xx)_1 & (n_y)_1 & (n_yy)_1 & (n_z)_1 & (n_zz-z_0)_1 & (n_yx)_1 & (n_x(z-z_0))_1 & (n_y(z-z_0))_1 \\ \vdots & \vdots & \vdots & \vdots & \vdots & \vdots & \vdots & \vdots & \vdots \\ (n_x)_N & (n_xx)_N & (n_y)_N & (n_yy)_N & (n_z)_N & (n_zz-z_0)_N & (n_yx)_N & (n_x(z-z_0))_N & (n_y(z-z_0))_N \end{bmatrix}}_{A} \underbrace{\begin{bmatrix} u_0 \\ \frac{\partial u}{\partial x} \\ v_0 \\ \frac{\partial v}{\partial y} \\ w_0 \\ \frac{\partial w}{\partial z} \\ \frac{\partial v}{\partial x} + \frac{\partial u}{\partial y} \\ \frac{\partial u}{\partial z} \\ \frac{\partial v}{\partial z} \end{bmatrix}}_{x} \tag{9}
$$

3. Results

This section presents the results of the data analysis. First, the lidar data is compared with the three dimensional ultra sonic anemometers (*USA*) data to determine if there is a directional dependency. For this comparison, the 3-D velocity information from the *USA* are reduced to the line-of-sight velocity by applying a projection on to the direction of the laser beam. Furthermore, in a second step the results of the three wind field reconstruction methods are compared with the met mast data. For the comparison between lidar and met mast data the method of the ordinary least squares (OLS) is used. The data from the Galion lidar were used for further investigations. Furthermore, the influence of the opening angle on the correlation with the met mast was examined and whether the scan duration has an influence on the results. In order to compare the different methods with each other, the determined confidence intervals are compared with each other.

3.1. Directional Dependency

To find out if there is a directional dependency at this test site, the wind vector from the *USA* $\vec{v}_{usa}$ at 98 m was projected onto the normal vectors $\vec{n}_{lidar}$ of the laser beams of the SWE Scanner. The projection is done using Equation (1).

The relative and absolute differences depending on the wind direction and the horizontal wind speed for each azimuthal position of the laser beam are shown in Figure 7. When looking at these plots, it should be noted that the sector disturbed by turbine wakes is from 341° to 171° (Figure 1). Figure 7 clearly shows increases in differences between the lidar and ultrasonic measurements in this sector. A closer inspection of the beam in the 288° direction shows that the positive differences (red dots) are mainly in the direction of the escarpment. There also appears to be an effect of the flow direction on the sign of the difference between the projected lidar wind speed and the value from the met mast in some sectors, whereby flow towards the lidar has a positive difference and flow away has a negative difference. This effect can be seen clearly in the beams at 288° and 216° but less so in the other directions. The particularly small differences do not appear to depend on direction or wind speed and also appear to have random distribution.

When the wind is perpendicular to the laser direction (e.g., from 18° or 198° for the azimuthal beam from 288°), the lidar can only observe the existing vertical component w and the horizontal components should be zero if the alignment is perfect. This arises from Equation (3). From Figure 7 it can be seen that the points on the axes orthogonal to the beam do not have zero magnitude. However, because of different sampling rates, the chance that the lidar and met mast were not perfectly aligned (despite using best practice), and the possibility that the wind direction was not exactly the same at the lidar and mast, it is impossible to say that this difference is solely the vertical wind component w.

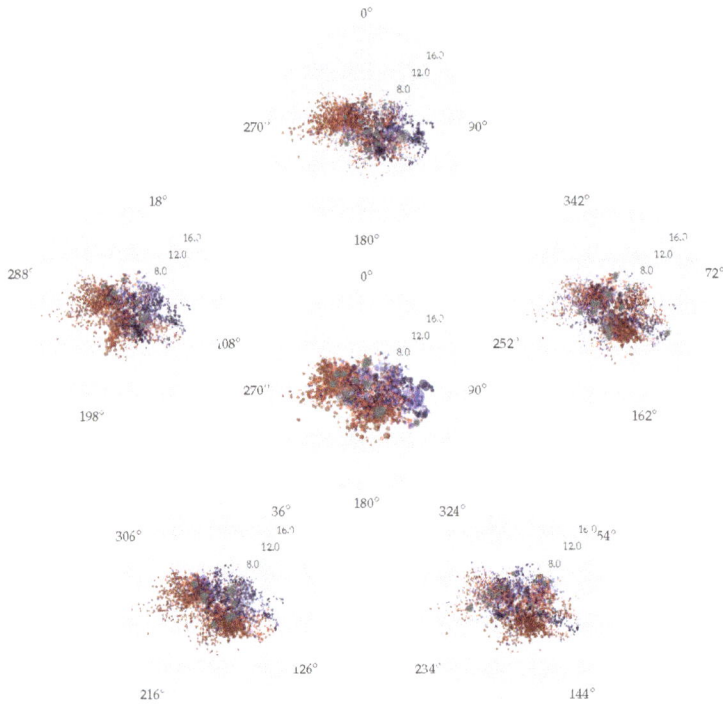

Figure 7. Results of the projection of the *USA* onto the five lidar beam normal vectors. Each polar plot represents a combination of azimuth-elevation angles of the laser beam. The positions of the plots correspond to the azimuth angle of the scanner laser beam: the upper left graph is equivalent to the azimuth angle at 288°, the lower left graph is equivalent to 216°, the middle graphic is the vertical measurement, etc. The axes of the graphs are rotated in the respective angles of the laser beam direction. The azimuth is the wind direction from the met mast. The distance of the point from the origin corresponds to horizontal wind speed (8, 12, 16mper s). The absolute differences between met mast and lidar ($\vec{v}_{usa} \cdot \vec{n}, i_{lidar} - \overline{v}_{los,i}$) are represented in colour and the size of points shows the relative differences between met mast and lidar. Positive absolute differences are marked in red, very small differences (magnitude < 0.001) are shown in grey, and negative differences in blue.

3.2. Wind Speed Comparisons

In this subsection the results of the different wind field reconstruction method compared to the cup anemometer are presented. The results of the statistical parameters of the OLS are summarized in Table 2.

SWE Scanner: Figure 8 shows the results of the comparison between the SWE Scanner and the cup anemometer installed at 100 m using the *CLS* and *PRESS* methods.

The results show very similar results for the methods and locations which are investigated. The parameter offset differs by up to 0.4 m·s^{-1} between the results and the slope changes by 0.04 between the lowest and the highest value. R^2 is similar with values between 0.94 and 0.95. It is noticeable that the results with *PRESS* (Figure 8b,d) have larger offsets and slopes than the *CLS* method (Figure 8a,c). These 10 min lidar data contain the point combinations that had the highest R^2 in time step t, which should reduce the effect of outliers. However, this does not lead to an improvement in the agreement between lidar and mast.

The investigated wind speed range in this case was range 0–20 m·s^{-1}. Data are most frequent in the lower range (up to 10 m·s^{-1}), where the 5–95th quantile range of the data is large compared to that seen in the more sporadic wind speeds over 15 m·s^{-1}.

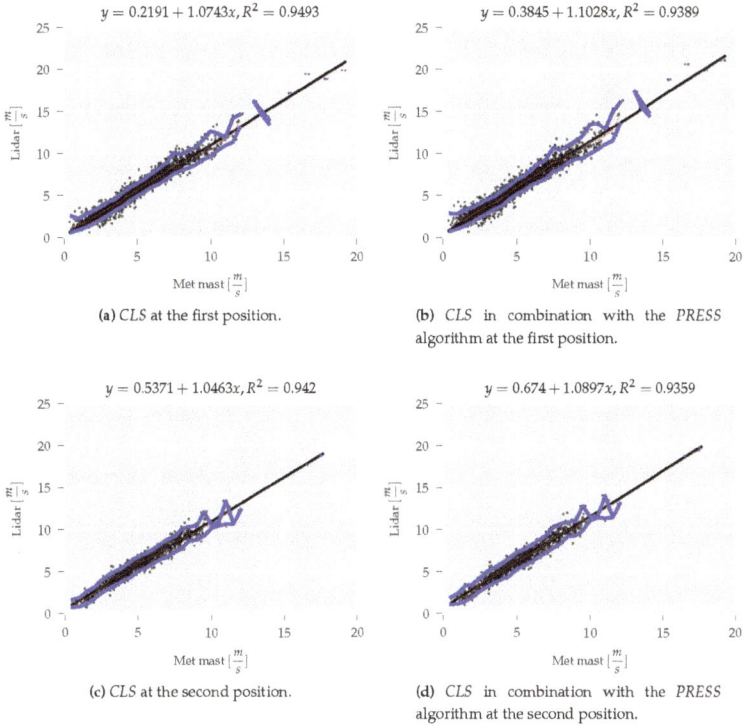

(a) *CLS* at the first position.

(b) *CLS* in combination with the *PRESS* algorithm at the first position.

(c) *CLS* at the second position.

(d) *CLS* in combination with the *PRESS* algorithm at the second position.

Figure 8. Correlation between SWE Scanner and the cup anemometer at 100 m above ground. The black dots represent the 10 min data points, the black line is the regression line, and the blue lines are the 5th and 95th quantiles.

The same comparison was also made using data from the SWE scanner at the second measurement location near the foot of the met mast. For brevity these results are not shown here. Results at this second (closer) site deviate more from the met mast at low wind speeds than at the first (further) site. At speeds greater than 8.8 m·s^{-1}, the deviations are smaller compared to position one. These results suggest that the improvement in agreement between the lidar and the met mast are due to the measurement method, not the different locations.

Galion: The results from the Galion system are shown in Figure 9. For the *CLS* method all 18 trajectory points were used for the evaluation. A multiple overdetermined system of equations has to be solved. This lidar and method shows a better agreement between the lidar measurements and the met mast than the SWE Scanner measurements (Figure 9a; offset 0.018 m·s^{-1}, slope 1.033, R^2 0.974).

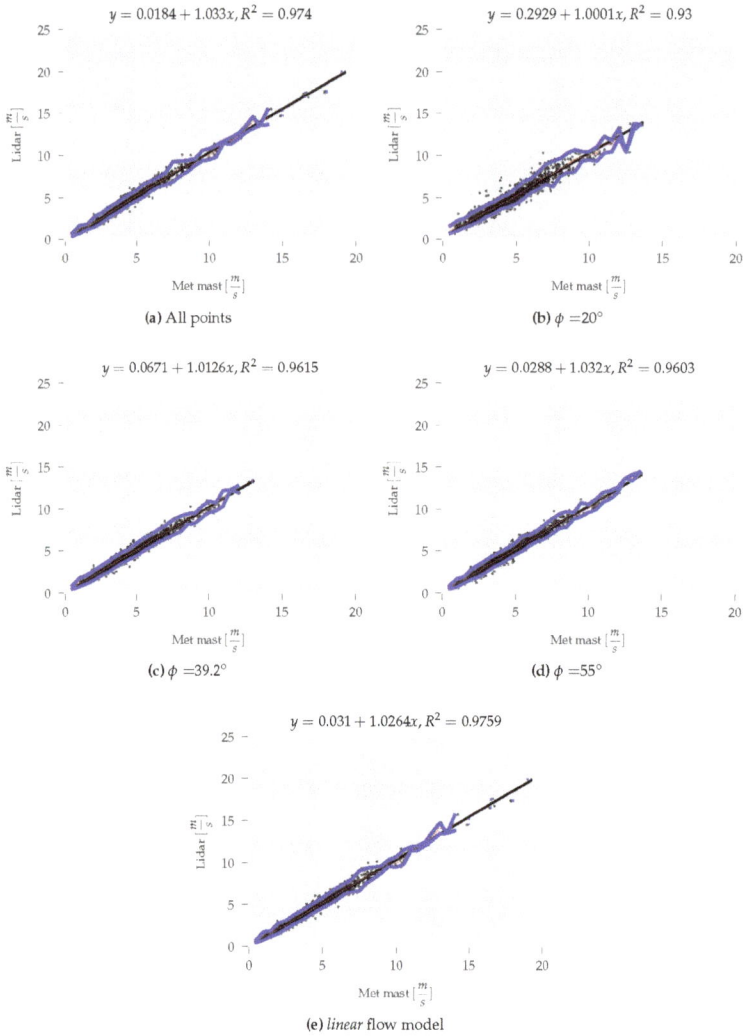

Figure 9. Correlation between Galion lidar and the cup anemometer at 100 m above ground. The black dots represent the 10 min data points, the black line is the regression line, and the blue lines are the 5th and 95th quantiles.

The trajectory of the Galion lidar consists of three identical trajectories with different half opening angles ϕ. In order to check whether there is an influence of the opening angle on the measuring quality, the three trajectories were examined and evaluated separately (Figure 9b–d). It can be seen that reducing the opening angle—implying a sharper cone which is narrower at a given height—leads to a worse correlation (i.e., increased offset, slope further from 1.0, or reduced R^2; details are in the Figures).

Results from the *linear* model are presented in Figure 9e. The offset is 0.03 and the slope (1.026) implies results that are very close to the met mast measurements. The dispersion of the data is small ($R^2 = 0.976$), the width of the 5–95th quantile is significantly reduced. Increased relative errors occur in the lower wind speed range (≤ 2 m·s^{-1}), likely due to the high turbulence intensity that reduces homogeneity across the measurement cone. However, this wind speed range is not relevant for the production of wind energy.

Table 2. Summary of the OLS (Ordinary least-squares) results.

Lidar System	Method	Slope [-]	Offset [m·s^{-1}]	R^2
SWE Scanner	CLS	1.074	0.219	0.949
SWE Scanner	PRESS	1.103	0.385	0.938
SWE Scanner 2nd position	CLS	1.046	0.537	0.942
SWE Scanner 2nd position	PRESS	1.09	0.674	0.936
Galion 18 Pts.	CLS	1.033	0.02	0.974
Galion 20°	CLS	1.00	0.293	0.93
Galion 39.2°	CLS	1.013	0.067	0.962
Galion 55°	CLS	1.032	0.03	0.960
Galion	linear	1.026	0.03	0.976

3.3. Comparison of the Statistical Parameters of Methods

Figure 10a shows the cumulative relative occurrence of R^2 of individual 10 min data for the different methods. The *PRESS* method has a very concentrated distribution with all R^2 greater than 0.85, and over 78% of the R^2 data greater than 0.95. The *CLS* method in combination with the SWE Scanner has the widest distribution of the systems and methods that was investigated. The results when using the Galion with an opening angle of $\phi = 20°$ are similarly wide, the differences of the shape of the distribution is very similar to the distribution of the SWE Scanner.

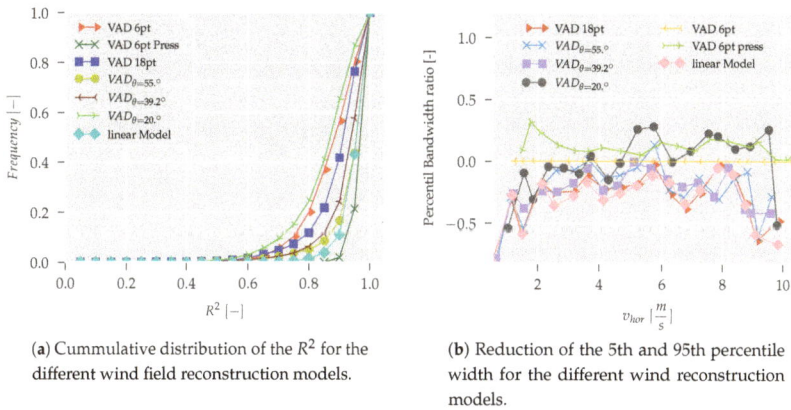

(a) Cummulative distribution of the R^2 for the different wind field reconstruction models.

(b) Reduction of the 5th and 95th percentile width for the different wind reconstruction models.

Figure 10. Overview of the distribution of R^2 and the reduction of the 5th and 95th percentile.

A comparison of the ranges of quantiles for the individual methods against a reference value gives a good impression of the relative dispersion of the individual methods. For this purpose the width of the 5–95th quantile is calculated for each method (Figure 10b). The results of the SWE Scanner with the *CLS* method were taken as reference because this trajectory is very similar to commercial lidar wind profilers. The narrowest quantile is observed using the *linear* model and the *CLS* model for the Galion Lidar with all 18 trajectory points. These procedures achieve a reduction between 20% and 60%. In comparison, the *PRESS* method raises the dispersion by up to 20%. The range is reduced for the other procedures, but not in order as for the *linear* model.

4. Discussion

The projection of the met mast *USA* data onto the normal vector of the lidar showed a directional dependence. This is in line with observations that the terrain significantly influences the inflow over the *Albtrauf* [16,21]. CFD analysis [16] does not give a value to quantify the difference and, the CFD analysis assumes neutral stratification. If stratification effects are added as in reality, this could explain the discrepancies between the lidar and the *USA*.

The difference in error between the met mast and wind field reconstruction using the *CLS* and *PRESS* methods is shown in Figure 11. The results are visualized using the frequency of $|v_{cup} - v_{CLS}| - |v_{cup} - v_{PRESS}|$. Consider a case where the reference or 'true' wind speed is 10 m·s^{-1}, and the *CLS* result is 9.7 m·s^{-1} while the *PRESS* result is 10.2 m·s^{-1}. In this case, the *PRESS* method has led to a better wind field reconstruction, and the result is 0.3 m·s^{-1} − 0.2 m·s^{-1} = 0.1 m·s^{-1}. In this study of 3820 data points, the *CLS* method resulted in a lower error in 91.1% of the measurement periods. One reason for this is, that the additional measuring point of the *CLS* method reduces the reconstruction error compared to the *PRESS* algorithm; the reconstruction error for the wind components is proportional to $1/\sqrt{N}$ where N as number of laser beams [30]. This is also surprising as it was hoped initially that the *PRESS* method would be more robust than the *CLS* method because of the rejection of outliers.

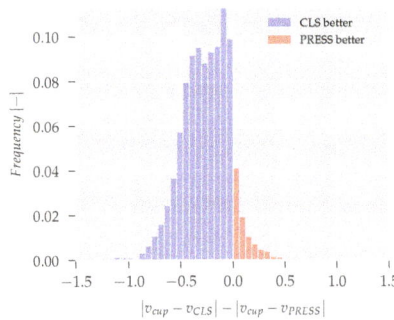

Figure 11. The error in the wind field reconstruction using the *CLS* method, versus the error due to the *PRESS* method. Negative values indicate that the *CLS* method had lower error compared to the met mast than the *PRESS* for a particular 10-minute period.

In the previous section, the correlation between the presented methods applied to lidar measurements and the horizontal wind speeds of in-situ sensors mounted on a met mast was presented. From those results it can be observed that a suitable measurement strategy can improve the agreement between the lidar and met mast.

In this study, using the *linear* model the agreement with the met mast was considerably improved. The results varies in the magnitude of the CFD simulation of ≈2% and the offset is nearly zero with 0.03 m·s^{-1}.

In addition, the width of the 5–95th quantile is significantly reduced in average of 26.96%. In summary, it is a clear improvement using the *linear* method compared to the *CLS* method. Due to the extended *linear* model, the horizontal wind speed in complex flow (at least at this location) can be determined much more precisely than with the standard VAD method. Whether this model delivers useful results in general for complex locations is still to be verified by further tests. For this location, the correlation is closer to unity and the offset is nearly zero. Furthermore, there are also optimization possibilities for this model in the trajectory configuration as well as in the selection of the angles ϕ and θ.

If this result is related to the distributions of R^2 (Figure 10a), then the influence of the used model is clearly visible. For the *PRESS* method, the R^2 distribution shows a very concentrated form with over 75% of the data with over $R^2 \geq 0.95$. The results of the correlation (Figure 8b) show, however, the worst matches with the measuring mast data. This has the highest offset, the highest slope and the smallest R^2, as well as the widest confidence interval. The results in Figure 10a shows a very good agreement between the data and the used model, however Figure 8b shows that the used model does not match the met mast measurements.

These results show that increasing the half opening angle ϕ leads to better correlations between the lidar and mast. The reason for this is that the deterministic error is reduced. This aspect has been shown by the determination of the second statistical moment by lidar data using an analytical model [23]. The optimum angle of a four point VAD method was estimated with an analytical solution to $\phi = 54.7356°$ [30]. This correlates very well with the result in this study. The condition number κ of the **A** solution matrix illustrates this. The κ describes the sensitivity of the solution to changes in the data. This means that in worst case, the input error is amplified by this κ.

The condition number κ of the different solutions tested here is shown in Table 3. The largest κ occurs in the *CLS* procedure for the SWE Scanner with a $\phi = 15°$. The smallest κ is found using the *CLS* method with a $\phi = 55°$. The ratio between largest and smallest κ (4.6) is not directly reflected in difference between the results, but the ability of κ to identify improved scan geometry for the same analysis methods is confirmed by the results.

The question that naturally arises is, how representative are measurements with a $\phi = 55°$ at a measuring height of 100 m? After all, the radius of the circle for this configuration is 76.5 m. This question cannot be answered without further investigation. For load simulations of wind turbines according to IEC 61400-01 [31], coherent wind fields are created using the exponential coherence model. This model depends on the average wind speed, distance of the observed points and the considered frequency. The validity of the coherence model is related to structures in the wind fields and the wind evolution over time and space, and should be investigated in future studies.

Table 3. Characteristics for the different configurations.

	SWE Scanner		Galion			
	VAD 5+1	VAD 18	ϕ_{20}	$\phi_{39.2}$	ϕ_{55}	*linear*
ϕ [°]	15	15	20, 39.2, 55	20	39.2	55	20, 39.2, 55
scan duration [m·s^{-1}]	8.8	4.78	49	16.3	16.3	16.3	49
κ [-]	5.78	5.78	2.02	4.30	2.00	1.26	120.66
Nb. of Points [-]	6	6	18	6	6	6	18
Location [-]	1	2	1	1	1	1	1

Another aspect that needs to be addressed here is the influence of the different scan durations on the measurement results. An overview of the detailed scan duration for the different trajectories is listed in Table 3. The SWE Scanner measures 5.57 times more data in 10 min than the Galion. This should lead to a reduction of the random error. This can be seen to a lesser extent in the comparison of Figures 8a and 9b which measured with a very similar ϕ. The characteristic of the different trajectories are listed in Table 3. The differences between these two half opening angles are marginal (the offset differs by 0.074 m·s^{-1}, the slope by 0.074 and R^2 by 0.019). At the second position the scan duration was reduced by a factor of 1.84, but there are no noticeable improvements. The averaged error over the wind speed range between 0 m·s^{-1} and 20 m·s^{-1} is for *CLS* at the first position 1.034 m·s^{-1} and at the second position is a bit less with 1.00 m·s^{-1}. Unfortunately, the question of the influence of the scan duration cannot be answered on the basis of the results. It is quite clear that the random error is reduced by a higher data availability, but other effects (e.g., opening angle) have a greater influence. Conducting additional experiments in flat terrain with homogeneous flow conditions would be a possible solution there to answer the question on the scan duration.

Further investigations are also necessary into the suitability of the *linear* model. It should be used at more sites to see if the results observed here apply for different complex terrains types. Furthermore, the gradients should be examined to see whether the assumed values are plausible and realistic. This could be done with the help of CFD simulations, or with the help of gradients the numerical simulation can be validated more precisely. So far, only the 10 min statistics have been compared; investigations with high-resolution data and in the frequency domain should also be carried out.

5. Conclusions

In this study, different methods for lidar-based wind field reconstruction in complex terrain were investigated. The main objective was to reduce uncertainties in the measurements by using optimized methods for wind field reconstruction. As a side product, the influence of the opening angle and the scanning duration could also be investigated. Two different methods were investigated and these were additionally coupled with the statistical *PRESS* procedure. The result of the *CLS* method in combination with *PRESS* shows a very skewed distribution of high values of R^2 based on the 10 min data, but this is not reflected in the correlation comparison with the in-situ measurements at the met mast. The results of the correlation are worse than the results of the *CLS* method without *PRESS* statistics. It could be shown that with the *linear* model for wind field reconstruction the agreement with the measurements of the in-situ sensors can reach the level of measurements with homogeneous flow. Furthermore, the range of the 5–95th quantile could be reduced by 32.2% on average compared to the reference VAD method. Furthermore, the influence of the opening angle on the measurement results could be shown. The agreement with the mast measurements and the width of the 5–95th quantile was reduced with larger half opening angles ϕ. The influence of the scan duration on the uncertainties reduction could not be verified with the current data set and measurement set up. In future measurement campaigns with complex flow, the *linear* model with an optimized trajectory should be used. However, this naturally also requires an appropriate lidar system in which the azimuth and elevation angle can be adjusted because for the application of the *linear* model, three different half-opening angles ϕ are necessary. This is currently not possible with the standard lidar wind profilers and it requires so-called all sky scanning devices.

Author Contributions: M.H. and A.C. conceived and designed the experiments; M.H. performed the experiments; M.H. and A.C. analyzed the data; M.H. wrote the manuscript; A.C. and P.W.C. read, reviewed, and approved the final manuscript and provided valuable input.

Funding: This research was funded by by the German Federal Ministry for Economic Affairs and Energy (BMWi) grant number [0325519].

Acknowledgments: The authors would like to thanks all the people who contributed to the measurement campaigns.

Conflicts of Interest: The authors declare no conflict of interest.

References

1. Weitkamp, C. *Lidar: Range-Resolved Optical Remote Sensing of the Atmosphere*; Springer Science+Business Media Inc.: New York, NY, USA, 2005.
2. Clifton, A.; Clive, P.; Gottschall, J.; Schlipf, D.; Simley, E.; Simmons, L.; Stein, D.; Trabucchi, D.; Vasiljevic, N.; Würth, I. IEA Wind Task 32: Wind Lidar Identifying and Mitigating Barriers to the Adoption of Wind Lidar. *Remote Sens.* **2018**, *10*, 406. [CrossRef]
3. Courtney, M.; Wagner, R.; Lindelöw, P. Testing and comparison of lidars for profile and turbulence measurements in wind energy. *IOP Conf. Ser. Earth Environ. Sci.* **2008**, *1*, 012021. [CrossRef]

4.   Alberts, A.; Janssen, A.; Mander, J. German Test Station for Remote Wind Sensing Devices. EWEC, Marseille, 2009. Available online: https://www.researchgate.net/profile/Axel_Albers/publication/237616810_German_Test_Station_for_Remote_Wind_Sensing_Devices/links/568e2aee08ae78cc0514b121.pdf (accessed on 10 September 2018).

5.   Clifton, A.; Boquet, M.; Burin Des Roziers, E.; Westerhellweg, A.; Hofsäß, M.; Klaas, T.; Vogstad, K.; Clive, P.; Harris, M.; Wylie, S.; et al. *Remote Sensing of Complex Flows by Doppler Wind Lidar: Issues and Preliminary Recommendations*; Technical Report; NREL (National Renewable Energy Laboratory (NREL): Golden, CO, USA, 2015.

6.   Clive, P.J.M. Compensation of bias in Lidar wind resource assessment. *Wind Eng.* **2008**, *32*, 415–432. [CrossRef]

7.   Bingöl, F.; Mann, J.; Foussekis, D. Conically scanning lidar error in complex terrain. *Meteorol. Z.* **2009**, *18*, 189–195. [CrossRef]

8.   Mikkelsen, T.; Mann, J.; Courtney, M.; Sjöholm, M. Windscanner: 3-D wind and turbulence measurements from three steerable doppler lidars. *IOP Conf. Ser. Earth Environ. Sci.* **2008**, *1*, 012018. [CrossRef]

9.   Pauscher, L.; Vasiljevic, N.; Callies, D.; Lea, G.; Mann, J.; Klaas, T.; Hieronimus, J.; Gottschall, J.; Schwesig, A.; Kühn, M.; et al. An Inter-Comparison Study of Multi- and DBS Lidar Measurements in Complex Terrain. *Remote Sens.* **2016**, *8*, 782. [CrossRef]

10.   Vasiljević, N.; Palma, J.M.; Angelou, N.; Carlos Matos, J.; Menke, R.; Lea, G.; Mann, J.; Courtney, M.; Frölen Ribeiro, L.; Gomes, V.M. Perdigão 2015: Methodology for atmospheric multi-Doppler lidar experiments. *Atmos. Meas. Tech.* **2017**, *10*, 3463–3483. [CrossRef]

11.   Mann, J.; Angelou, N.; Arnqvist, J.; Callies, D.; Cantero, E.; Arroyo, R.C.; Courtney, M.; Cuxart, J.; Dellwik, E.; Gottschall, J.; et al. Complex terrain experiments in the New European Wind Atlas. *Philos. Trans. R. Soc. A Math. Phys. Eng. Sci.* **2017**, *375*, 20160101. [CrossRef] [PubMed]

12.   Schlipf, D.; Rettenmeier, A.; Haizmann, F.; Hofsäß, M.; Courtney, M.; Cheng, P.W. Model based wind vector field reconstruction from lidar data. In Proceedings of the 11th German Wind Energy Conference DEWEK 2012, Bremen, Germany, 7–8 November 2012. [CrossRef]

13.   Bradley, S.; Strehz, A.; Emeis, S. Remote sensing winds in complex terrain—A review. *Meteorol. Z.* **2015**, *24*, 547–555. [CrossRef]

14.   Klaas, T.; Pauscher, L.; Callies, D. LiDAR-mast deviations in complex terrain and their simulation using CFD. *Meteorol. Z.* **2015**, *24*, 591–603. [CrossRef]

15.   Schulz, C.; Hofsäß, M.; Anger, J.; Rautenberg, A.; Lutz, T.; Cheng, P.W.; Bange, J. Comparison of Different Measurement Techniques and a CFD Simulation in Complex Terrain. *J. Phys. Conf. Ser.* **2016**, *753*, 082017. [CrossRef]

16.   Schulz, C. Numerische Untersuchung des Verhaltens von Windenergieanlagen in Komplexem Gelände unter Turbulenter AtmosphäRischer Zuströmung. Ph.D Thesis, Universität Stuttgart, Stuttgart, Germany, 2018.

17.   Rettenmeier, A.; Bischoff, O.; Hofsäß, M.; Schlipf, D.; Trujillo, J. Wind Field Analysis Using A Nacelle-Based LiDAR System. In Proceedings of the EWEC, Warsaw, Poland, 20–23 April 2010.

18.   Wildmann, N.; Bernard, S.; Bange, J. Measuring the local wind field at an escarpment using small remotely-piloted aircraft. *Renew. Energy* **2017**, *103*, 613–619. [CrossRef]

19.   Schulz, C.; Klein, L.; Weihing, P.; Lutz, T. Investigations into the Interaction of a Wind Turbine with Atmospheric Turbulence in Complex Terrain. *J. Phys. Conf. Ser.* **2016**, *753*, 032016. [CrossRef]

20.   Hofsäß, M.; Bergmann, D.; Bischoff, O.; Denzel, J.; Cheng, P.W.; Lutz, T.; Peters, B.; Schulz, C. *Lidar Complex*; Technical Report; Universität Stuttgart, Fakultät 6 Luft- und Raumfahrttechnik und Geodäsie, Institut für Flugzeugbau (IFB), Stuttgarter Lehrstuhl für Windenergie: Stuttgart, Germany, 2017. [CrossRef]

21.   Anger, J.; Bange, J.; Blick, C.; Brosz, F.; Emeis, S.; Fallmann, J. *Erstellung einer Konzeption eines Windenergie-Testgeländes in bergig komplexem Terrain : Kurztitel : KonTest : Abschlussbericht des Forschungsprojektes: ein Vorhaben des WindForS Windenergie Forschungscluster Forschungsnetzwerks*; Technical Report; Universität Stuttgart–Stuttgarter Lehrstuhl für Windenergie (SWE) am Institut für Flugzeugbau (IFB): Stuttgart, Germany, 2015. [CrossRef]

22.   IEC (International Electrotechnical Commission). *Wind Turbines. Part 12-1, Power Performance Measurements of Electricity Producing Wind Turbines*; Number 12, 1; IEC: Geneva, Switzerland, 2005.

23.   Sathe, A.; Mann, J.; Vasiljevic, N.; Lea, G. A six-beam method to measure turbulence statistics using ground-based wind lidars. *Atmos. Meas. Tech.* **2015**, *8*, 729–740. [CrossRef]

24. Frisch, A.S. On the measurement of second moments of turbulent wind velocity with a single Doppler radar over non-homogeneous terrain. *Bound.-Layer Meteorol.* **1991**, *54*, 29–39. [CrossRef]

25. Hofsäß, M.; Bischoff, O.; Cheng, P.W. Comparison of second moments between remote sensing devices in complex errain. In Proceedings of the 18th International Symposium for the Advancement of Boundary-Layer Remote Sensing, Varna, Bulgaria, 6–9 June 2016.

26. Peña, A.; Hasager, C.B.; Badger, M.; Barthelmie, R.J.; Bingöl, F.; Cariou, J.P.; Emeis, S.; Frandsen, S.T.; Harris, M.; Karagali, I.; et al. *Remote Sensing for Wind Energy*; DTU Wind Energy: Roskilde, Denmark, 2015.

27. Allen, D.M. Mean Square Error of Prediction as a Criterion for Selecting Variables. *Technometrics* **1971**, *13*, 469–475. [CrossRef]

28. Allen, D.M. The Relationship Between Variable Selection and Data Agumentation and a Method for Prediction. *Technometrics* **1974**, *16*, 125–127. [CrossRef]

29. Draper, N.R.; Smith, H. *Applied Regression Analysis*, 3rd ed ed.; Wiley Series In Probability and Statistics; Wiley: New York, NY, USA, 1998.

30. Teschke, G.; Lehmann, V. Mean wind vector estimation using the velocity–azimuth display (VAD) method: An explicit algebraic solution. *Atmos. Meas. Tech.* **2017**, *10*, 3265–3271. [CrossRef]

31. IEC (International Electrotechnical Commission). *Wind turbines. Part 1, Design Requirements*; IEC: Geneva, Switzerland, 2014.

remote sensing

MDPI

Article

Wind in Complex Terrain—Lidar Measurements for Evaluation of CFD Simulations

Andrea Risan [1,*], John Amund Lund [2], Chi-Yao Chang [3] and Lars Sætran [1]

[1]   Department of Energy and Process Engineering, Norwegian University of Science and Technology,
     7491 Trondheim, Norway; andrea.ris@hotmail.com or lars.satran@ntnu.no
[2]   Meventus AS, 4630 Kristiansand, Norway; john@meventus.com
[3]   Fraunhofer-Institute for Wind Energy and Energy System Technology, IWES, 26129 Oldenburg, Germany;
     chi-yao.chang@gmx.de
*   Correspondence: andrea.ris@hotmail.com

Received: 13 November 2017; Accepted: 28 December 2017; Published: 4 January 2018

Abstract: Computational Fluid Dynamics (CFD) is widely used to predict wind conditions for wind energy production purposes. However, as wind power development expands into areas of even more complex terrain and challenging flow conditions, more research is needed to investigate the ability of such models to describe turbulent flow features. In this study, the performance of a hybrid Reynolds-Averaged Navier-Stokes (RANS)/Large Eddy Simulation (LES) model in highly complex terrain has been investigated. The model was compared with measurements from a long range pulsed Lidar, which first were validated with sonic anemometer data. The accuracy of the Lidar was considered to be sufficient for validation of flow model turbulence estimates. By reducing the range gate length of the Lidar a slight additional improvement in accuracy was obtained, but the availability of measurements was reduced due to the increased noise floor in the returned signal. The DES model was able to capture the variations of velocity and turbulence along the line-of-sight of the Lidar beam but overestimated the turbulence level in regions of complex flow.

Keywords: detached eddy simulation; turbulence; Lidar; range gate length

1. Introduction

In recent years, computational fluid dynamics (CFD) have frequently been applied for predicting wind conditions in the wind energy industry. Such flow models can provide a three-dimensional description of the flow field in a large area using input data from point measurements or meso-scale meteorological models. However, although CFD models have become increasingly advanced, the challenge of accurately describing turbulent flow, e.g., in complex terrain, remains. For a large three-dimensional area, the requirement of spatial and temporal resolution to accurately resolve turbulent structures is simply not computationally affordable. An approach that has proved to yield valuable results for turbulence prediction is the Large Eddy Simulation (LES) method, which separates the flow in large and small scale eddies to save computational effort [1].

Research has been done regarding the performance of various LES models for describing turbulent wind conditions in complex terrain. A comprehensive blind test including several models, called the Bolund experiment, has been conducted by Bechmann et al. [1], where the accuracy of these models across an isolated hill was tested. The performance of the LES models included in the analysis yielded somewhat disappointing results with significant speed-up errors over the Bolund Hill. One reason for the large deviations might be the challenge of obtaining the correct free stream boundary condition, which the LES models failed to do in this study [1]. A similar experiment has been done by Bechmann and Sørensen [2], where a hybrid Reynolds-Averaged Navier-Stokes (RANS)/Large Eddy Simulation (LES) model was tested over the Askervein Hill in Scotland. In this model, the near-wall regions are resolved in a Reynolds-Averaging manner with the two-equation k-Epsilon turbulence model. The model was able to predict the high turbulence level in the complex wake region downwind of the hill reasonably well, but underestimated the mean velocity [2].

Experiments like Askervein Hill and Bolund Hill have provided invaluable insights into flow model performance and provided a benchmark for further flow model development. However, as wind power development expands into areas of even more complex terrain and challenging flow conditions, there is a need for full-scale validation cases, which reflects the challenges the wind industry meets today. This paper presents a validation case in highly complex terrain, using a pulsed Doppler Lidar. Lidars are particularly useful for this purpose as they can measure the spatial distribution of the wind along the Lidar beam. However, there are limitations in the Lidar technology that need to be addressed in order to make the measurements useful for flow model validation purposes.

One of the main limitations of the Lidar technology is that a horizontally homogeneous velocity field is assumed when deriving the three-dimensional wind field, which is not a valid assumption in complex terrain. Several studies regarding the performance of Lidars in complex terrain have been conducted, among others by Guillén et al. [3] and Vogstad et al. [4]. Guillén et al. found that the deviation between ten-minute averaged Lidar and cup anemometer measurements was significantly larger when the wind direction was such that the complex terrain features were most prominent. A greater discrepancy was also observed for higher turbulence intensities, and for higher vertical velocities [3]. Vogstad et al. [4] tested the performance of three different Lidars—WindCube V1, ZephIR 300 and Galion/StreamLine in complex terrain by comparing measurements with cup anemometer data. A numerical flow model was used to correct for the inhomogeneities of the terrain when deriving the three-dimensional velocity field. They found that the uncertainty of the ten-minute averaged velocities from all Lidars were in the order of 2.5% when applying the appropriate numerical corrections, which is comparable to the uncertainty of cup anemometers [4]. However, none of the Lidar instruments were found to predict the ten-minute averaged horizontal turbulence intensity accurately.

Lidar systems are also limited by spatial averaging along the Lidar beam. This effect is most prominent in the accuracy of turbulence estimations when small fluctuations are of vital importance. Several studies have investigated the ability of Lidar systems to provide accurate one-dimensional turbulence statistics, and methods such as the one described by Lenschow et al. [5] can be applied to improve the accuracy of these estimates under the assumption of isotropic turbulence. Bonin et al. [6] shows how measurement noise in Lidar data can be treated to improve the turbulence estimates. They investigated the performance of a StreamLine and WindCube Lidar compared to a sonic anemometer, and applied the methods of Lenschow to correct for white noise and limitations of spatiotemporal averaging of the measurements. Sjöholm et al. [7] investigated the spatial averaging effect for a ZephIR continuous-wave Lidar by comparing one-dimensional velocities with sonic measurements projected onto the line-of-sight (LOS) of the Lidar. Two periods with different atmospheric conditions were investigated—one with low clouds and high backscattering, and one with clear conditions. The power density spectra were almost identical for low frequencies, but the Lidar spectrum fell off more rapidly than the sonic spectrum for higher frequencies in both cases, proving that the Lidar did not capture the small-scale turbulent features of the wind as accurately as the sonic anemometer. The spectra deviated at approximately 0.02 Hz in the clear conditions case and 0.05 Hz in the low cloud case with stronger backscattering. Cañadillas et al. [8] investigated the same effect for a WindCube pulsed Lidar with a range gate length of 20 m on an offshore site. In this case, the power density spectra for line-of-sight velocities from the Lidar and the sonic anemometer were only comparable up to a frequency of 0.21 Hz due to the scanning pattern of the Lidar. The spectra showed a good compliance, and it was concluded that the spatial averaging along the Lidar beam had a negligible effect for this range of frequencies.

Although Lidars have become widely used in wind energy applications [9], they can only provide a relatively accurate estimate of turbulence along the line of sight. Multi-Lidar systems are capable of addressing this shortcoming [10,11]; however, these systems can still only measure the three-dimensional wind in one point at a time.

The lack of ability to provide a three-dimensional estimate of atmospheric turbulence limits the applicability of Lidar systems for evaluation of wind turbine structural integrity in complex terrain, where flow homogeneity and isotropic turbulence cannot be assumed. Therefore, CFD flow models are widely used to evaluate the spatial distribution of turbulence in complex terrain. However, such models are often only validated against point measurements, being meteorological masts or Lidars operating in Velocity azimuth display (VAD) mode. Using the method proposed in this paper, we believe that a pulsed Lidar, capable of measuring higher-order statistics of the wind along the line of sight, can be used efficiently to improve the accuracy of the flow models used in the industry today.

In this study, we have used a free-scanning Lidar operating in fixed direction stare-mode aligned with the mean flow direction. This way, accurate estimations of the wind fluctuations along the line-of-sight may be derived, which can be used to validate how the flow model predicts the variations of turbulence in the mean flow along the same line. The results can also give an indication of where the model succeeds or fails in predicting turbulence transport, production or dissipation.

The main objectives of this study are to (1) evaluate the accuracy of 1D Lidar measurements with a focus on turbulence estimation; (2) investigate how reduced gate length may increase the accuracy of turbulence estimations with the Lidar; and (3) use the Lidar data to validate a computational flow model in highly complex terrain. Lidar line-of-sight measurements will be validated with sonic anemometer data projected onto the Lidar beam, and the effect of spatial averaging will be investigated by changing the range gate length of the Lidar. For the flow model validation, a hybrid RANS/LES (DES) model will be applied along a horizontal Lidar beam parallel to the mean flow.

2. Materials and Methods

The measurement campaign was carried out by Meventus in Roan in Sør-Trøndelag, Norway, with a ground-based StreamLine XR pulsed Lidar in the proximity of a meteorological mast. The Lidar system is capable of recording raw data, making it possible to reprocess the data to provide time series with different gate lengths and frequencies. The system is operated with a collimated output beam, and the full width at half maximum (FWHM) is about 35 m. The instrument is described in more detail in Pearson et al. [12]. In the following sections, a brief description of the site will be provided before the setup of the instruments and the methodology of the analysis will be explained.

2.1. Site Description

The site is located in central Norway, approximately 3 km from the coastline. The terrain at the site is complex, with rocky, mountainous and open topography. A steep ridge located 1300 m west of the Lidar and mast is expected to generate complex flow with large-scale turbulent eddies. The positions of the Lidar and mast, and the surrounding terrain are shown in Figure 1.

2.2. Experimental Setup for Lidar Validation

The triangular lattice mast is located at 366 meters above sea level (m a.s.l.). The instrumentation installed on the mast are summarized in Table 1. The StreamLine XR v14-8 Lidar is located 344 m north of the mast at 370 m a.s.l. The range gate length of the Lidar was set to 18 m.

In order to avoid movement of the Lidar in strong winds, the Lidar was bolted to the ground using rock anchors. The instrument was leveled and oriented towards north using binoculars with compass. Lidar roll and tilt was logged throughout the campaign. The Lidar bearing was determined by scanning towards the mast and identifying at what azimuth backscatter from the mast was observed. The Lidar bearing was confirmed at the end of the measurement program, in order to confirm that no drift occurred.

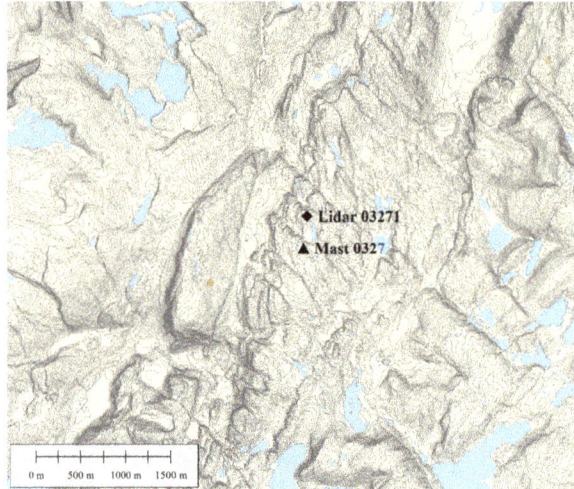

Figure 1. Map of the site with the position of the Lidar (64°08′20.7″ N 10°19′04.0″ E) and the meteorological mast (64°08′09.7″ N 10°19′00.7″ E). The height contours represent 5 m height difference.

Table 1. Instrumentation installed on the measurement mast. All cup anemometers are mounted on booms pointing southwest from the mast. At 100.5 m height, there is an additional cup anemometer on a boom pointing northeast.

Parameter	Type	Height (m)	Boom Direction (°) from Mast	Boom Length (m)
Wind speed	Thies First ClassAdvanced	100.5	225	1.015
Wind speed	Thies First ClassAdvanced	100.5	45	1.015
Wind speed	Thies First ClassAdvanced	80.0	225	5.028
Wind speed	Thies First ClassAdvanced	60.0	225	5.028
Wind speed	Thies First ClassAdvanced	40.0	225	5.028
Wind speed	Thies First ClassAdvanced	20.7	225	5.028
Wind speed	3D UltrasonicThies	98.0	225	5.028
Wind direction	3D UltrasonicThies	98.0	225	5.028
Temperature	3D UltrasonicThies	98.0	225	5.028
Vertical wind speed	3D UltrasonicThies	98.0	225	5.028
Wind direction	Thies First ClassTMR	98.0	44	5.028
Wind direction	Thies First ClassTMR	40.0	44	5.028
Wind direction	NRG IceFree3	80.0	42	5.028
Temperature	Galltec Mess KP	97.0	20	0.305
Relative humidity	Galltec Mess KP	97.0	20	0.305
Temperature	Galltec Mess TP	3.0	20	0.305
Barometric pressure	Ammonit AB 60	3.0	-	-
Aviation lights	ObeluxLI-32+IR-DCW-F	97.6	-	-

Line-of-sight velocities from the Lidar were collected with a sampling frequency of 1 Hz throughout a two-month period from 13 April 2015 to 11 June 2015. Corresponding horizontal and vertical velocities and wind directions were collected with the sonic anemometer at 98 m height for the same period. Due to inaccuracies in alignment tools for the sonic anemometer, an offset in the alignment of −7° was detected and corrected for. The offset was determined through a correlation analysis determining the offset required to obtain the highest correlation between the Lidar and the corresponding decomposed sonic signal. Figure 2 illustrates the setup of the two instruments.

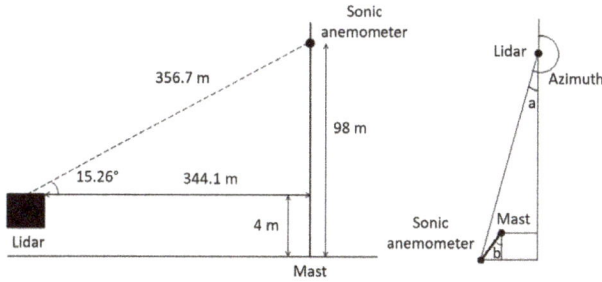

Figure 2. Schematic of the Lidar and mast positions. **Left**: side view of the setup. The elevation angle is 15.26°; **Right**: top view of the setup. Azimuth angle = 180° + a, a = 6.86°. To avoid backscatter from the sonic anemometer, the Lidar azimuth was set to 186.94°, pointing approximately 0.5 m west of the sonic anemometer. The sonic anemometer is mounted on a 5 m long boom at an angle b = 45°.

Data Analysis

The sonic anemometer measurements were projected onto the line-of-sight of the Lidar to allow for a comparison of one-dimensional velocities along the Lidar beam. As the elevation angle is fairly large, a 3D projection was applied to account for the vertical velocity component.

To provide information about the quality of the Lidar data, these include values for the pitch and roll angles and the signal-to-noise ratio (SNR) of the backscattered signal. The pitch and roll angles are the forward/backward and sideways tilt angles of the Lidar. During the measurement period, these were checked remotely to be within a 0.25° threshold. This yields a maximum deviation of 1.55 m of the beam from the sonic anemometer. When validating the Lidar data, a stronger filtration of these values was performed to ensure that the beam deviation was within 1 m, corresponding to a maximum 0.16° pitch and roll angle. An analysis was carried out to determine how the correlation of the Lidar and sonic anemometer data was affected by filtration of SNR values. Although recent studies [6,13] have shown that the spatial averaging effects of the Lidar can be corrected to some extent, none of these methods were applied. The aim of this study was to verify that the Lidar is sufficiently accurate to validate numerical models in complex terrain. As the uncertainty involved in numerical models is generally much higher than the Lidar uncertainty, additional corrections were considered to be outside the scope of this study.

Velocity measurements and turbulence estimates were compared using standard linear regression analysis. The coefficient of determination R^2, defined by Equation (1) [14] for data sets x and y, was used as a measure of the correlation between the Lidar and sonic anemometer measurements, respectively:

$$R^2 = \frac{(\sum(x-\bar{x})(y-\bar{y}))^2}{\sum(x-\bar{x})^2 \sum(y-\bar{y})^2}. \tag{1}$$

A spectral analysis was performed for a more detailed comparison of the data. Power density spectra illustrate how the energy is transferred from larger to smaller eddies. As a reference, the spectra were compared with the theoretical Kolmogorov slope of $-5/3$ in the inertial subrange [15].

The effect of changing the range gate length of the Lidar was investigated by reprocessing raw Lidar data in the program Raw Data Processor v14 developed by Halo Photonics (Worcestershire, UK). A 24 h period on 30 April 2015 with large variations in wind velocity and direction was chosen for this analysis to challenge the Lidar with varying wind conditions. The wind velocity and direction measured by the sonic anemometer during this period are illustrated in Figure 3. Data were reprocessed with range gate lengths of 9 m and 30 m.

Throughout the entire analysis, only coinciding data from the Lidar and the sonic anemometer were used. For periods when Lidar data were not available due to filtration requirements, the sonic anemometer data were also removed. This data removal causes the spectra to appear with certain artifacts, as the resulting time series does not represent the physical processes in the atmosphere. However, the spectra will be suitable for evaluating the ability of the Lidar system to represent high-frequency wind speed variations.

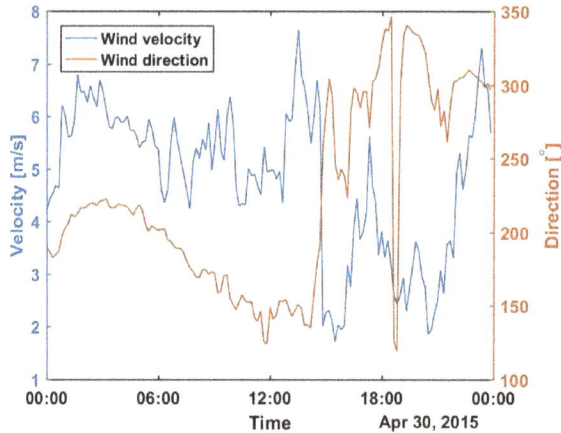

Figure 3. Wind velocity and direction during 30 April 2015 measured by the sonic anemometer. Times are local (UTC + 1:00).

2.3. Experimental Setup for the Flow Model Validation

The large-scale turbulent structures of interest in this study are expected to be generated downwind of the ridge located approximately 1300 m southwest of the Lidar. The ridge has a steep vertical cliff with an elevation of 150 m facing westward. For the purpose of validating the flow models, the Lidar was operating in stare-mode towards this ridge, with an azimuth of 262°, parallel to the mean flow. This corresponds to an azimuth of 259.70° relative to grid north (UTM zone 32, WGS 84). (The use of grid north as the reference is chosen in order to avoid differences between the coordinate system of the flow model and the Lidar.) The velocity was measured along the length of the beam, and the locations (a)–(c) shown in Figure 4 were used for a more detailed spectral analysis.

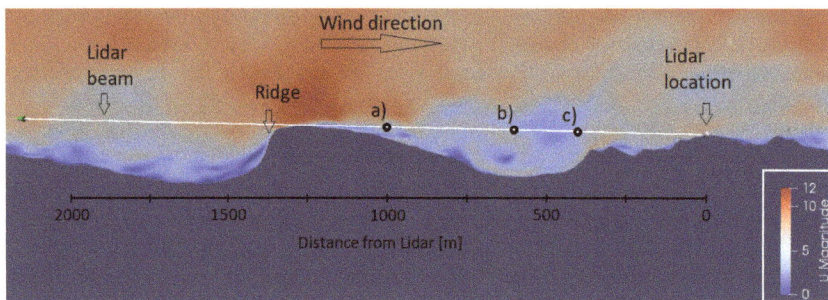

Figure 4. Schematic describing the location of the Lidar and meteorological mast, and the ridge of interest to the study. Flow model results were compared with Lidar measurements along the line-of-sight parallel to the mean flow, and points at distances (**a**) 1000 m; (**b**) 600 m; and (**c**) 400 m from the Lidar were used for further evaluations.

A detailed assessment of wind data from the mast was used to identify a period with steady wind speeds and wind direction approximately orthogonal to the ridge. The estimated Monin–Obukhov length scale [16] was found to be near-neutral ($|MOL| > 450$) for the entire period, implying that the vertical heat flux is close to zero. The conditions during the selected time period are presented in Table 2.

Table 2. Wind conditions observed during the selected time period. Times are local (UTC + 1:00).

Time Period	Wind Speed @100 m Mean (min, max) (m/s)	Wind Direction @100 m Mean (min, max) (°)	Monin–Obhukov Length Scale (min) (m)
14 June 2015 1:00 p.m. to 10:30 p.m.	8.3 (7, 10)	260 (250, 290)	(450)

Lidar alignment was verified by pointing the Lidar beam towards the hill at different elevations, obtaining solid target return values where the Lidar beam hit the terrain. The Lidar beam was aligned to pass approximately 5 m over the crest of the hill, with an elevation angle of 2°. Analysis of the variations in pitch and roll during the experiment showed that the variation in elevation of the Lidar beam was on the order of 0.2 degrees. This corresponds to ±4.4 m at the crest of the hill.

Results from the flow models were extracted along a line that started at the position of the Lidar and passed through a point 5 m above the digital terrain elevation at the crest of the hill. The azimuth was kept at 259.7° for the evaluations of the mean flow field.

Simulation Procedure

Classical methodology regarding wind farm modeling using computational fluid dynamics (CFD) refers to two general strategies: the Reynolds–Averaged Navier–Stokes (RANS) and Large Eddy Simulation (LES) methods. In the RANS method, the simulation is executed aiming for a steady state solution, for which the turbulent properties are modeled in the framework of applied transport models. As a result, the transient behavior of turbulent flows is suppressed. The general principle of the LES method is to resolve the turbulent structures in the mean flow, i.e., the large eddies, and model the effect of the smaller eddies. Although this approach will yield more realistic results, the computational cost is much higher [17].

Due to the filtering of large and small eddies, LES models encounter severe difficulties in the near-wall region, for which the correct physical characteristics cannot be reproduced due to insufficient mesh resolution. This is a significant problem for atmospheric boundary layer flows with high Reynolds numbers, as the mesh requirement becomes computationally unaffordable. To solve this issue, a Detached Eddy Simulation (DES) is used in this study. A DES is a hybrid RANS/LES method, which compensates this shortcoming by applying the RANS model in the near-wall region. The switch from LES to RANS is based on the to-wall distance, as well as the modeling length scale and cell size [2]. The hybrid RANS/LES approach selected for this numerical study was first proposed by Spalart [17]. Details about the transport models, as well as the filtering strategies are described by Bechmann et al. [2]. The traditional RANS model is also included in the analysis for comparison.

The computational domain was constructed by Fraunhofer Institute for Wind Energy and Energy System Technology's (IWES) terrainMesher (Oldenburg, Germany), and the simulations were performed by Fraunhofer IWES in OpenFOAM (version 2.3.1, OpenCFD Ltd, Bracknell, UK). A region covering a 9.5 km × 7.3 km area orthogonal to the wind direction was meshed by ~56 million degrees of freedom, with increasing mesh resolution near the surface. The governing equation system is the Navier–Stokes equations, and the RANS and DES flow models are applied without heat transfer (neutral conditions).

The inlet condition with a mean bulk velocity of 8.3 m/s was first estimated by a RANS simulation using the k-Epsilon turbulence model. This model is widely used in the wind energy industry today, and serves as a baseline for the evaluation of the DES model. The RANS model also served as the starting point for the DES simulation. However, to obtain the required eddy structures in the inlet profile for the DES simulation, a prolonged inlet with circulating flow was used. The velocity fluctuations were triggered by surface shear, which caused an instability in the momentum equation. The domain with the prolonged inlet is shown in Figure 5.

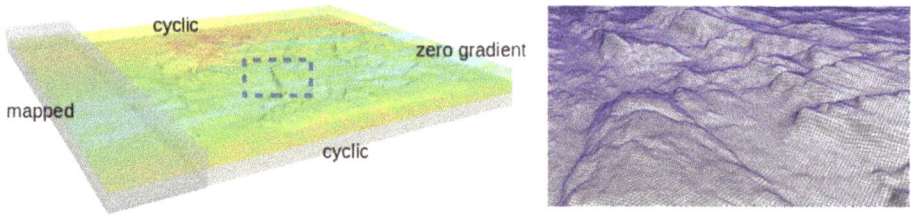

Figure 5. Right: illustration of the computational domain with the prolonged inlet. The boundary conditions used in the simulation are illustrated. The ridge of interest can be seen in the center of the domain. The color coding represent velocity magnitude; **Left**: close up of the structured grid at the ridge.

The DES simulation was executed over 25,000 physical seconds, which corresponds to approximately 20 flow through times (FTT). The results were averaged over 15 FTT, which can demonstrate a plausible statistical representation.

A spectral analysis was performed for a few locations along the beam. The spectra were compared with the predicted Kaimal spectrum for the longitudinal wind speed, which is given by Equation (2) [18]:

$$\frac{nS_u(n)}{u_*^2} = \frac{105f}{(1+33f)^{5/3}}. \tag{2}$$

u_* is the friction velocity, S_u is the power density for the longitudinal wind and f is the normalized frequency, related to frequency n, height z and velocity $U(z)$ by relation (3):

$$f = \frac{nz}{U(z)}. \tag{3}$$

The friction velocity is estimated using the logarithmic law, given by Equation (4) [19]:

$$u(z) = \frac{u_*}{K} ln\left(\frac{z}{z_0}\right). \tag{4}$$

K is the von Karman constant (=0.4) [19], and z_0 is the surface roughness, which is approximately 0.03 m for bare mountains [20]. The friction velocity is estimated at 100 m height at the measurement mast. Using the mean velocity of 8.3 m/s yields a friction velocity of 0.41 m/s.

3. Results

In this section, a validation of the Lidar will be presented by comparison with sonic anemometer data. Next, the effect of changing the range gate length of the Lidar will be analyzed. At last, the Lidar measurements will be used to validate the performance of a RANS and DES flow model.

3.1. Validation of Lidar Measurements

For evaluating the accuracy of turbulence estimates with the Lidar, the data were compared to sonic anemometer measurements with a sampling frequency of 1 Hz. As an initial analysis, the effect of the level of filtration of Lidar data was investigated. Noisy measurements were removed by increasing the filtering limit on the Signal-to-noise ratio (SNR) value of the Lidar signals, and the correlation between the Lidar and sonic anemometer data were evaluated by incrementally increasing the SNR filter value by 0.01 for each step. Figure 6 shows how the coefficient of determination changes with the limiting SNR value, and how the availability of measurements is affected.

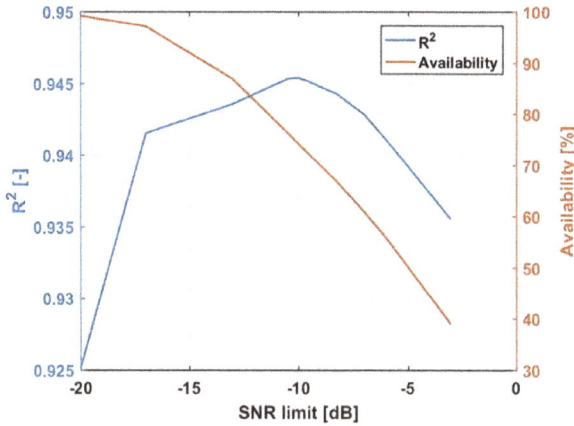

Figure 6. Coefficient of determination R^2 and availability versus the lower Signal-to-noise ratio (SNR) limit.

The figure shows that the coefficient of determination has a maximum value of $R^2 = 0.9454$ with an optimal SNR limit of -10 dB. The slope of the linear regression line for this limit is 0.9705. With a higher limit, the reduction of availability will dominate and cause a decrease in the coefficient of determination. This effect occurs because the differences between the Lidar and sonic anemometer measurements are amplified for large and rapid changes in velocity, which typically occurs across long time gaps.

When increasing the SNR limit, periods with clear conditions are also filtered out. Figure 7 illustrates how the relative humidity and SNR are related. In clear conditions with low humidity, the SNR is low due to a weak backscattering. When the humidity increases, the aerosols swell and the backscattering is increased [21]. With the optimal SNR limit of -10 dB, the availability is reduced to 74.2% due to a loss of such periods. The figure only shows a time period of 24 h on 30 April 2015, but the illustrated phenomenon is applicable for all times and causes a loss of clear-condition periods with higher SNR limits.

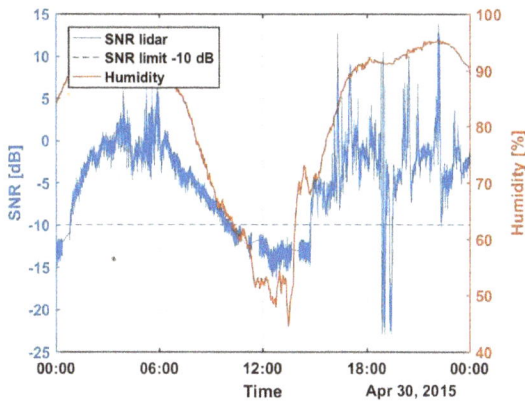

Figure 7. Temperature, relative humidity and SNR versus time for a 24 h period on 30 April 2015. The optimal SNR limit of -10 dB is included as a reference. Times are local (UTC+1:00).

In Figure 8, the one-dimensional wind velocity measured with the Lidar and the sonic anemometer with a sampling frequency of 1 Hz and the optimal SNR limit of −10 dB are plotted as a function of time. Due to clear conditions during the first four days, these data have been filtered out with an SNR limit of −10 dB. The Lidar measurements follow the sonic measurements very well, although the sonic anemometer captures more fluctuations than the Lidar. As explained in Section 1, the spatial averaging along the Lidar beam removes some of the small-scale features of the wind, and hence the small fluctuations are not described as accurately as with the sonic anemometer. Similar results were observed by Sjöholm et al. [7].

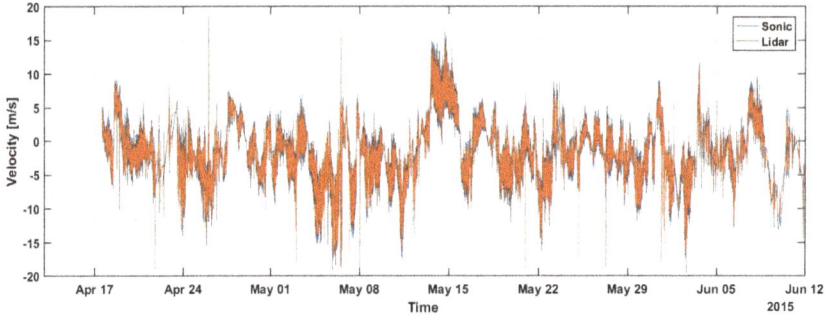

Figure 8. Time series of one-dimensional 1 Hz velocities from 13 April 2015–11 June 2015 measured with the Lidar and the sonic anemometer. Times are local (UTC + 1:00).

Figure 9 shows a correlation plot of one-dimensional ten-minute averaged velocities from the Lidar and the sonic anemometer. The coefficient of determination during this period is $R^2 = 0.9972$, and the linear regression slope is 1.0043. The figure confirms the Lidars' ability to obtain accurate mean velocities. The ten-minute averaged velocities correlate almost perfectly and are comparable with results from similar experiments done in previous research [4,22].

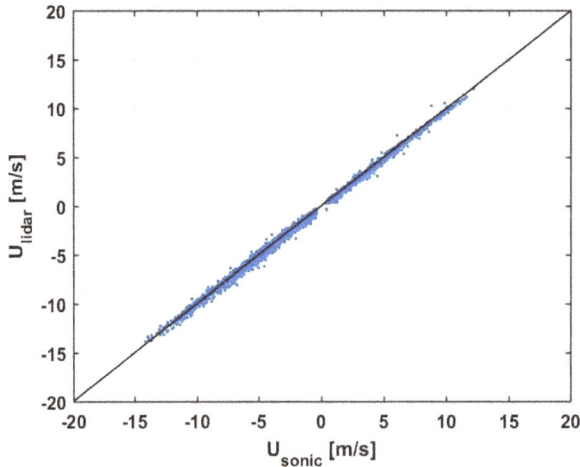

Figure 9. Correlation plot of ten-minute averaged velocities from 13 April 2015–11 June 2015. The coefficient of determination is $R^2 = 0.9972$ and the linear regression slope is 1.0043.

Figure 10 shows a correlation plot with the optimal SNR limit of −10 dB and a sampling frequency of 1 Hz. There is a larger spread in this case than for the ten-minute averaged velocities in Figure 9, which is also reflected in the lower coefficient of determination $R^2 = 0.9454$ and slope of the linear regression line 0.9705. The spread does not necessarily represent a flaw in the instruments, but rather a physically different approach to wind speed estimation. While the sonic anemometer is close to a point measurement, the Lidar measures the wind over a larger volume.

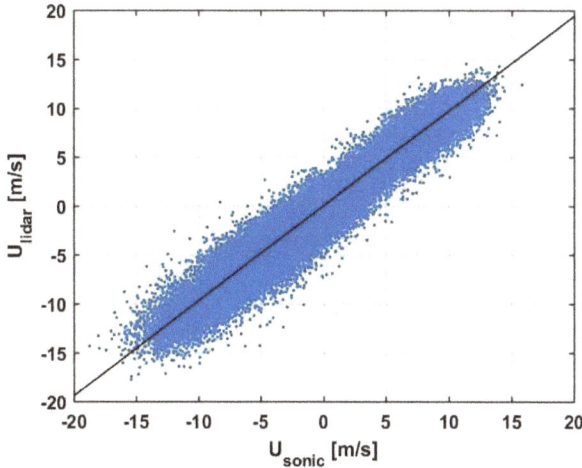

Figure 10. Correlation plot of one-second velocities from 13 April 2015–11 June 2015. The coefficient of determination is $R^2 = 0.9454$ and the linear regression slope is 0.9705.

In Figure 11, the ten-minute mean standard deviations along the line-of-sight estimated by the Lidar and the sonic anemometer are plotted as a function of velocity. There is a slightly larger spread between the instruments for low velocities, when the standard deviation is also low. On a mean level, the Lidar slightly underestimates the turbulence intensity.

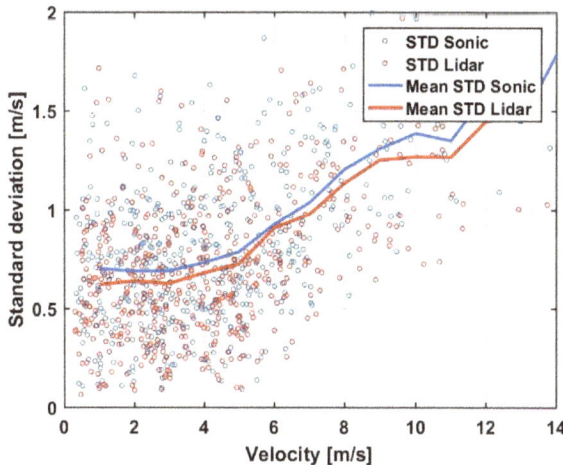

Figure 11. Ten-minute mean turbulence intensity versus velocity for the Lidar and the sonic anemometer.

Figure 12 shows the power density spectra for the Lidar and the sonic anemometer as well as the theoretical Kolmogorov $-5/3$ slope for the inertial subrange [15].

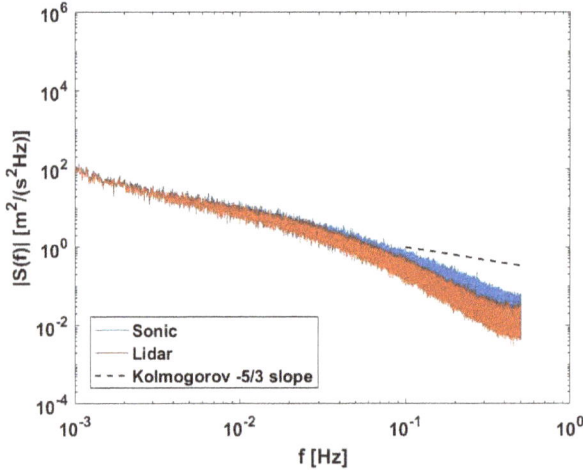

Figure 12. Power density spectra for the Lidar and the sonic anemometer, and the theoretical Kolmogorov slope.

The spectra does not fall in line with the $-5/3$ slope for the higher frequencies. This is probably due to the fact that the dataset is structured so that the longest complete sequential measurement series covers a period of 600 s. However, with data gaps in between, the majority of the complete sequential measurement series are shorter. As data in between these measurement series are removed, data gaps will lead to an increased power spectral density in the region below 2×10^{-3} Hz, as is clearly visible in the figure. As we approach the higher frequencies, the majority of the sequential periods of corresponding shorter time periods will fall inside a longer sequential measurement series. Hence, the power spectral density falls with a steeper slope for higher frequencies compared to the Kolmogorov slope.

The figure also shows that the Lidar spectrum has a lower power density for higher frequencies than the sonic anemometer spectrum. Hence, the Lidar does not capture the small-scale turbulent features of the wind as accurately as the sonic anemometer due to the spatial averaging along the Lidar beam. The deviation between the spectra occurs at approximately 0.03–0.07 Hz. Similar results were observed by Sjöholm et al., with a deviation in the spectra at 0.02–0.05 Hz [7]. As mentioned in Section 1, Cañadillas et al. concluded that the spatial averaging had a negligible effect up to 0.21 Hz [8]. In their analysis, the availability was over 90%, which might be a reason for the favorable results. Bonin et al. [6] reports that the deviations are apparent for frequencies higher than 0.1 Hz, but are able to obtain better estimates by applying the methods of Lenchow [5]. However, as the measurements in this study are performed in the presence of large-scale recirculation, isotropic turbulence cannot be expected. Therefore, these methods are not applied.

3.2. Effect of Range Gate Length

Figure 13 shows how the coefficient of determination changes with the SNR limit for the different range gate lengths during the 24 h analysis period on 30 April 2015. The same methodology as for Figure 6 of calculating the coefficient of determination for increasing values of the SNR limit was used. It can be seen that R^2 is higher for longer range gate lengths with low SNR limits (<-10 dB), and higher for shorter range gate lengths with higher limits (>-10 dB). This is because the uncertainty of a line-of-sight measurement increases for a given SNR value when the range gate length is reduced. With a short range gate length, the amount of noisy measurements with a high uncertainty is larger, contributing to a lower correlation. However, when the SNR limit is increased, many of the uncertain measurements are filtered out and a shorter range gate length with a smaller spatial averaging effect gives a better correlation than a longer range gate length.

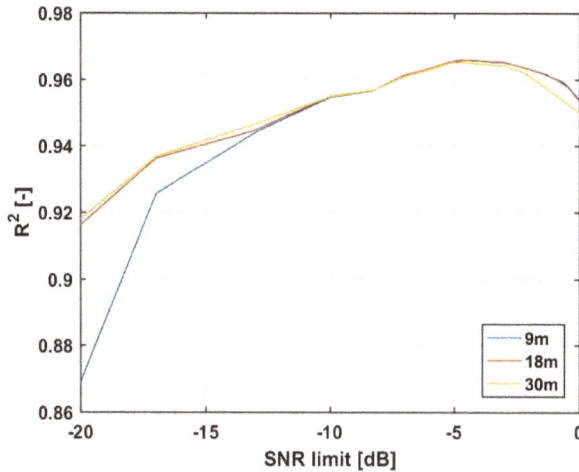

Figure 13. Coefficient of determination versus lower SNR limit for the three different range gate lengths on 30 April 2015.

Table 3 summarizes the optimal case for each range gate length. The table shows that the optimal SNR limit is higher for shorter range gate lengths. This appears to be because the smaller spatial averaging effect with shorter range gate lengths dominates the effect of reduced availability. Thus, the SNR limit can be increased further before the correlation decreases, notwithstanding the reduced availability.

Table 3. Optimal R^2 values and the corresponding Signal-to-noise ratio (SNR) limit and availability for the different range gate lengths

Range Gate Length (m)	Opt. R^2 (-)	Opt. SNR Limit (dB)	Availability (%)
9	0.9657	−4.44	55.4
18	0.9657	−4.69	57.2
30	0.9650	−4.95	58.0

Note that the optimal SNR limit is higher with the 18 m range gate length for this period (−4.69 dB) than for the longer period discussed in Section 3.1 (−10 dB). This difference is due to differences in availability and atmospheric conditions during the two different time periods.

In Figure 14, the power density spectrum for the sonic anemometer is plotted together with the Lidar spectra using different range gate lengths. In order to avoid any effects of time gaps, the spectra are averaged for the ten-minute stare mode periods with a 100% availability only. Additionally, the data were filtered with an SNR limit of −4.95 dB to obtain the best possible correlation for all range gate lengths, before concurrent measurements were plotted.

The Lidar spectra have a steeper slope than the sonic spectrum, and the deviation occurs at approximately 0.05 Hz for all three range gate lengths. However, a higher power density is obtained with a range gate length of 9 m for the very highest frequencies. As the noise floor is raised when the gate length is reduced, it is difficult to state that the increased power level is caused by better representation of the flow turbulence. A more probable cause might be that an increasing noise level in the signal is causing the energy level to rise. With the FWHM significantly larger than the gate length, this is assumed to be the probable cause.

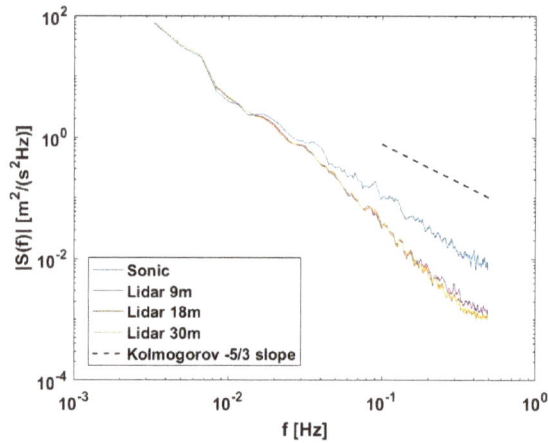

Figure 14. Power density spectra for the Lidar with different range gate lengths, the sonic anemometer and the theoretical Kolmogorov slope.

3.3. Validation of Numerical Model

Figure 15 shows the flow model results compared to the Lidar measurements averaged for the time period 14 June 2015 1:00 p.m. to 10:30 p.m. along the Lidar beam parallel to the mean flow. The DES model is averaged over 15 FTT. Velocity estimates for both flow models are normalized by the upstream free flow velocity at 250 m from the Lidar.

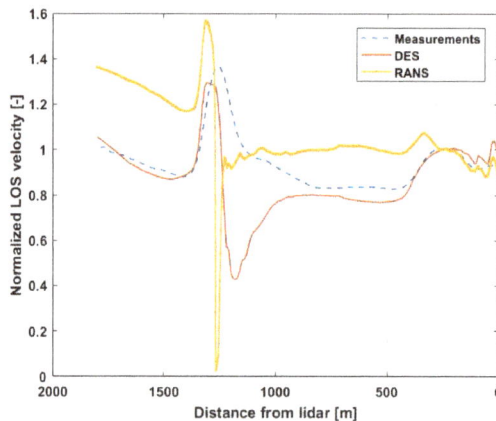

Figure 15. Velocity along the Lidar beam normalized by the upstream free flow velocity averaged for 14 June 2015 1:00 p.m. to 10:30 p.m. measured by the Lidar and computed with the Reynolds-Averaged Navier-Stokes (RANS) and Detached Eddy Simulation (DES) model.

The numerical results are found to predict the speed-up on the crest of the hill. However, both models predict a much higher adverse pressure gradient on the lee-side of the hill, resulting in a lower wind speed compared to the measurements. The RANS model is found to predict a full recirculation, with zero velocity at the crest of the hill. For the DES model, the largest deviations are found at approximately 1300 m from the Lidar. In the region close to the Lidar location, from approximately 0–900 m from the Lidar, the DES model provides a very good estimate for the wind speed variations along the line-of-sight. The RANS model over predicts the velocity in most of the region between the Lidar and the crest, but falls in agreement for the closest 300 m.

Figure 16 shows the measured and computational results for the one-dimensional standard deviation along the Lidar beam. The RANS model fails in predicting the variations in turbulence along the line of sight, predicting a large increase in turbulence at the peak of the crest, where both the measurements and DES-model indicates a reduction. This is expected according to theory, as wind acceleration should cause a relative reduction in turbulence at this point.

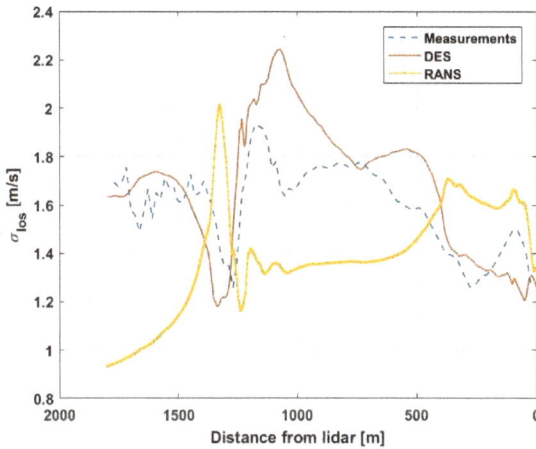

Figure 16. Standard deviation along the Lidar beam averaged for 14 June 2015 1:00 p.m. to 10:30 p.m. estimated by the Lidar and computed with the RANS and DES model.

It is clear that the RANS model fails to predict the expected behavior of turbulence production and advection over the crest. The DES model provides a qualitative improvement even though the erroneous prediction of flow separation on the lee-side of the hill causes a relatively large discrepancy in the estimates.

A spectral analysis was performed at three different distances from the Lidar ((a) 1000 m, (b) 600 m and (c) 400 m). The results were compared to the predicted Kaimal spectrum for the longitudinal wind based on the measured wind speed at 100 m at the measurement mast. The power density spectra for the Lidar and the DES model at the three locations are plotted in Figure 17 together with the predicted Kaimal spectrum. The RANS model is not included in this analysis as it is not a transient model.

The figure shows how the model manages to predict the low-frequency part of the spectra for all locations. There is a significant cut-off in the higher frequencies at $\sim 4 \times 10^{-3}$ Hz most probably due to insufficient mesh and temporal resolution to capture the small-scale fluctuations.

For the Lidar measurements, the spectrum for location (a) has a higher power density than for location (b) and (c) for the highest frequencies. This is expected, as the measurement location is very close to the ground, where turbulence production by high shear is present.

Concerning the numerical results, the expected $-5/3$ slope can be observed in the frequency region between 5×10^{-3} and 5×10^{-2}, where quantitative agreement with the measurements is achieved. For the higher frequency region (5×10^{-2}–5×10^{-1}), the numerical simulation under-predicts the turbulent kinetic energy, denoting less fine flow structures are captured at all 3 positions than for the measured data. Abnormal evolutions can be seen in highest frequency region

$(f > 5 \times 10^{-2})$. This denotes that the temporal resolution (time step size) is incapable of adopting the mesh resolution (cell size) applied in the simulation, and only numerical noise without any physical meaning is obtained.

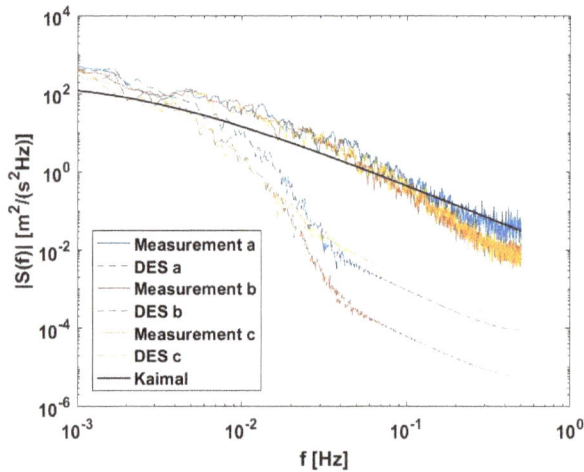

Figure 17. Power density spectra for the Lidar and the DES model at distances (**a**) 1000 m, (**b**) 600 m and (**c**) 400 m from the Lidar, and the estimated Kaimal spectrum during 14 June 2015 1:00 p.m. to 10:30 p.m.

The predicted Kaimal spectrum underestimates the power density for most frequencies ($<10^{-1}$ Hz). The model is based on shear-introduced turbulence [18], and these results suggest that the model is not able to capture the additional mechanically induced turbulence due to the complexity of the terrain. For higher frequencies, the model is a good fit for location (a) with a higher turbulence level. Nonetheless, as the model is based on the measured speed at 100 m height, the model is not expected to be a highly accurate fit.

Note that there is a difference in the azimuth angle of 0.3° and elevation angle of ±0.2° between the measured and modeled results. The resulting deviations in the horizontal and vertical directions at the three locations along the beam are presented in Table 4.

Table 4. Horizontal and vertical deviations at the three locations along the beam caused by the difference in azimuth and elevation angle between the measured and modeled results.

Location	Horizontal Deviation (m)	Vertical Deviation (m)
(a)	5.2	3.5
(b)	3.1	2.1
(c)	2.1	1.4

Even though the error introduced is non-negligible, it can be disregarded considering the resolution of the mesh and the uncertainties involved in predicting the wind regime in this terrain.

4. Conclusions

In this study, the performance of a hybrid RANS/LES (DES) flow model for turbulence estimation has been evaluated by comparison with Lidar measurements in highly complex terrain.

First, the accuracy of Lidar turbulence estimates were evaluated by validation with sonic anemometer data. The analysis proved the Lidar to be very accurate in prediction of 10-min mean velocities, while the accuracy was lower for one-second measurements. This is expected, as the Lidar is affected by both spatial averaging and presence of white noise in the measurements. The possibility of increasing the accuracy of the Lidar was investigated by reducing the range gate length. With a shorter range gate length, the spatial averaging is done over a smaller volume, and the system is less affected by the spatial averaging effect. However, as the range gate is reduced, the noise level is increased, and only a slight improvement was observed when reducing the gate length from 30 m to 18 m, which is about half the FWHM scale of the Lidar. As the uncertainties related to the spatial averaging of the Lidar are relatively small compared to uncertainties in the computational model, the Lidar is assumed to be suitable for validation of flow model turbulence estimates.

The transient numerical simulation using a DES modeling strategy for turbulence estimation was tested by comparison with Lidar measurements along a beam aligned with the mean flow, pointing towards a steep ridge. This way the variation of velocity and turbulence along the beam can be determined, providing a significant improvement compared to a traditional meteorological mast measurement campaign, where only point measurements can be used for flow model validation. The DES simulation in this study overestimated the mean turbulence level somewhat, but outperformed a more traditional RANS model approach that failed to describe the turbulence behavior over the ridge.

The study shows the applicability and viability of using Lidar systems in stare mode for validating numerical models over larger distances, without a need for a significant number of point measurements. The study also shows the need for better and more reliable numerical models for predicting turbulence in highly complex terrain.

Acknowledgments: The study was funded by the ENERGIX-programme of the Research Council of Norway. We thank Statkraft AS, Meventus AS, Fraunhofer Institute for Wind Energy and Energy System Technology and the Department of Energy and Process Engineering at the Norwegian University of Science and Technology for making this research work possible.

Author Contributions: J.A.L. and Meventus conducted the experiments and collected the Lidar data used in the analysis. C.Y.C. performed the CFD simulations. A.R. performed the analysis with close help from J.A.L. and L.S. for the Lidar validation, and from C.Y.C. for the numerical analysis.

Conflicts of Interest: The authors declare no conflict of interest.

References

1. Bechmann, A.; Sørensen, N.N.; Berg, J.; Mann, J.; Réthoré, P.E. The Bolund Experiment, Part II: Blind Comparison of Microscale Flow Models. *Bound. Layer Meteorol.* **2011**, *141*, 245–271.
2. Bechmann, A.; Sørensen, N.N. Hybrid RANS/LES Method for Wind Flow Over Complex Terrain. *Wind Energy* **2010**, *13*, 36–50.
3. Guillén, B.F.; Gómez, P.; Rodrigo, S.J.; Courtney, M.S.; Cuerva, A. Investigation of Sources for Lidar Uncertainty in Flat and Complex Terrain. In Proceedings of the EWEC-11, Marseille, France, 16–19 March 2009.
4. Vogstad, K.; Simonsen, A.H.; Brennan, K.J.; Lund, J.A. Uncertainty of Lidars in Complex Terrain. In Proceedings of the EWEA 2013, Vienna, Austria, 4–7 February 2013.
5. Lenschow, D.H.; Wulfmeyer, V. Measuring Second- Through Fourth-Order Moments in Noisy Data. *J. Atmos. Ocean. Technol.* **2000**, *17*, 1330–1347.
6. Bonin, T.A.; Newman, J.E.; Klein, P.M.; Chilson, P.B.; Wharton, S. Improvement of Vertical Velocity Statistics Measured by a Doppler Lidar Through Comparison With Sonic Anemometer Observations. *Atmos. Meas. Tech.* **2016**, *9*, 5833–5852.
7. Sjöholm, M.; Mikkelsen, T.; Mann, J.; Enevoldsen, K.; Courtney, M. Spatial Averaging-Effects on Turbulence Measured by a Continuous-Wave Coherent Lidar. *Meteorol. Z.* **2009**, *18*, 281–287.
8. Cañadillas, B.; Westerhellweg, A.; Neumann, T. Testing the Performance of a Ground-Based Wind LiDAR System: One Year Intercomparison at the Offshore Platform FINO1. *Dewi Mag.* **2011**, *38*, 58–64.

9.   Clifton, A.; Boquet, M.; Roziers, E.B.; Westerhellweg, A.; Hofsass, M.; Klaas, T.; Vogstad, K.; Clive, P.; Harris, M.; Wylie, S.; et al. Remote Sensing of Complex Flows by Doppler Wind Lidar: Issues and Preliminary Recommendations. *Natl. Renew. Energy Lab.* **2015**, doi:10.2172/1351595

10.   Courtney, M.; Mann, J.; Mikkelsen, T.; Sjöholm, M. Windscanner: 3-D Wind and Turbulence Measurements from Three Steerable Doppler Lidars. In *IOP Conference Series: Earth and Environmental Science (1:1)*; IOP Publishing: Bristol, UK, 2008.

11.   Fuertes, F.C.; Iungo, G.V.; Porté-Angel, F. 3D Turbulence Measurements Using Three Synchronous Wind Lidars: Validation against Sonic Anemometry. *J. Atmos. Ocean. Technol.* **2014**, doi:10.1175/JTECH-D-13-00206.1.

12.   Pearson, G.; Davies, F.; Collier, C. An Analysis of the Performance of the UFAM Pulsed Doppler Lidar for Observing the Boundary Layer. *Remote Sens.* **2009**, doi:10.1175/2008JTECHA1128.1.

13.   Smalikho, I.N.; Banakh, V.A. Measurements of Wind Turbulence Parameters by a Conically Scanning Coherent Doppler Lidar in the Atmospheric Boundary Layer. *Atmos. Meas. Tech.* **2017**, *10*, 4191–4208.

14.   Walpole, R.E.; Ye, K.; Myers, S.L.; Myers, R.H. *Probability and Statistics for Engineers and Scientists*, 8th ed.; Pearson Education International: Upper Saddle River, NJ, USA, 2007.

15.   Pope, S.B. *Turbulent Flows*; Cornell University: Ithaca, NY, USA, 2000; pp. 182–191.

16.   Monin, A.S.; Obukhov, A.M. Basic Laws of Turbulent Mixing in the Surface Layer of the Atmosphere. *Tr. Akad. Nauk SSSR Geophiz. Inst.* **1954**, *24*, 163–187.

17.   Spalart, P.R. Strategies for Turbulence Modelling and Simulations. *Int. J. Heat Fluid Flow* **2000**, *21*, 252–263.

18.   Kaimal, J.; Wyngaard, J.; Izumi, Y.; Coté, O. Spectral Characteristics of Surface-Layer Turbulence. *Q. J. R. Meteorol. Soc.* **1972**, *98*, 563–589.

19.   Beaupuits, J.P.P.; Otárola, A.; Rantakyrö, F.; Rivera, R.; Radford, S.; Nyman, L. *Analysis of Wind Data Gathered at Chajnantor*; ALMA Memo: Charlottesville, VA, USA, 2004; Volume 497.

20.   Weir, D.E. *Vindkraft—Produksjon i 2013*; Norges Vassdrags- og Energidirektorat: Oslo, Norway, 2014.

21.   Quaas, J.; Stevens, B.; Stier, P.; Lohmann, U. Interpreting the Cloud Cover—Aerosol Optical Depth Relationship Found in Satellite Data Using a General Circulation Model. *Atmos. Chem. Phys.* **2010**, *10*, 6129–6135.

22.   Sathe, A.; Mann, J.; Gottschall, J.; Courtney, M. Can Wind Lidars Measure Turbulence? *J. Atmos. Ocean. Technol.* **2011**, *28*, 853–868.

remote sensing

MDPI

Article

Airborne Doppler Wind Lidar Observations of the Tropical Cyclone Boundary Layer

Jun A. Zhang [1,2,*], **Robert Atlas** [3], **G. David Emmitt** [4], **Lisa Bucci** [1,2] **and Kelly Ryan** [1,2]

[1] Hurricane Research Division, Atlantic Oceanographic and Meteorological Laboratory, NOAA, 4301 Rickenbacker Causeway, Miami, FL 33149, USA; lisa.r.bucci@noaa.gov (L.B.); kelly.ryan@noaa.gov (K.R.)
[2] Cooperative Institute for Marine and Atmospheric Studies, University of Miami, Miami, FL 33149, USA
[3] Atlantic Oceanographic and Meteorological Laboratory, NOAA, Miami, FL 33149, USA; Robert.Atlas@noaa.gov
[4] Simpson Weather Associates, Charlottesville, VA 22920, USA; gde@swa.com
* Correspondence: jun.zhang@noaa.gov; Tel.: +1-305-361-4557

Received: 16 April 2018; Accepted: 23 May 2018; Published: 25 May 2018

☑ check for updates

Abstract: This study presents a verification and an analysis of wind profile data collected during Tropical Storm Erika (2015) by a Doppler Wind Lidar (DWL) instrument aboard a P3 Hurricane Hunter aircraft of the National Oceanic and Atmospheric Administration (NOAA). DWL-measured winds are compared to those from nearly collocated GPS dropsondes, and show good agreement in terms of both the wind magnitude and asymmetric distribution of the wind field. A comparison of the DWL-measured wind speeds versus dropsonde-measured wind speeds yields a reasonably good correlation ($r^2 = 0.95$), with a root mean square error (RMSE) of 1.58 m s^{-1} and a bias of -0.023 m s^{-1}. Our analysis shows that the DWL complements the existing P3 Doppler radar, in that it collects wind data in rain-free and low-rain regions where Doppler radar is limited for wind observations. The DWL observations also complement dropsonde measurements by significantly enlarging the sampling size and spatial coverage of the boundary layer winds. An analysis of the DWL wind data shows that the boundary layer of Erika was much deeper than that of a typical hurricane-strength storm. Streamline and vorticity analyses based on DWL wind observations explain why Erika maintained intensity in a sheared environment. This study suggests that DWL wind data are valuable for real-time intensity forecasts, basic understanding of the boundary layer structure and dynamics, and offshore wind energy applications under tropical cyclone conditions.

Keywords: tropical cyclones; Doppler Wind Lidar; atmospheric boundary layer; wind structure

1. Introduction

Although substantial progress has been made in the accuracy of tropical cyclone (TC) track forecasts, progress to improve intensity forecasts has lagged, especially for TCs undergoing rapid intensity (RI) change [1]. The difficulty in forecasting intensity change is due mainly to the complicated nature of TC intensification, which has been neither well understood nor correctly represented in forecast models. The atmospheric boundary layer that connects the ocean with the upper level TC vortex is a critical region for intensity change, because it governs both the energy distribution and dynamics required for TC intensification [2–4]. Numerical studies have also emphasized the critical role of the boundary layer parameterization in simulations of TC intensity and structure [5–9]. However, the TC boundary layer has been the least observed part of a storm until now.

The routine collection of kinematic and thermodynamic observations in the TC boundary layer remains limited [10]. Currently, boundary-layer observations are scarce, due to the danger involved in manned aircraft gathering direct wind and humidity measurements in this turbulent region of the storm. The use of unmanned aircraft is a promising tool for TC boundary layer observations, but the technology is not yet advanced enough to collect fast response wind and thermal data (e.g., the rate of data transfer through a satellite). Other in-situ observing platforms such as research buoys suffer the same plight as manned aircraft, due to the likelihood of damage to instrumentation. Even if a research buoy does occasionally survive in a strong TC [11], it must be located in the eyewall to obtain hurricane-force wind measurements. The probability of this occurring is small, due to uncertainties in the track forecast at the time of the buoy's pre-storm deployment. This lack of observational data is believed to be one of the primary reasons why boundary layer processes remain poorly represented in operational TC models [12,13], which limits their ability to improve intensity forecasts.

Our understanding of the mean boundary layer structure has improved since the advent of the Global Positioning System (GPS) dropsonde in 1997 [14]. Due to limited resources, however, the sampling size of GPS dropsondes in individual storms is generally quite small (<20 for a 12 h observational period). In rare cases where multiple research aircraft are flown simultaneously, such as in Hurricanes Earl (2010) and Edouard (2014), more than 40 dropsondes can be collected in a 12-h window. Such composite analyses can present a radius-height view of the boundary layer [15,16]. Previous studies have used this type of composite method to analyze a large number of dropsonde data collected in multiple storms, in order to characterize the mean climatological boundary-layer structure [17,18]. Of note, these studies focused on the boundary layer of TCs with hurricane-force (33 m s^{-1}) and stronger-strength winds. The differences in the boundary layer structure between that of a tropical storm and a hurricane are not well documented.

Doppler radar onboard research aircraft provides extensive wind observations in hurricanes, but its vertical resolution is generally too coarse for boundary layer studies. When a TC experiences strong environmental vertical wind shear, its convective structure is usually asymmetric, which makes the distribution of precipitation asymmetric. Under such a scenario, Doppler radar wind measurements are limited by the lack of backscattering from precipitation. Doppler Wind Lidar (DWL) observations complement Doppler radar observations in regions of little to no precipitation in TCs. Additionally, the DWL provides a much larger data coverage area for wind profiles than GPS dropsondes.

Baker et al. [19] provided an excellent review of previous impact studies that used both simulated satellite and real-world, aircraft-based DWL data to demonstrate the DWL's ability to measure winds in TCs. Impact experiments with real data, termed Observing System Experiments, have been conducted with and without DWL data to show the positive impact of this observing system on TC track and intensity forecasts [20]. Similar experiments with simulated DWL data, termed Observing System Simulation Experiments, have also shown positive impacts on track forecasts [21–23]. The present study further illustrates the usefulness of the DWL for TC studies, with a focus on understanding TC structure by analyzing wind profiles collected in Tropical Storm (TS) Erika (2015). More recently, a coherent DWL was flown in 2017 on the National Aeronautics and Space Administration (NASA)'s DC8 aircraft during the Convective Processes Experiment (CPEX), which has provided a new perspective on tropical convective systems [24]. DWLs continue to mature as airborne systems, based on their ability to derive wind measurements from molecular motions through direct detection, providing wind data in aerosol-sparse areas.

2. Material and Methods

The DWL aboard the National Oceanic and Atmospheric Administration (NOAA) P3 aircraft is a coherent system (1.6-micron wavelength) that depends on atmospheric aerosols for its return signal, such that vertical coverage varies from one storm to the next [25]. In general, however, the convection and high winds associated with TCs provide ample aerosols and thus profiles from the flight level (usually 3 km) down to the ocean surface. The three main components of the DWL are the transceiver, scanner, and data processing system. Table 1 summarizes general information about the DWL.

Table 1. National Oceanic and Atmospheric Administration (NOAA) P3 DWL system and data product generation parameters.

Parameter (units)	Value	Comments
Wavelength (nm)	1600	Eyesafe for NOAA P3 DWL configuration
Pulse energy (Joules)	0.0015	0.0023 maximum
Pulse repetition frequency (Hz)	500	Due to data processing limitations, the effective pulse repetition frequency is 166 Hz
Pulse full width half maximum (m)	90	Full width half maximum of Gaussian pulse; duration is 320 ns
Telescope diameter (m)	0.10	
Scanner		Biaxial conical scanner side mounted starboard on P3
Digitization rate (MHz)	250	
Line of sight range gate (m)	~90	Sliding gate provides 45 m line of sight product
Shot integration, nominal (seconds)	1	Nominal scan consists of 12 point step and stares with 1-s dwells
Time between u,v,w profiles (seconds)	~25	Assumes 1 s dwells
Distance between u,v,w profiles (km)	3.75	Assumes 150 m/s P3 ground velocity

The latest version of the coherent Doppler transceiver developed by Lockheed Martin Coherent Technologies was used aboard the NOAA P3 aircraft. The scanner shown in Figure 1 is a bi-axial scanner that has a scanning range of 30 degrees in azimuth and 120 degrees in elevation. The scanner can be programmed to change scanning modes during flight. The standard scanning mode employs 12 step-stares at 20 degrees off nadir, with a 1-s duration at each stare and a 1-s transition between stares, followed by a 5-s dwell at nadir. During the mission into TS Erika, however, the scanner was set to scan forward at 30 degrees off nadir and ±30 degrees azimuth.

Figure 1. P3 Doppler Wind Lidar (DWL) bi-axis scanner.

The DWL was operated at 166 Hz, such that each 1-s line of sight (LOS) integrated product contained 166 laser shots. The shot pulses were approximately Gaussian in shape with a full width half maximum of about 90 m and a diameter of about 10 cm. During the Erika mission, the DWL operated within a range of 4000 m. A "sliding range gate" approach was used in generating a 50 m vertical resolution wind profile based on 100 m basic range gates. The closest usable signals were about 300 m below or above the aircraft. The aircraft's position, speed, and attitude (pitch, roll, and yaw) were obtained from the DWL's dedicated Global Positioning System (GPS) and Inertial Navigation System (INS). The LOS winds were navigated in space and converted into vertical profiles of the full three-dimensional (3D) mean wind vectors. At nominal P3 cruising speeds (~100 m s^{-1}), the wind profiles are representative of the mean flow over approximately 3 km. The data processing system, along with other instruments that help cool the system, is shown in Figure 2.

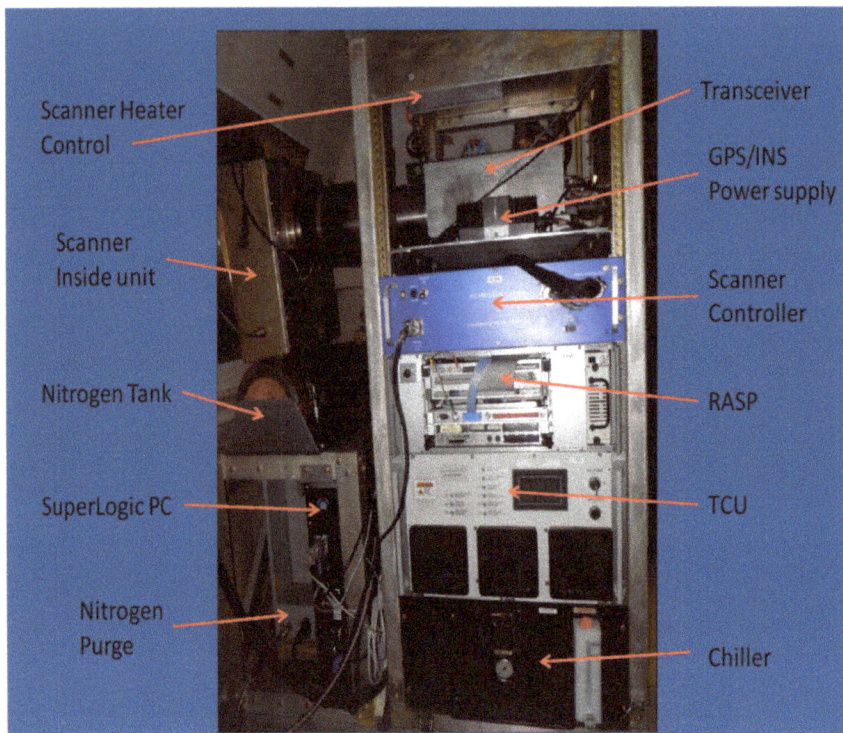

Figure 2. The Doppler Wind Lidar system inside the P3 aircraft, showing the laser (transceiver), data processing system, and cooling system from top to bottom. GPS represents the Global Positioning System, INS represents the Inertial Navigation System, RASP represents the Real-time Advanced Signal Processor, and TCU represents the Transceiver Control Unit.

The DWL has been successfully flown on one of NOAA's P3 Hurricane Hunter aircraft since the 2015 hurricane season. DWL wind profile data were first collected in TS Erika (2015). Erika began as a tropical wave west of Africa on 21 August and became a tropical storm in the Atlantic on 24 August (Figure 3a). The system remained nearly steady-state throughout its life cycle, weakening by 5 kt (~2.5 m s^{-1}) in the first two days, and then intensifying by 10 kt (~5 m s^{-1}) until 28 August (Figure 3b).

The P3 mission into TS Erika was conducted on 26 August, when the storm's intensity was 40 kt, just before its weak intensification. During the period of P3 observations, Erika was under strong northwesterly environmental wind shear, and convective activity mostly appeared in the downshear side of the storm (Figure 4). Despite the detrimental influence of the shear, Erika maintained steady state and slightly intensified after the period of DWL measurements.

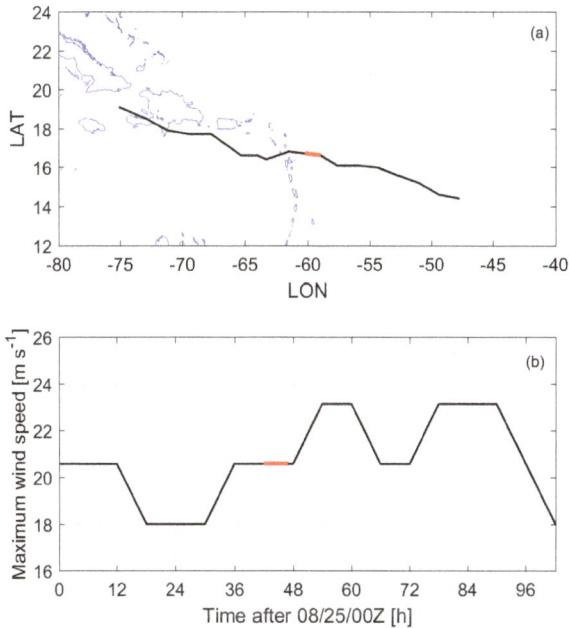

Figure 3. Plots of the track (**a**) and intensity (**b**) of Tropical Storm Erika (2015) from the National Hurricane Center's best track. The red line indicates the period of P3 observations.

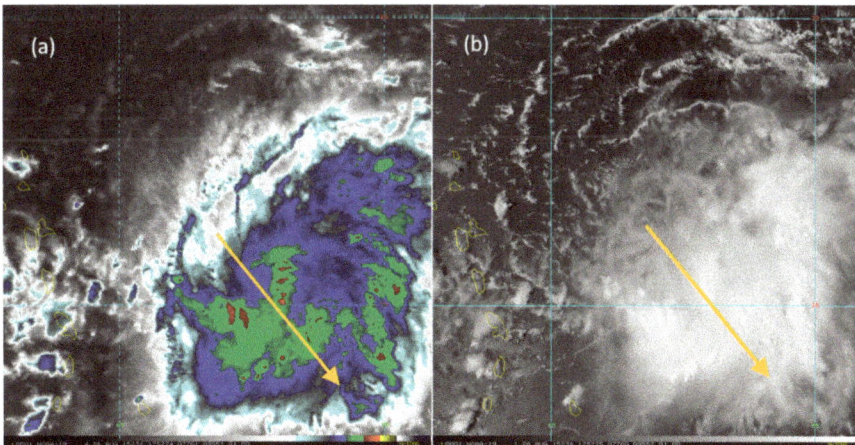

Figure 4. (**a**) GOES satellite Infrared (IR) image and (**b**) visible image of Tropical Storm Erika (2015) on 26 August during the time of the P3 observations. The yellow arrow represents the shear direction. The colors in the IR image show the cloud top brightness temperature, with red and green colors indicating convective activity. In the visible image, the cloudy region indicates convection.

The earth-relative flight track of the P3 aircraft flown in TS Erika is shown in Figure 5a, and followed a standard Figure 4 flight pattern that is routinely flown as part of the NOAA Hurricane Research Division's annual operational hurricane field program. After taking into account the storm's motion, the flight track was plotted in a storm-relative framework, using center fixes based on flight level wind observations (Figure 5b). Of note, when calculating the storm-relative track, the storm centers from both the best track of the National Hurricane Center (NHC) and fixes based on flight-level winds are used, following the method of Willoughby and Chelmow [26]. The dropsondes released in TS Erika were mostly located in the storm center and turn points, as shown in Figure 5b by red circles. As mentioned earlier, the number of dropsondes deployed in a P3 mission is limited by available resources, and typically only the eye, eyewall, and turn points at the end of each radial penetration leg are sampled, as was the case with Erika. With this type of limited coverage, dropsonde data can only provide isolated snapshots of the mean boundary layer structure. The DWL, on the other hand, can provide extensive wind observations of the boundary layer, as shown in next section.

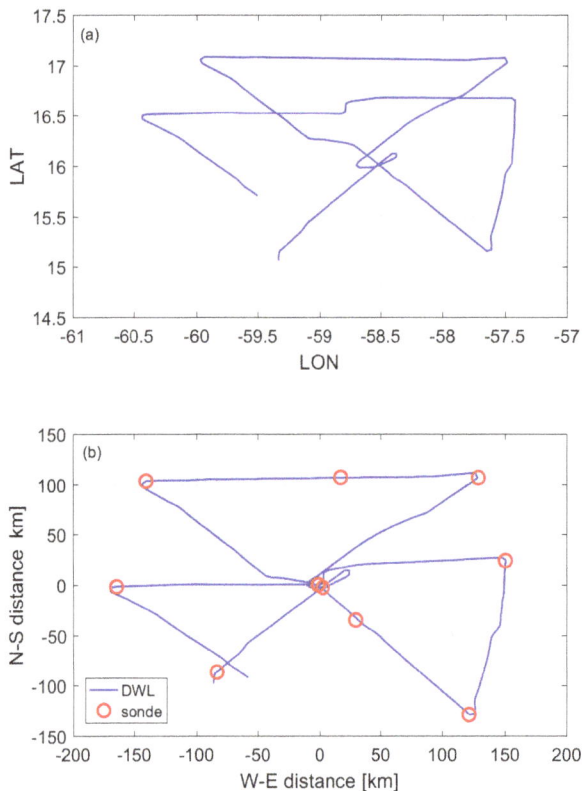

Figure 5. Plots of the earth-relative (**a**) and storm-relative (**b**) P3 aircraft track into Tropical Storm Erika (2015). Red circles represent the location of dropsonde deployments.

3. Results

We first compared GPS dropsonde data to DWL wind profiles. Of note, dropsondes provide an inherently different measure of the wind from that of DWL. Dropsondes usually follow a drifting trajectory, and take approximately 2–3 min to reach the ocean surface. Depending on the relative direction of the flight path and mean wind, dropsondes generally sample a very different part of the mean flow. In a strongly sheared and turbulent boundary layer, such as is found in tropical cyclones, dropsonde profiles are best considered to be single realizations of the wind. Furthermore, since dropsondes follow a slanted path in conditions of both strong horizontal and vertical shear, the near-surface portion of the dropsonde profile may not be representative of what the near-surface profile would be below the portion of the profile that was sampled higher in the atmosphere. In contrast, the DWL is closer to the mean wind measurement within a volume of the atmosphere.

Figure 6 shows vertical wind profiles from the DWL measurements that are collocated with the dropsondes at each dropsonde location shown in Figure 5b. The wind comparison in Figure 6 is displayed in a storm-relative sense, in that the eye sounding is placed at the center of the figure. It is evident from Figure 6 that in the storm center, the wind speed is weak (<5 m/s); this wind feature is captured by both the DWL and dropsonde instruments. Furthermore, both the DWL and dropsonde observations show that the surface wind speed is strongest in the right-front quadrant, on the order of 20 m s^{-1}. The DWL data also captured a similar asymmetric distribution (front versus back, and left versus right) of the wind field that was observed by the dropsondes. The difference between the dropsonde and DWL wind speed measurements near the surface (<100 m altitude) is mainly due to the dropsonde drift effect mentioned above.

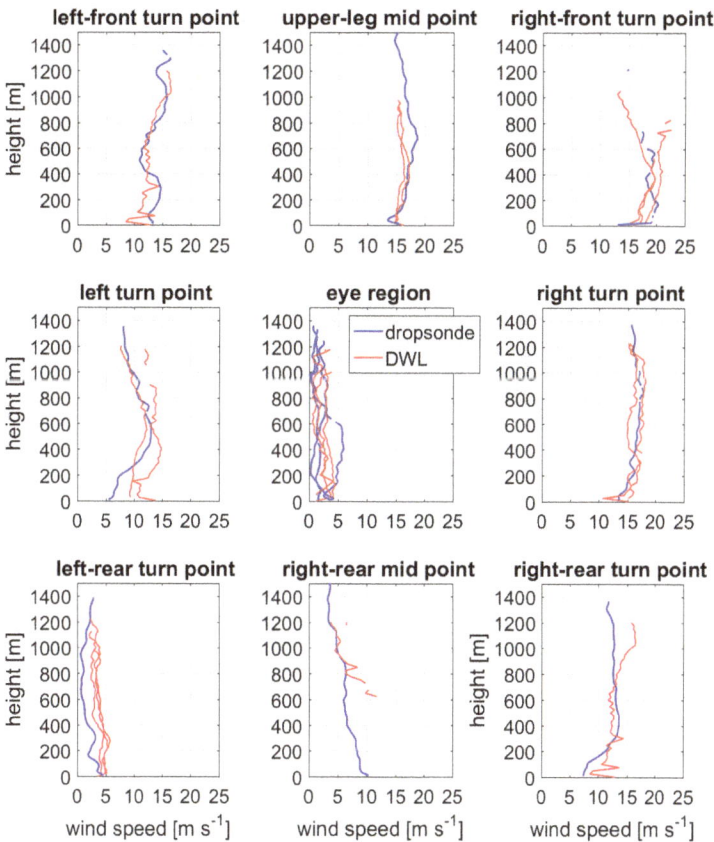

Figure 6. Plots of vertical wind profiles from the DWL (red) and GPS dropsonde (blue) observations. The wind comparison is plotted in a storm-relative framework, with the location of the observations shown in the title of each panel.

A comparison of wind speeds measured by the DWL within a 10-km distance from the dropsonde data and within a 2-min time interval of the two types of observations shows good agreement (Figure 7). There is also a reasonably good correlation ($r^2 = 0.95$) with the root mean square error (RMSE) of 1.58 m s^{-1} compared to the dropsonde data. The bias of the DWL-measured winds appears to be small (-0.023 m s^{-1}). This result is consistent with previous verifications of the DWL-measured wind speeds with dropsonde data [20,27]. Other studies [28–31] have shown comparisons of DWL data to dropsonde data that are in better agreement than with our study (i.e., smaller RMSE). This is most likely due to the fact that TCs have a highly variable wind field, and the different measurement volumes had different wind speeds in the comparison. Of note, the RMSE of the DWL-measured wind speed is much smaller (~4 m s^{-1}) than that of the wind speed measured by the Stepped Frequency Microwave Radiometer (SFMR), both compared to dropsonde data [32,33]. Given that the SFMR wind measurements have been routinely used for real-time TC intensity forecasts, the DWL wind data have great potential to assist forecasters with intensity estimates.

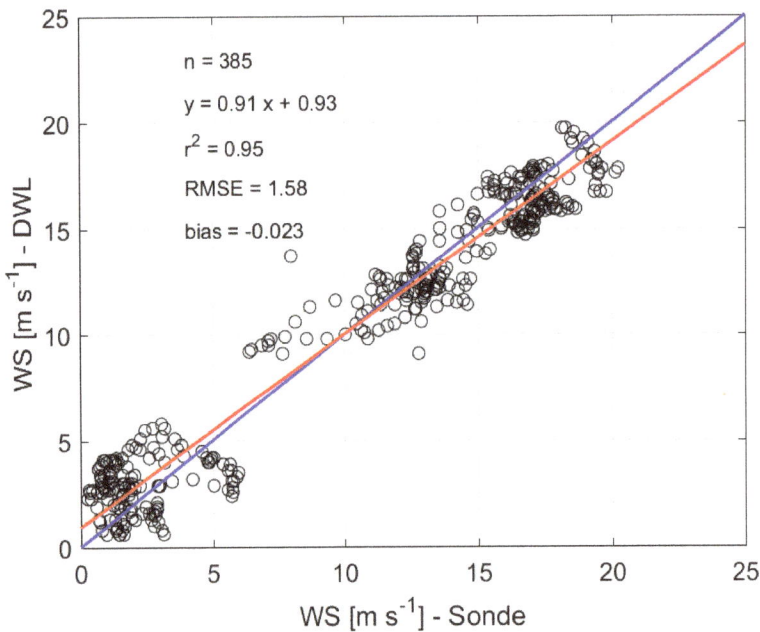

Figure 7. Scatterplot of the DWL measured wind speed (WS) versus dropsonde wind speed and the linear regress (red line). The blue line shows the 1:1 ratio, which is the line of perfect correlation. The regression equation, correlation coefficient (r), root mean square error (RMSE), and bias are also shown.

To further evaluate the usefulness of the DWL wind observations in TCs, we conducted a two-dimensional (2D) analysis of wind speeds measured by the DWL, at 500 m and 1 km altitudes, and compared these observations to Tail Doppler radar observations at the same altitudes (Figure 8). As mentioned earlier, Doppler radar has been used to routinely measure 3D wind velocities during NOAA P3 missions before the DWL was installed on the P3. We used a piece-wise cubic spline method for the 2D wind analysis, following Zhang et al. [34]. This method preserves original data (i.e., along the flight track) and only interpolates data at locations where no observations are available.

It is evident from Figure 8 that wind speeds measured by the DWL at the two altitudes of interest, i.e., 500 m and 1 km, generally agreed with those measured by the Doppler radar in terms of wind asymmetry. For instance, the Doppler radar measured the strongest and weakest wind speeds on the northeast and southwest sides of the storm, respectively, which were captured by the DWL wind observations. Since Doppler radar can only measure wind speed when there is precipitation, winds in almost the entire northwest quadrant were not measured well, due to there being little precipitation in this region. The strong winds (~20 m/s) to the left of the storm at ~100 m radius were measured by the DWL, but these winds were not observed by Doppler radar (c.f., Figure 6). This result suggests that the DWL wind data had better spatial coverage, clouds permitting, than the Tail Doppler radar data at the two altitudes of the Doppler radar observations.

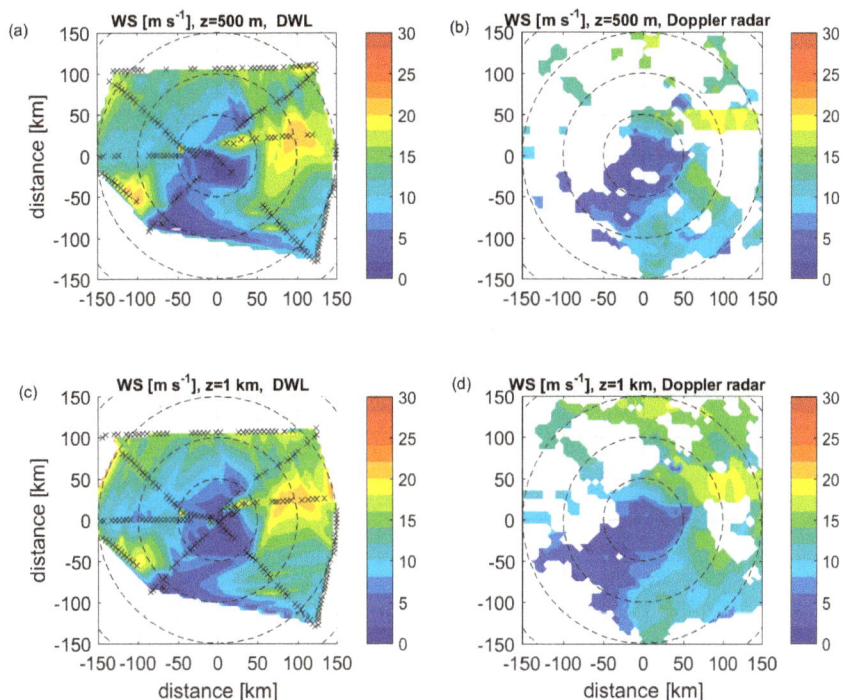

Figure 8. Plots of the wind speed at 500 m (**a,b**) and 1 km (**c,d**) altitudes from the DWL two-dimensional (2D) analysis (**a,c**) and Doppler radar observations (**b,d**). Black crosses in the left panels indicate the location of the DWL wind observations used in the analysis, while black dashed lines indicate the radial distance from the center every 50 km.

Furthermore, Doppler radar is limited by its vertical resolution (500 m), and the swath data have no observations below an altitude of 500 m. On the other hand, the DWL measures the wind to 25 m above the sea surface with a vertical resolution of ~50 m [35]. Figure 9 shows the wind speed measured by the DWL at altitudes as low as 25 m (middle of 50 m height gate). This suggests the DWL is capable of measuring the near-surface maximum wind speed, which is close to a tropical cyclone's intensity.

The maximum wind speed measured by the DWL at 25 m in Erika (2015) was 23 m s^{-1}, which is quite close to the storm's intensity, based on the NHC's best track (c.f., Figure 3b). These highly accurate DWL wind measurements not only complement dropsonde observations by significantly enlarging the sampling size, but also provide useful information for validating SFMR surface wind observations. Large biases in SFMR wind measurements can be identified and corrected by using the collocated DWL wind profiles. This process can improve NHC forecasts by providing better real-time intensity estimates.

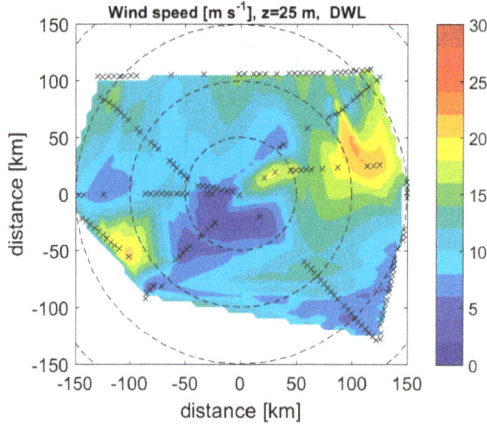

Figure 9. Plot of the wind speed measured by the DWL at 25 m in Tropical Storm Erika (2015). The black crosses are the location of the DWL winds used in the 2D analysis. The black dashed lines indicate the radial distance from the center every 50 km.

4. Discussion

Given the excellent data coverage provided by the DWL wind observations, storm-relative tangential and radial velocities can be studied. Due to a lack of observations, no previous study has shown the detailed inflow layer structure of an individual TC, to the authors' knowledge. Figure 10 shows the boundary layer inflow and outflow structure of Erika at four vertical levels (25 m, 100 m, 500 m, and 1000 m).

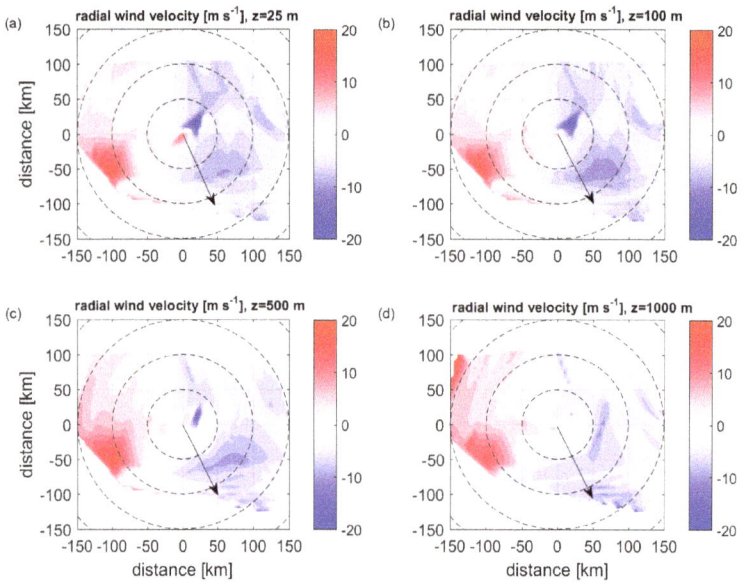

Figure 10. Plots of the radial wind velocity at 25 m, 100 m, 500 m, and 1000 m altitudes, respectively. The black arrow represents the shear direction, while black dashed lines indicate the radial distance from the center every 50 km.

It is evident that inflow is much stronger on the right side of the storm than on the left side. Note that the strongest inflow is located along the shear direction, as indicated by the black arrow in Figure 10. The inflow layer being deeper on the downshear side rather than on the upshear side in Erika is consistent with the result of the dropsonde composite analysis given by Zhang et al. [18].

An analysis of the height of the maximum tangential wind speed also shows that Erika's boundary layer is deeper on the downshear side of the storm (Figure 11). Of note, both the inflow layer depth and height of the maximum tangential wind speed are found to decrease toward the storm center, consistent with a previous dropsonde composite [17]. However, the kinematic boundary layer heights on average are much larger than that of a typical hurricane-strength TC, which suggests that the boundary layer structure of a tropical storm is different from that of a hurricane. Forecast models should consider this difference.

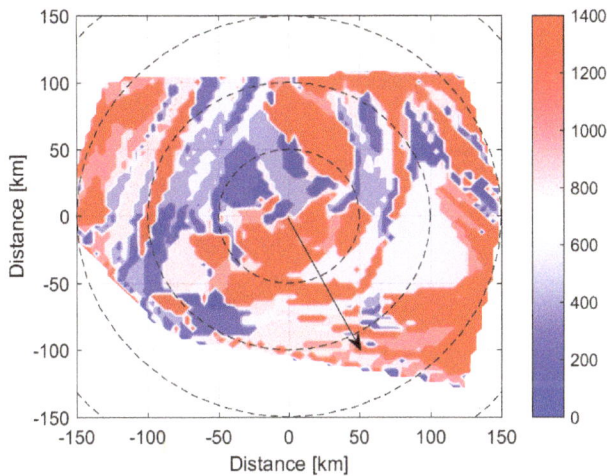

Figure 11. Height of the maximum tangential wind speed based on the DWL data in Tropical Storm Erika (2015). The black arrow represents the shear direction, while black dashed lines indicate the radial distance from the center every 50 km.

The streamline pattern based on storm-relative winds is shown in Figure 12, at the four altitude levels in Figure 10, along with the relative vorticity shown in shading. The closed circulation and large vorticity near the storm center suggest that Erika was able to maintain tropical storm strength despite strong wind shear, likely due to vorticity development in the boundary layer. It is evident from Figure 12 that absolute vorticity is maximized in the storm center, with a broad region of relatively large values of vorticity ($>4 \times 10^{-4}$ s^{-1}) located in the downshear side of the storm in the boundary layer. Interestingly, the circulation center of the vortex of Erika varies with height, showing a weak tilt of the vortex in the upper levels (>750 m) toward the downshear direction, and implying a vortex tilt signature even in the boundary layer. This type of structure can only be detected by high-resolution wind measurements from an instrument like the DWL.

Of note, there is no distinct eyewall in a tropical storm like Erika, so the radius of the maximum wind speed is not well defined. The maximum azimuthally-averaged tangential wind speed is found to be located at a radius of ~100 km, which is nearly twice the size of a typical hurricane. Streamline analysis shows that the circulation of the vortex near the surface is closed. The vortex center is slightly tilted to the shear direction in the boundary layer. Despite this vortex tilt feature, the largest vorticity is located in the storm center, suggesting the development of circulation occurs within the boundary layer. This is consistent with the hypothesis of a progressive boundary layer control of the spin-up process, as suggested by previous theoretical studies [3,36].

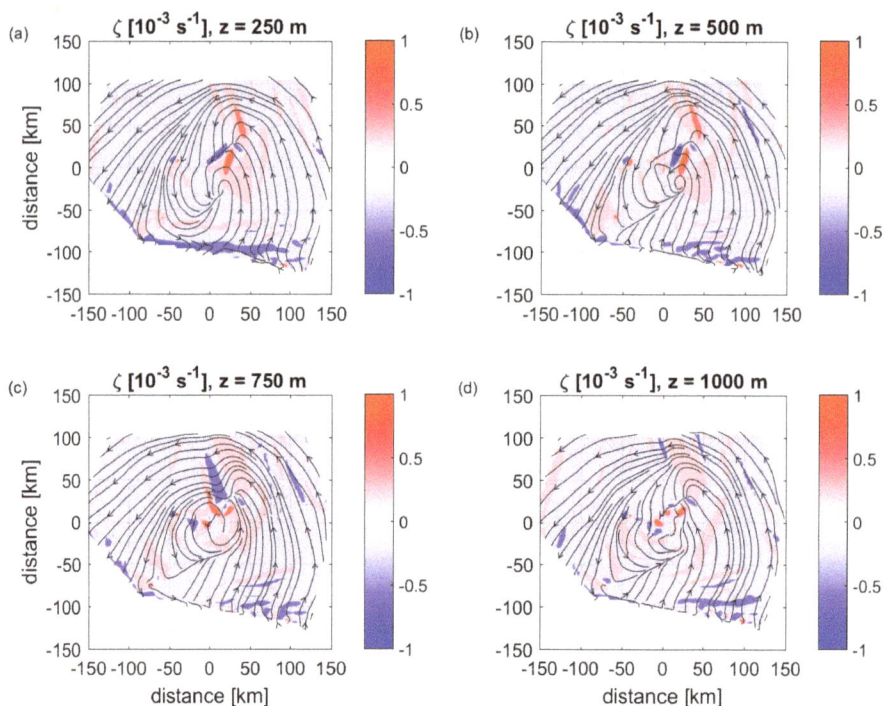

Figure 12. Plots of the relative vorticity (shading) and streamlines (contour) at (**a**) 250 m, (**b**) 500 m, (**c**) 750 m, and (**d**) 1000 m altitudes based on the DWL measured winds.

5. Conclusions

This study presents an analysis of wind profile data collected by the DWL onboard NOAA's P3 aircraft in TS Erika (2015). The DWL observations complement the existing P3 Doppler radar observations, in that the DWL collects wind data in rain-free and low-rain regions where Doppler radar is limited by its inability to capture the backscatter. In addition, the DWL wind data have a much higher vertical resolution (50 m) than the Doppler radar wind data (500 m). The DWL observations also complement the dropsonde measurements by significantly enlarging the sampling size of the wind profiles. Observations of wind speeds down to ~25 m can provide valuable intensity information in real time to NHC forecasters when a NOAA P3 Hurricane Hunter mission is flown with a DWL onboard.

Our analysis shows good agreement when DWL-measured wind profiles are compared to those from collocated dropsondes (i.e., high correlation, small bias, and relatively small RMSE). A comparison of DWL-measured wind speeds at the same altitudes as the Doppler radar observations also show good agreement in regions where there are extensive Doppler data (i.e., rain regions) in terms of wind asymmetry. To the authors' knowledge, the DWL data collected in TS Erika provide the best data coverage of the boundary layer of any given tropical cyclone. The DWL data presented in this study will be invaluable for evaluating the boundary layer structure and physics in TC forecast models.

The kinematic boundary layer height as depicted by both the inflow layer depth and height of the maximum tangential wind speed is found to decrease with decreasing distance from the storm center, consistent with the structure of a typical hurricane. However, the boundary layer in TS Erika is much deeper than in a hurricane based on climatology. The shear-relative analysis of the DWL data shows that the strength of the inflow is larger in the downshear-side quadrants than in the upshear-side quadrants, again mimicking the structure of a hurricane. On the other hand, the extent of the asymmetry of both the tangential and radial winds was larger in TS Erika than that observed in the hurricane composites.

Future work will analyze DWL data collected in recent hurricanes, including landfall cases, in order to further explore wind structure in the boundary layer for both science and engineering applications. For instance, DWL wind data can be used to improve our understanding of the boundary layer structure of TCs close to US coastal regions where offshore wind energy development has been conducted under the guidelines of the Department of Energy. Wind turbines built offshore for power generation are usually affected by tropical cyclone winds, especially strong wind gusts [37]. However, the impact of tropical cyclone winds on the structural integrity of turbines is poorly understood, due to a lack of observations at typical turbine heights (<200 m above sea level). Dropsondes provide single slant profiles of wind observations that are inadequate for high temporal or spatial analysis across the rotor layer at the turbine height. Offshore tower and buoy observations usually only collect wind data near the ocean surface with a measurement altitude of <50 m. Airborne DWL provides a unique tool for 3D wind observations around offshore wind turbines, which can be used in future investigations to explore the impact of tropical cyclones on turbine loads.

Author Contributions: J.A.Z. led this paper and wrote the paper with R.A. and G.D.E. J.A.Z. performed the data analysis and produced the results. D.E. installed the Doppler Wind Lidar on NOAA's P3 aircraft and conducted data processing. L.B. and K.R. participated the P3 aircraft mission into Tropical Storm Erika (2015) for data collection. L.B. also assisted in data verification. R.A. supervised the overall investigation of this project.

Acknowledgments: This work was mainly supported by NOAA's Sandy Supplemental Award. Jun Zhang also acknowledges supports from NSF Grants AGS1822128 and AGS 1654831, NOAA Grant NA14NWS4680030, and NASA Grant NNX14AM69G. We thank three reviewers for their comments that substantially improved the paper. We are grateful to Gail Derr for editorial assistance. We also acknowledge scientists from the Hurricane Research Division, P3 crew and engineers from the Aircraft Operational Center, who helped with the DWL installation and P3 mission into Tropical Storm Erika (2015).

Conflicts of Interest: The authors declare no conflict of interest. The founding sponsors had no role in the design of the study; in the collection, analysis, or interpretation of data; in the writing of the manuscript, and in the decision to publish the results.

References

1. Kaplan, J.; Rozoff, C.M.; DeMaria, M.; Sampson, C.R.; Kossin, J.P.; Velden, C.S.; Cione, J.J.; Dunion, J.P.; Knaff, J.A.; Zhang, J.A.; et al. Evaluating environmental impacts on tropical cyclone rapid intensification predictability utilizing statistical models. *Weather Forecast.* **2015**, *30*, 1374–1396. [CrossRef]
2. Emanuel, K.A. Sensitivity of tropical cyclones to surface exchange coefficients and a revised steady-state model incorporating eye dynamics. *J. Atmos. Sci.* **1995**, *52*, 3969–3976. [CrossRef]
3. Smith, R.K.; Montgomery, M.T.; Nguyen, S.V. Tropical cyclone spin-up revisited. *Q. J. R. Meteorol. Soc.* **2009**, *135*, 1321–1335. [CrossRef]
4. Montgomery, M.T.; Smith, R.K. Recent developments in the fluid dynamics of tropical cyclones. *Ann. Rev. Fluid Mech.* **2017**, *49*, 541–574. [CrossRef]
5. Braun, S.A.; Tao, W.-K. Sensitivity of high-resolution simulations of Hurricane Bob (1991) to planetary boundary layer parameterizations. *Mon. Weather Rev.* **2000**, *128*, 3941–3961. [CrossRef]
6. Nolan, D.S.; Zhang, J.A.; Stern, D.P. Evaluation of planetary boundary layer parameterizations in tropical cyclones by comparison of in-situ data and high-resolution simulations of Hurricane Isabel (2003). Part I: Initialization, maximum winds, and outer core boundary layer structure. *Mon. Weather Rev.* **2009**, *137*, 3651–3674. [CrossRef]
7. Smith, R.K.; Thomsen, G.L. Dependence of tropical-cyclone intensification on the boundary layer representation in a numerical model. *Q. J. R. Meteorol. Soc.* **2010**, *136*, 1671–1685. [CrossRef]
8. Kepert, J. Choosing a boundary layer parameterization for tropical cyclone modeling. *Mon. Weather Rev.* **2012**, *140*, 1427–1445. [CrossRef]
9. Bu, Y.P.; Fovell, R.G.; Corbosiero, K.L. The influences of boundary layer vertical mixing and cloud-radiative forcing on tropical cyclone size. *J. Atmos. Sci.* **2017**, *74*, 1273–1292. [CrossRef]

10. Black, P.G.; D'Asaro, E.A.; Drennan, W.M.; French, J.R.; Niiler, P.P.; Sanford, T.B.; Terrill, E.J.; Walsh, E.J.; Zhang, J.A. Air-sea exchange in hurricanes: Synthesis of observations from the Coupled Boundary Layer Air-Sea Transfer Experiment. *Bull. Am. Meteorol. Soc.* **2007**, *88*, 357–374. [CrossRef]

11. Potter, H.; Graber, H.C.; Williams, N.J.; Collins, C.O., III; Ramos, R.J.; Drennan, W.M. In situ measurements of momentum fluxes in typhoons. *J. Atmos. Sci.* **2015**, *72*, 104–118. [CrossRef]

12. Zhang, J.A.; Gopalakrishnan, S.G.; Marks, F.D.; Rogers, R.F.; Tallapragada, V. A developmental framework for improving hurricane model physical parameterization using aircraft observations. *Trop. Cyclone Res. Rev.* **2012**, *1*, 419–429.

13. Zhang, J.A.; Nolan, D.S.; Rogers, R.F.; Tallapragada, V. Evaluating the impact of improvements in the boundary layer parameterization on hurricane intensity and structure forecasts in HWRF. *Mon. Weather Rev.* **2015**, *143*, 3136–3155. [CrossRef]

14. Franklin, J.L.; Black, M.L.; Valde, K. GPS dropwindsonde wind profiles in hurricanes and their operational implications. *Weather Forecast.* **2003**, *18*, 32–44. [CrossRef]

15. Rogers, R.F.; Reasor, P.D.; Zhang, J.A. Multiscale structure and evolution of Hurricane Earl (2010) during rapid intensification. *Mon. Weather Rev.* **2015**, *143*, 536–562. [CrossRef]

16. Rogers, R.F.; Zhang, J.A.; Zawislak, J.; Jiang, H.; Alvey, G.R.; Zipser, E.J.; Stevenson, S.N. Observations of the structure and evolution of Hurricane Edouard (2014) during intensity change, Part II: Kinematic structure and the distribution of deep convection. *Mon. Weather Rev.* **2016**, *144*, 3355–3376. [CrossRef]

17. Zhang, J.A.; Rogers, R.F.; Nolan, D.S.; Marks, F.D. On the characteristic height scales of the hurricane boundary layer. *Mon. Weather Rev.* **2011**, *139*, 2523–2535. [CrossRef]

18. Zhang, J.A.; Rogers, R.F.; Reasor, P.D.; Uhlhorn, E.W.; Marks, F.D. Asymmetric hurricane boundary layer structure from dropsonde composites in relation to the environmental vertical wind shear. *Mon. Weather Rev.* **2013**, *141*, 3968–3984. [CrossRef]

19. Baker, W.E.; Atlas, R.; Cardinali, C.; Clement, A.; Emmitt, G.D.; Gentry, B.M.; Hardesty, R.M.; Kallen, E.; Kavaya, M.J.; Langland, R.; et al. Lidar-measured wind profiles: The missing link in the global observing system. *Bull. Am. Meteorol. Soc.* **2014**, *95*, 543–564. [CrossRef]

20. Pu, Z.; Zhang, L.; Emmitt, G.D. Impact of airborne Doppler Wind Lidar data on numerical simulation of a tropical cyclone. *Geophys. Res. Lett.* **2010**, *37*, L05801. [CrossRef]

21. Atlas, R. Atmospheric observations and experiments to assess their usefulness in data assimilation. *J. Meteorol. Soc. Jpn.* **1997**, *75*, 111–130. [CrossRef]

22. Atlas, R.; Hoffman, R.N.; Ma, Z.; Emmitt, G.D.; Wood, S.A.; Greco, S.; Tucker, S.; Bucci, L.; Annane, B.; Murillo, S. Observing system simulation experiments (OSSEs) to evaluate the potential impact of an optical autocovariance wind lidar (OAWL) on numerical weather prediction. *J. Atmos. Ocean. Technol.* **2015**, *32*, 1593–1613. [CrossRef]

23. Atlas, R.; Zhang, J.A.; Emmitt, G.D.; Bucci, L.; Ryan, K. Application of Doppler wind lidar observations to hurricane analysis and prediction. In Proceedings of the 2017 Symposium on Lidar Remote Sensing for Environmental Monitoring, San Diego, CA, USA, 6–10 August 2017; Singh, U.N., Ed.; International Society for Optical Engineering: Bellingham, WA, USA, 2017; p. 10406. [CrossRef]

24. Emmitt, G.D.; Greco, S.; Garstang, M.; Beaubien, M. CPEX 2017: Utilizing the airborne Doppler aerosol wind lidar and dropsondes for convective process studies. In Proceedings of the 22nd Conference on Integrated Observing and Assimilation Systems for the Atmosphere, Oceans, and Land Surface, Austin, TX, USA, 7–11 January 2018; American Meteorological Society: Boston, MA, USA, 2018.

25. Emmitt, G.D. Hybrid technology Doppler wind lidar: Assessment of simulated data products for a space-based system concept. In Proceedings of the A Symposium on Lidar Remote Sensing for Industry and Environment Monitoring, Sendai, Japan, 9–12 October 2000; Singh, U.N., Asai, J., Ogawa, T., Itabe, T., Sugimoto, N., Eds.; International Society for Optical Engineering: Bellingham, WA, USA, 2000; p. 4153. [CrossRef]

26. Willoughby, H.E.; Chelmow, M.B. Objective determination of hurricane tracks from aircraft observations. *Mon. Weather Rev.* **1982**, *110*, 1298–1305. [CrossRef]

27. Weissmann, M.; Busen, R.; Dörnbrack, A.; Rahm, S.; Reitebuch, O. Targeted observations with an airborne wind lidar. *J. Atmos. Ocean. Technol.* **2015**, *22*, 1706–1719. [CrossRef]

28. Chouza, F.; Reitebuch, O.; Groß, S.; Rahm, S.; Freudenthaler, V.; Toledano, C.; Weinzierl, B. Retrieval of aerosol backscatter and extinction from airborne coherent Doppler wind lidar measurements. *Atmos. Meas. Technol.* **2015**, *8*, 2909–2926. [CrossRef]

29. Chouza, F.; Reitebuch, O.; Jähn, M.; Rahm, S.; Weinzierl, B. Vertical wind retrieved by airborne lidar and analysis of island induced gravity waves in combination with numerical models and in situ particle measurements. *Atmos. Chem. Phys.* **2016**, *16*, 4675–4692. [CrossRef]

30. Witschas, B.; Rahm, S.; Dörnbrack, A.; Wagner, J.; Rapp, M. Airborne wind lidar measurements of vertical and horizontal winds for the investigation of orographically induced gravity waves. *J. Atmos. Ocean. Technol.* **2017**, *34*, 1371–1386. [CrossRef]

31. Lux, O.; Lemmerz, C.; Weiler, F.; Marksteiner, U.; Witschas, B.; Rahm, S.; Schafler, A.; Reitebuch, O. Airborne wind lidar observations over the North Atlantic in 2016 for the pre-launch validation of the satellite mission Aeolus. *Atmos. Meas. Technol.* **2018**. [CrossRef]

32. Uhlhorn, E.W.; Black, P.G.; Franklin, J.L.; Goodberlet, M.; Carswell, J.; Goldstein, A.S. Hurricane surface wind measurements from an operational stepped frequency microwave radiometer. *Mon. Weather Rev.* **2007**, *135*, 3070–3085. [CrossRef]

33. Klotz, B.W.; Uhlhorn, E.W. Improved stepped frequency microwave radiometer tropical cyclone surface winds in heavy precipitation. *J. Atmos. Ocean. Technol.* **2014**, *31*, 2392–2408. [CrossRef]

34. Zhang, J.A.; Cione, J.J.; Kalina, E.A.; Uhlhorn, E.W.; Hock, T.; Smith, J.A. Observations of infrared sea surface temperature and air-sea interaction in Hurricane Edouard (2014) using GPS dropsondes. *J. Atmos. Ocean. Technol.* **2017**, *34*, 1333–1349. [CrossRef]

35. Emmitt, G.D. Airborne Doppler wind lidar atmospheric boundary layer research. In Proceedings of the Workshop on the Future of Boundary Layer Observing, Warrenton, VA, USA, 24–26 October 2017; National Academy of Sciences: Washington, DC, USA, 2017.

36. Lussier, L.L.; Montgomery, M.T.; Bell, M.M. The genesis of Typhoon Nuri as observed during the Tropical Cyclone Structure 2008 (TCS-08) field experiment–Part 3: Dynamics of low-level spin-up during the genesis. *Atmos. Chem. Phys.* **2014**, *14*, 8795–8812. [CrossRef]

37. Worsnop, R.; Bryan, G.H.; Lundquist, J.K.; Zhang, J.A. Using large-eddy simulations to define spectral and coherence characteristics of the hurricane boundary layer for wind energy applications. *Bound. Layer Meteorol.* **2017**, *165*, 55–86. [CrossRef]

remote sensing

MDPI

Article

On the Use of Dual-Doppler Radar Measurements for Very Short-Term Wind Power Forecasts

Laura Valldecabres [1],*, Nicolai Gayle Nygaard [2], Luis Vera-Tudela [1], Lueder von Bremen [1,3] and Martin Kühn [1]

[1]   Institute of Physics, ForWind—University of Oldenburg, Küpkersweg 70, 26129 Oldenburg, Germany; luis@veratudela.com (L.V.-T.); lueder.von.bremen@dlr.de (L.v.B.); martin.kuehn@forwind.de (M.K.)
[2]   Ørsted Wind Power, Kraftværksvej 53, 7000 Fredericia, Denmark; nicny@orsted.dk
[3]   DLR Institute of Networked Energy Systems, Carl von Ossietzky Straße 15, 26129 Oldenburg, Germany
*   Correspondence: laura.valldecabres@forwind.de

Received: 30 August 2018 ; Accepted: 24 October 2018; Published: 29 October 2018

check for updates

Abstract: Very short-term forecasts of wind power provide electricity market participants with extremely valuable information, especially in power systems with high penetration of wind energy. In very short-term horizons, statistical methods based on historical data are frequently used. This paper explores the use of dual-Doppler radar observations of wind speed and direction to derive five-minute ahead deterministic and probabilistic forecasts of wind power. An advection-based technique is introduced, which estimates the predictive densities of wind speed at the target wind turbine. In a case study, the proposed methodology is used to forecast the power generated by seven turbines in the North Sea with a temporal resolution of one minute. The radar-based forecast outperforms the persistence and climatology benchmarks in terms of overall forecasting skill. Results indicate that when a large spatial coverage of the inflow of the wind turbine is available, the proposed methodology is also able to generate reliable density forecasts. Future perspectives on the application of Doppler radar observations for very short-term wind power forecasting are discussed in this paper.

Keywords: Doppler radar; five-minute ahead wind power forecasting; probabilistic forecasting; remote sensing forecasting; offshore wind speed forecasting

---

1. Introduction

The increasing participation of offshore wind power in electricity markets continuously poses challenges for ensuring grid stability and power quality, especially due to the enhanced variability of offshore wind power in short scales [1]. A recent project in Germany allows wind farms to downregulate their power to participate in the reserve market, where reserves are calculated in one-minute intervals [2]. During the pilot phase of the project, the standard deviation of the prediction error of the possible power is required to be less than ±5%. These changes in grid regulations, driven by faster temporal responses of the grid, show the importance of improving wind power forecasts with shorter horizons and higher temporal resolutions in order to reduce the costs associated with operating the grid and minimize balancing reserves.

Reliable very short-term forecasts of wind power also benefit electricity market participants since they allow them to be more competitive in intraday markets [3]. As electricity markets are becoming more flexible with the use of intraday gate closure times as short as five minutes [4], the technical challenge arises on how to generate accurate very short-term wind power forecasts.

Very short-term or minute-scale forecasts of wind power (<60 min) are usually based on statistical time-series models. For larger horizons, numerical weather prediction (NWP) models are more accurate [5]. Examples of very short-term forecasts of wind speed and power using time series methods can be found in [6–8]. Very short-term forecasts of wind speed and power are often based on machine learning methods such as artificial neural networks [9] or Markov chain models [10,11], which are trained with historical data. Alternatively, combinations of different models (known as hybrid models) have been proposed to overcome the deficiencies of single models [12]. Parallel to the development of single point or deterministic forecasting models, the use of probabilistic forecasts, which include information about the uncertainty associated with the predicted events, has increased. Examples of very short-term probabilistic forecasts of wind power using statistical methods can be found in [13,14]. These probabilistic forecasts are known as predictive densities or probability distributions and provide important information for making risk-based decisions [15].

Over the last two decades, the use of remote sensing measurements such as long-range lidars [16] has been extended in the wind industry. These systems are capable of measuring wind speed and direction (under certain assumptions) up to 30 km [17]. Unlike conventional wind measurements from met-mast or satellites, they present an adequate trade-off between temporal and spatial resolution for wind farm applications. However, the prediction horizon of a remote sensing-based forecasting model is limited by the maximum range of the remote sensing measurements and also influenced by meteorological conditions [18]. Indeed, publications on the use of long-range lidar measurements for wind energy applications have reported measurements with a maximum range of less than ten kilometers [19–21]. Despite those limitations, a recent contribution showed that a lidar-based forecasting technique can provide better results than conventional statistical benchmarks when forecasting near-coastal winds with lead times of five minutes [22].

Alternatively, the use of weather radars has also been investigated for short-term (minutes to few hours) forecasting of wind power, due to its appropriate spatial and temporal resolution. Trombe et al. [23] introduced the use of ground-based weather radars, which measure precipitation reflectivity, for predicting large power fluctuations in the offshore wind farm Horns Rev. As precipitation fields are highly correlated to strong wind speed fluctuations, weather radar systems show a high potential in anticipating strong power fluctuations at offshore sites [24].

The use of wind speed and direction observations from Doppler radars for wind energy applications has recently been explored. Wind farm operational data was coupled with wind fields derived from dual-Doppler (two synchronized Doppler radars) measurements to further investigate wind farm wake effects and evaluate wake modelling [25]. Dual-Doppler (DD) radar measurements of the wake behind an offshore wind farm were also reported in [26]. Power performance studies [27] and ramp events detection [28] were also documented using DD radar measurements.

In [29], the authors introduced a methodology that uses DD radar measurements as input for predicting the average aggregated power of seven offshore wind turbines in a probabilistic framework. This paper extends the methodology presented in [29] to predict the power generated by individual wind turbines and analyses the characteristics of DD radar measurements for very short-term forecasts of wind power. Thus, the main goals of this paper are (i) the description of a methodology that uses fully resolved DD radar measurements to create probabilistic forecasts of wind power, (ii) its evaluation on seven wind turbines in an offshore wind farm with a very short-term horizon of five minutes and (iii) the assessment of the measurement characteristics and quality aspects of DD radar observations for very short-term forecast of wind power.

Our paper is organised as follows: first, we introduce the DD radar system along with the measurement campaign in Section 2. The forecasting methodology is detailed in Section 3. We investigate the use of probabilistic and deterministic forecasts of power for individual wind turbines and a row of wind turbines in Section 4. A discussion on the use of DD radar observations for wind power forecasting is given in Section 5. Conclusions are drawn in Section 6.

2. Data Description

The data used in this analysis was collected during the BEACon research and development (R&D) project conducted by Ørsted [26]. The BEACon measurement campaign started in July 2016, lasting until the spring of 2018. Two Doppler radar units scanned the flow within and surrounding the Westermost Rough offshore wind farm in the North Sea (Figure 1a,b). Westermost Rough is composed of 35 turbines with a hub height of 106 m, a rotor diameter (D) of 154 m and a rated power of 6 MW. The two Doppler radars (Figure 2) are located at the shoreline 8 km from the closest wind turbines (Figure 1b). The last row of turbines is 14 km from the coast.

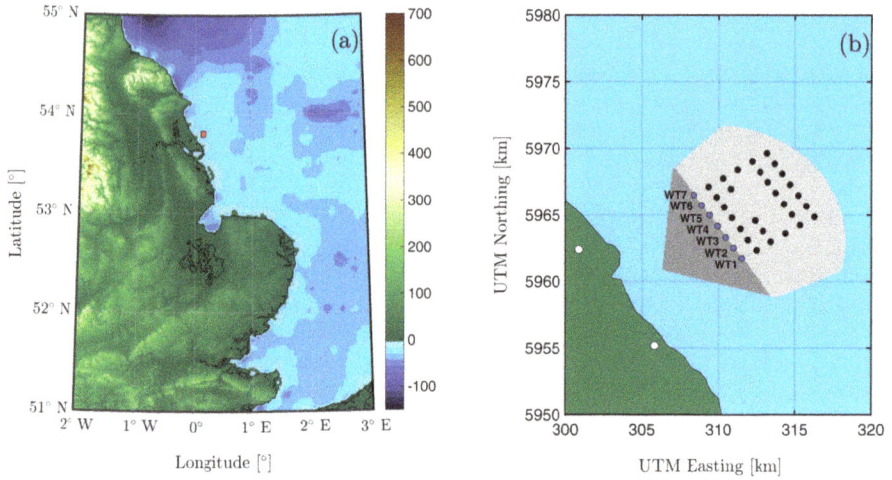

Figure 1. (**a**) location of the Westermost Rough wind farm (■), 8 km off the Holderness coast, in the North Sea. The colourbar indicates the height above mean sea level in meters; (**b**) layout of the wind farm showing the position of the radars (○) and the wind turbines (● and ●). Wind turbines used for this analysis (●) are labeled. The dark and light gray shadowed areas indicate the overlapping dual-Doppler measurement area.

Figure 2. Doppler radar unit deployed on the shore of the Westermost Rough wind farm.

The DD radar system [26] was configured to obtain volumetric wind field measurements with a temporal resolution of roughly one minute. Each radar scanned a 60° sector at 13 different elevation tilts, ranging from 0.2 to 1.4°. During the performance of a one-minute volumetric measurement cycle, flow homogeneity is assumed. The radars measured the flow with a spatial resolution of 0.5° in the azimuthal direction, given by the beam width, and 15 m along the beam direction. To retrieve the two horizontal wind speed components, the line-of-sight (radial) velocity measurements from the two radars are interpolated into a three-dimensional Cartesian grid over the overlapping region measured by the radars (Figure 1b). It is assumed that no vertical flow component exists. The maximum range of the radar measurements is 32 km. After interpolation into the Cartesian grid, the final DD volumetric wind fields have a spatial horizontal resolution of 50 m and vertical resolution of 25 m. The accuracy of the radial velocity measurements of one of the Doppler radars has been validated in a measurement campaign using a scanning lidar, which had been previously calibrated with an anemometer. Results indicate a good correlation between the Doppler radar and the scanning lidar measurements [30]. The uncertainty in the dual-Doppler wind speed data is currently being quantified and will be reported elsewhere.

A sample of 2795 one-minute DD radar measurements during south-westerly winds was made available for this study. Those periods were continuously collected during several hours on 11 days between November 2016 and February 2017 and correspond to free-flow conditions (DD wind direction: 191.7–281.7°) for the first row of wind turbines (Figure 1b). We focus our analysis only on these wind turbines to avoid the additional complication of wake effects. The horizontal spacing of those wind turbines ranges from 5.92D to 6.43D. The DD radar images were filtered for periods with spatial availability greater than 90% in the area corresponding to the inflow zone of the first row of wind turbines (dark gray shadowed area in Figure 1b). Only continuous periods of at least 20 min were considered. With these criterion applied, a total number of 1134 one-minute DD radar wind fields, with a mean spatial availability in the inflow area of 98.4%, were further investigated. During these periods, mean wind speeds at 100 m height (averaged over the radar domain) in the range of 5–14 m/s with a prevailing south-southwesterly direction were observed (Figure 3).

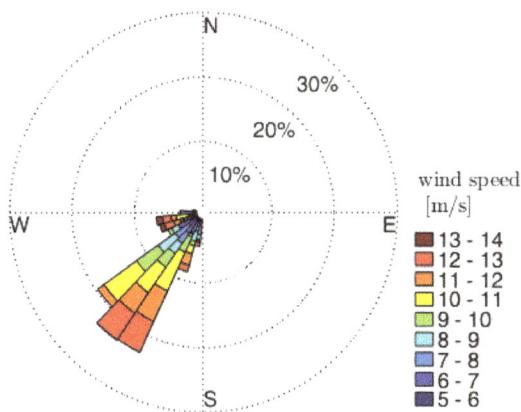

Figure 3. Wind rose of one-minute mean wind speeds at the height of 100 m (averaged over the radar domain) observed by the dual-Doppler radar system during the period covered in this analysis.

In addition to the DD wind field data, power data from the wind turbines' supervisory control and data acquisition (SCADA) system is used to derive the wind turbine power curve and to validate the performance of the forecast. The power data with a frequency of 1 Hz is averaged every minute and temporally synchronized with the DD data. Periods with start-ups, shut-downs and abnormal performance such as power curtailment, power boosting and downtimes have been filtered.

3. Forecasting Methodology

The forecasting methodology presented here uses fully-resolved DD wind field measurements as input to derive probabilistic forecasts of wind power. Figure 4 outlines the proposed methodology. To calculate a remote sensing-based forecast (RF) of power for the ith wind turbine P_i^{RF} at a future time $t + k$, we need first to estimate the density forecast of the hub height wind speed ws_i^{RF} at instant $t + k$. Then, the predictive density of wind speed at each rotor is transformed to power density by means of a probabilistic power curve. Due to the prevailing wind direction of the available data set and the particular measurement setup of the BEACon project focusing on wake effects, we forecast wind power output with a five-minute prediction horizon, i.e., five-steps ahead ($k = 5$). Here, we first introduce how the predictive densities of wind speed are estimated.

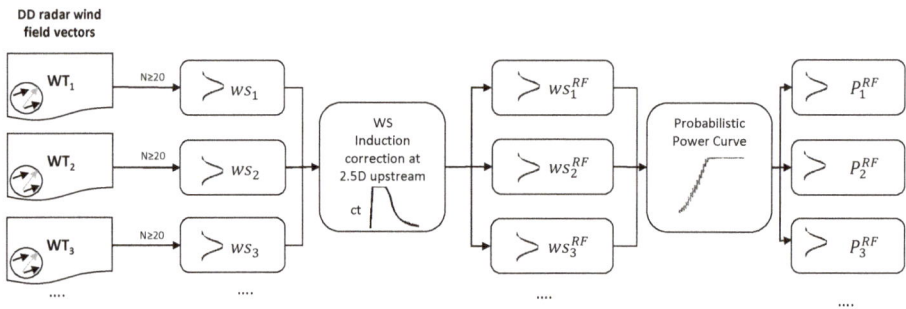

Figure 4. Scheme of the remote sensing probabilistic forecasting model (RF) showing the unmodified wind speed predictive densities (ws_i), the wind speed densities after correcting for induction effects (ws_i^{RF}) and the predictive densities of power (P_i^{RF}).

3.1. Predictive Wind Speed Densities

The probabilistic forecast of wind speed is based on a Lagrangian persistence technique widely applied in probabilistic forecasts of precipitation [31,32]. The principle underlying this method is the persistence of moving radar precipitation patterns. Our model uses wind speed information measured by the DD radar system at 100 m height to create the probabilistic forecast of wind speed. Thus, given a DD radar observed wind field at any given time t, the model propagates the horizontal wind field vectors with their respective trajectories defined by their local wind speed and direction for a duration k. The probabilistic wind speed advection forecast is constructed under the following premises: During the prediction horizon, (i) the observed DD radar wind field vectors maintain a constant horizontal trajectory (ii) mass conservation, vorticity and diffusion are neglected.

A simple approach to generate the predictive wind speed densities ws_i at time $t + k$ at the target wind turbine is to search in the surroundings of the wind turbine for the velocity vectors advected during the forecast horizon. The ensemble of the magnitude of the advected velocity vectors comprises the distribution of wind speeds or predictive densities of wind speed.

The forecasting location or point of interest is defined as an area of influence A_i encompassing the target wind turbine. We only consider the wind vectors from the DD scan at 100 m that will reach the area of influence within a time window of $\tau = 60$ s centred around the forecast horizon. Thus, the basis of our probabilistic very short-term forecast of wind speed is the cloud of points or observations that fall inside the defined spatio-temporal window. For a forecasting horizon of five minutes, as it is the case here, we consider the wind field vectors found within the area of influence between 4.5 min and 5.5 min after the forecast is issued, under the Lagrangian persistence hypothesis.

To evaluate the predictive wind speed densities, we compare them with the DD wind speeds observed in front of the rotor (at 100 m height) of the considered wind turbines. Close to the rotor, a reduction of wind speed is experienced due to the extraction of axial momentum of the flow. The axial induction factor a expresses the velocity reduction at the rotor and can be defined in terms of the thrust coefficient C_T by:

$$a = \frac{1}{2}\left(1 - \sqrt{1 - C_T}\right).$$ (1)

The International Electrotechnical Commission (IEC) standard for power curve measurements [33] recommends rotor wind speed observations to be measured $2.5D$ upstream of the rotor. At this distance, the flow is assumed to be "outside" the induction zone. The power curve used in this analysis and introduced in Section 3.2 is based on DD measurements at this distance, at the height of 100 m. Wind tunnel measurements have shown, however, that wind speed deficits due to the rotor blockage effect extend to $3D$ and beyond upstream of the wind turbine [34]. In considering the velocity reduction due to the induction zone in front of the rotor, we correct our wind speed distributions ws_i using Equation (2), [34], at the distance of $x = -2.5D$ upstream of the wind turbine, where x is the spatial coordinate in the longitudinal flow direction. In our case, U_∞ refers to the predicted DD wind speed ws_i based on observations upstream (or in the undisturbed zone) and U (here ws_i^{RF}) denotes the wind speed at the distance of $2.5D$ upstream of the rotor. As mentioned before, ws_i^{RF} will be transformed into power by using a power curve based on measurements at this distance. The induction factor is obtained from Equation (1), using the C_T given by the manufacturer's thrust curve at each wind speed level. For an induction factor $a = \frac{1}{3}$, the correction factor at $x = -2.5D$ equals 0.994, which is close to unity:

$$\frac{ws_i^{RF}}{ws_i} = \frac{U}{U_\infty} = 1 - a\left(1 + \frac{2x}{D}\left(1 + \left(\frac{2x}{D}\right)^2\right)^{-0.5}\right).$$ (2)

3.1.1. Optimization of the Wind Turbine Area of Influence

To optimize the probabilistic forecasting methodology, we conducted a sensitivity analysis on the area of influence encompassing the target wind turbine, as introduced in [29]. The area of influence A_i is defined as a circle centred at the wind turbine (Figure 5c). The optimization criterion is the minimization of the average continuous ranked probability score (CRPS) of the wind speed predictions, after correcting for induction effects of the wind turbines in the first row. The continuos ranked probability score (crps) evaluates the spread of the predictive densities in regard to the observation [35] and is given by:

$$\text{crps}\,(F, x_0) = \int_{-\infty}^{\infty} [F(x) - \theta(x - x_0)]^2\, dx,$$ (3)

where F is the cumulative distribution function of the predictive density (in our case ws_i^{RF}), x_0 is the observation (here $ws_i^{2.5D}$) and θ is the Heaviside step function which takes the value 1 when $x \geq x_0$ and 0 otherwise. If the predictive density is reduced to a point forecast, then the crps can be understood as the absolute error. The lower the crps, the better the density forecast. For a skillful forecast, crps is close to zero, as F approximates the Heaviside step function of the observation. CRPS is given by:

$$\text{CRPS} = \frac{1}{T}\sum_{j=1}^{T} \text{crps}(F_j, x_{0,j}),$$ (4)

where T refers to the number of samples analysed. To optimize the area of influence, we use wind speed densities predicted one-minute ahead ($k = 1$), as we want to optimize the area of influence with predictions close to the real distribution of wind speeds in front of the rotor.

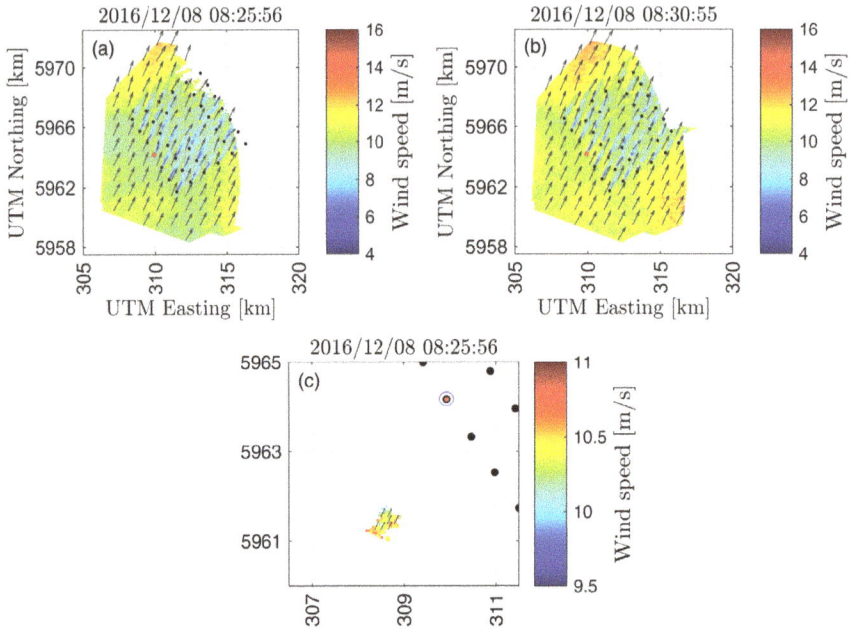

Figure 5. Wind speed forecast for *WT*4 (marked in red). (**a**) Dual-Doppler flow field at the time that the forecast is issued and (**b**) validated; (**c**) cloud of wind vectors used to derive the probabilistic forecast for *WT*4 and the respective area of influence (blue circle). Notice the different scales on Figure 5c.

Figure 6 depicts the CRPS for different areas of influence, in terms of number of rotor diameters *D*. Here, we consider samples with a minimum number of wind field vectors N_{min} = 20 to estimate the predictive densities, following [29]. A total of *T* = 10,447 predictions are considered for the sensitivity analysis. The figure also depicts the CRPS for a wind speed distribution estimated with the probabilistic extension of the forecasting model persistence. Persistence is the most often employed forecasting benchmark, which uses the last available measurement at time *t*, as the prediction at time *t* + *k*, and is known for being difficult to outperform for short horizons [5]. As we want to compare our model to a probabilistic prediction, we consider a persistence distribution using the persistence point and the 19 previous forecasting errors, as defined in [36]. The CRPS for the predictions based on the DD observations is smaller than for persistence. A decrease in the CRPS value can be observed when increasing the diameter of the area of influence up to 2*D*. Further increase of the area of influence shows no improvement. An area of influence larger than the optimum appears to extend the temporal and spatial characteristics of the wind speed distribution, which seem to be no longer representative of the sampling effect of the rotor. At the same time, a smaller area of influence than the optimum is not able to capture the distribution of wind speeds characteristic of the rotor.

Based on the results of the sensitivity analysis of the area of influence, we use an area of influence with a diameter of 2*D*, for the results presented below. Figure 5c shows an example of the cloud of wind field vectors that will reach the area of influence of the wind turbine *WT*4 (in red) in 5 min ± 30 s. The wind fields at the time that the prediction is issued at and validated are illustrated in Figure 5a,b, respectively. The predicted wind speed distribution after implementing the correction due to the induction zone, along with its mean and the observed DD wind speed 2.5*D* upstream of the rotor at the validation time are depicted in Figure 7a.

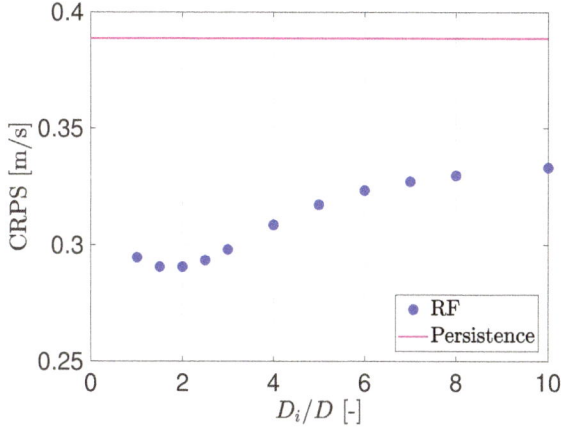

Figure 6. Average continuous ranked probability score (CRPS) for the one-minute ahead wind speed predictive densities for different areas of influence A_i with the remote-sensing forecasting (RF) model (blue dots) and for a probabilistic persistence method (magenta line). The area of influence is expressed in number of rotor diameters (D).

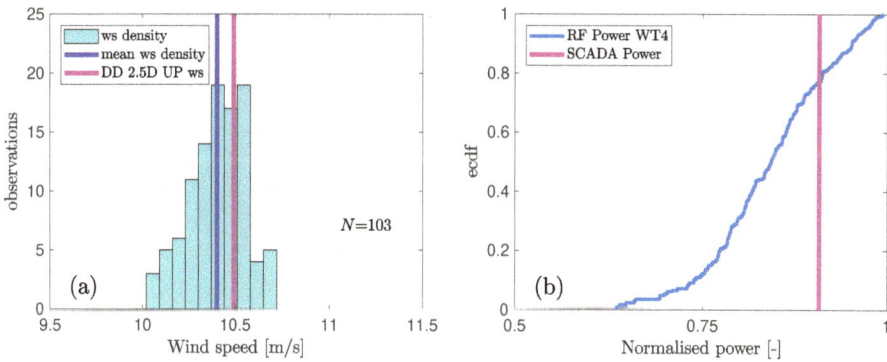

Figure 7. (**a**) predictive histogram distribution of wind speeds for *WT4* at the time shown in Figure 5b. The blue line represents the mean of the distribution and the magenta line the verifying dual-Doppler wind speed 2.5D upstream of the rotor. N indicates the number of wind field vectors; (**b**) predictive empirical cumulative distribution function of the normalised power for *WT4* at the same time. The magenta line indicates the observed power.

3.1.2. Evaluation of Predicted Wind Speeds

Five-minute ahead single point predicted wind speeds ws_i^{RF} are evaluated by comparison with the wind speeds observed 2.5D upstream of the turbine rotor $ws_i^{2.5D}$. Following the perspective of [36] and [13], the optimal single point predictor should be chosen from the predictive densities according to the target metric. Given that the focus of the paper is on evaluating the root-mean-square error (RMSE) of the predicted variables, the mean of the distribution is considered. Figure 8a–c compare the wind speeds observed 2.5D upstream of the turbine rotor with the mean of the predictive wind speed densities forecasted five minutes ahead, before and after applying the wind speed correction due to induction effects, and for the persistence method, respectively. The correction of wind speeds considering the induction effects improves the RMSE by 2%. Relative to persistence there is an improvement of 6%. In general, a high correlation is found between the predicted and the observed wind speeds. For wind speeds below 6 m/s, the DD predicted mean wind speeds (including the induction correction) overestimate the DD wind speeds $2.5D$ upstream by an average of 0.32 m/s. For wind speeds in the range 6–10 m/s, the predicted wind speeds exceed the DD wind speeds $2.5D$ upstream by an average of 0.15 m/s. Over 10 m/s, this difference is 0.12 m/s. Those differences are attributed, among others, to the assumption of the persistence of the wind field trajectories during the forecasting horizon and to the radar uncertainty.

Although this work focuses on predicting coastal winds, we are not considering the effects of the wind speed gradient present at the discontinuity between the land and the sea, which increases the uncertainty in the predictions. Studies on coastal gradients [37] have reported velocity reductions between 4% and 8% from 3 km to 1 km from the shore. Over 3 km from the coast, velocity gradients of 0.5%/km have been observed for different offshore sites [38]. Future work should include corrections for wind speeds due to coastal effects.

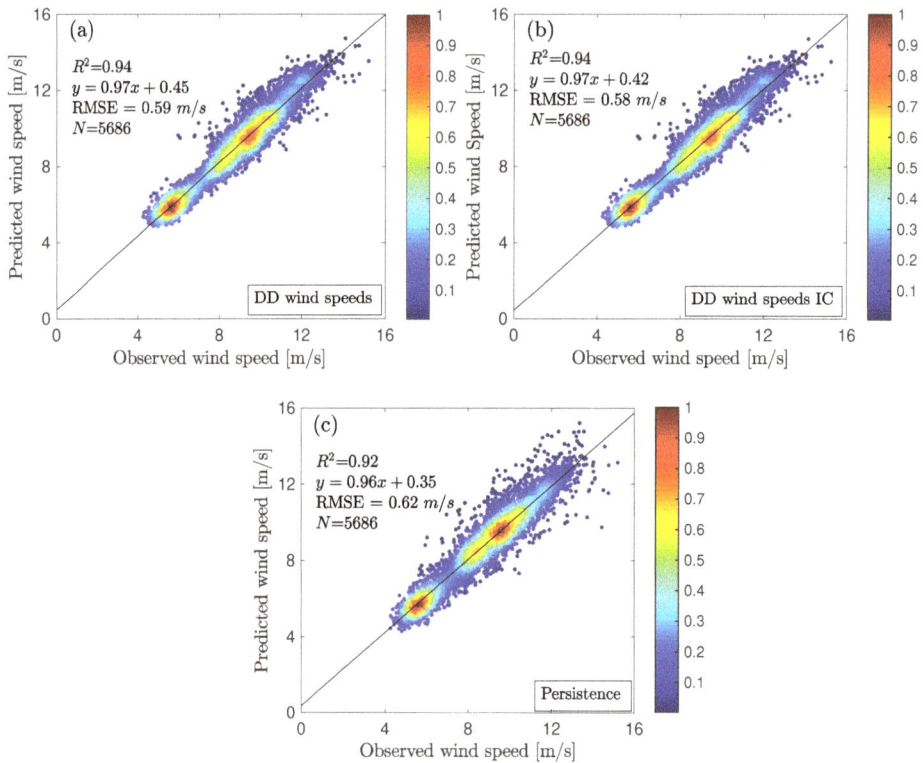

Figure 8. (**a**) density scatter plot of the dual-Doppler wind speeds $2.5D$ upstream of the wind turbine rotor (observed wind speed) and the mean of the five-minute ahead wind speeds distributions without velocity correction due to induction effects, (**b**) including the velocity correction due to induction effects and (**c**) for persistence.

3.2. Predictive Wind Power Densities

To estimate the predictive wind power densities P_i^{RF}, we transform the predictive wind speed densities ws_i^{RF} into power densities using a probabilistic wind turbine power curve. Wind turbine power curves are normally built using ten-minute averages of wind speed and power, following an IEC standard [33]. Since our goal is to forecast wind power with a frequency of one minute, we derive the power curve from the one-minute DD wind speeds observed $2.5D$ upstream of the rotor of the considered wind turbines (Figure 9). This power curve is representative of an undisturbed inflow as only free-flow sectors are considered. The resulting power curve presents an irregular degree of variance for different wind speed levels, as shown in Figure 9. This is characteristic of a power curve constructed with high frequency data [39]. To include the uncertainty of the power curve in our forecasting model, a probabilistic power curve is built. First, we collect the power data into wind speed bins of 0.5 m/s width. Next, we estimate the empirical cumulative distribution function (*ecdf*) of the wind turbine power for each wind speed bin. We apply a resampling technique with replacement (bootstrap) [40] to derive the wind power predictive distributions, as wind speed predictive densities are based on an irregular number of wind field vectors. Thus, for each wind speed distribution, a total of 10,000 random values, out of the original *ecdf* of the predicted wind speed distributions, are selected. With the resulting set of wind speed values, the predictive densities of power are derived by random selection from the power *ecdf* associated to each wind speed bin. The estimated *ecdf* of the predictive density of power for *WT*4, for the example introduced in Figure 5, is shown in Figure 7b.

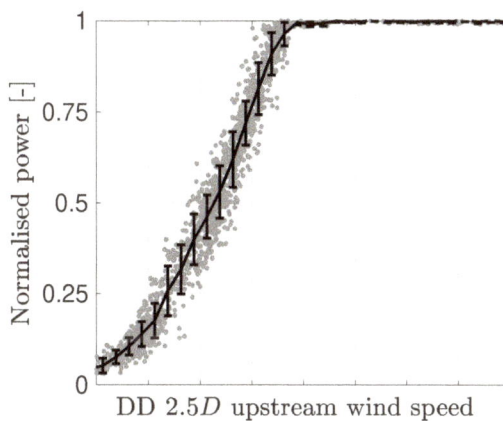

Figure 9. Normalised wind turbine power curve based on 1656 samples of dual-Doppler wind speeds 2.5*D* upstream of the rotor, at 100 m height (first row of wind turbines). The line is the binned mean power. The error bars represent the standard deviation in 0.5 m/s wind speed bins.

4. Results

We assess the predictive performance of the RF model in terms of single point and probabilistic forecasts. Five-minute ahead predictions of power from the wind turbines of the first row and its aggregation are evaluated. The forecasting skill of the RF model is compared with the benchmarks persistence and climatology, which are described below. An analysis of the influence of the radar spatial availability on the performance of the RF model is also included.

4.1. Probabilistic Forecast Evaluation

To evaluate our probabilistic forecast, we follow Gneiting guidelines [35]. A probabilistic forecast aims at maximizing the sharpness of the predictive distributions under the constraint of calibration. Sharpness refers to the spread of the predictive distributions and is only a property of the forecasting variable. Calibration, however, is a joint property of the forecasts and observations and indicates the statistical consistency between the predictive distributions and the observed values. To evaluate the skill of the RF model we compare its sharpness and calibration with the probabilistic version of persistence and the climatology benchmarks. Climatology is the most common benchmark to assess climatological variables and can be understood as the average of the variable during a long period. Here, we define the climatology distribution as the probability distribution of all available

SCADA power measurements. We derive the persistence distribution, defined in Section 3.1.1, using the persistence point forecast and the 19 most recent consecutive observed values of the persistence error, as described in [36].

4.1.1. Individual Wind Turbine Power

The overall performance of the predictive power densities of the seven wind turbines is assessed with the CRPS, previously defined in Section 3.1.1, but in this case using power. CRPS addresses both calibration and sharpness. Due to the reduced availability of time stamps with simultaneous measurements for the seven wind turbines, we first explore the results for all measurements available for each wind turbine. The respective results are given in % of nominal power P_n (Table 1, upper row). In general, the CRPS of the predictions with the RF model for individual wind turbines outperforms persistence and climatology. When evaluating the predictions made with the climatology and persistence models, a similar CRPS is found for each wind turbine, except for *WT*1 which shows a worse performance in general. We assume that these differences stem from the different operational behaviour of the wind turbines. In contrast to the benchmarks, the CRPS of the RF model shows higher variability among the wind turbines. Wind turbines *WT*1 and *WT*7 show higher CRPS values than the other wind turbines whilst *WT*3 performs best. Figure 10 depicts a one-hour episode of the RF model of power for *WT*3 together with the observed power. As stated before, the forecast is generated every minute. A strong decrease of power during the first half hour of the event is shown, which is properly captured by the RF model. To assess all wind turbines equally, we reduce the number of samples to the periods where all wind turbines operate simultaneously (Table 1, lower row) and there are available forecasts ($T = 343$). In general, RF is more skillful than persistence, except for *WT*1.

Table 1. Average continous ranked probability score (CRPS), in % of the nominal capacity (P_n), of the five-minute ahead power forecasts for the seven wind turbines evaluated. Results are shown for the remote sensing-based forecast (RF), persistence and climatology benchmarks. Upper row presents the results for all available measurements (T) of each wind turbine. Lower row provides the results for all simultaneous available measurements. Minimum values are shown in bold.

		*WT*1	*WT*2	*WT*3	*WT*4	*WT*5	*WT*6	*WT*7
	T	596	903	997	983	960	846	608
	RF	**5.11**	**4.06**	**3.92**	**4.57**	**4.50**	**4.29**	**4.82**
CRPS [%]	Persistence	5.37	4.98	5.04	5.07	4.96	4.88	4.85
	Climatology	17.93	17.45	17.32	17.15	17.07	16.96	16.79
	T	343	343	343	343	343	343	343
	RF	6.42	**5.03**	**4.83**	**5.27**	**4.76**	**4.39**	**4.76**
CRPS [%]	Persistence	**6.12**	5.79	5.63	5.76	5.76	5.85	5.91
	Climatology	16.78	16.31	15.58	15.37	15.32	15.03	14.94

When assessing calibration of predictive densities, the use of quantile–quantile reliability diagrams is recommended [41]. In a reliable probabilistic forecast model, *x*% of the observations should be below the xth percentile of the distributions, as in the diagonals of Figure 11. However, the sample size of the evaluated forecast strongly influences the reliability of the diagrams, and even if the forecasts are highly reliable, a reduced sample size can lead to a reliability diagram deviating from the diagonal. Therefore, 95% consistency bars are generated following the work of Bröcker and Smith [42]. Here, we evaluate the predictions in quantile intervals with steps of 5%. Climatology is not represented since it has perfect reliability when evaluated over the whole sample set, as predictive densities are directly derived from the observations.

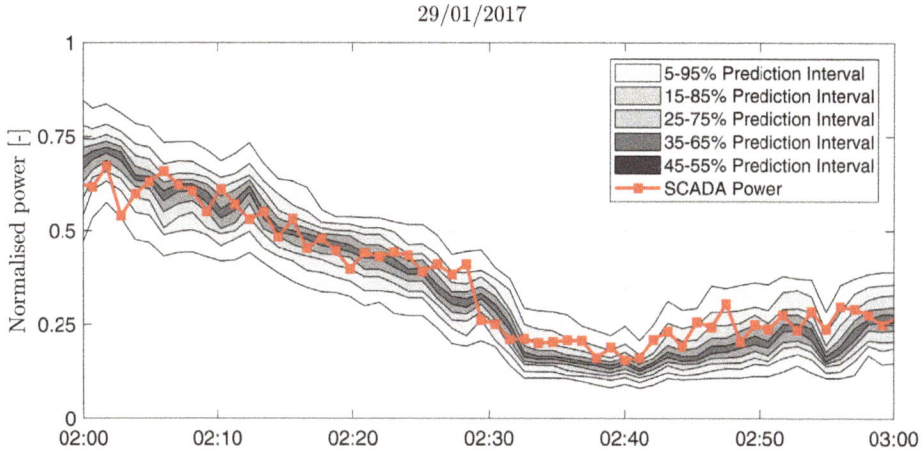

Figure 10. A 60 min episode of five-minute ahead predictions of normalised power for WT3 with the remote sensing-based forecasting model (RF). Prediction intervals are shown together with the observed power (red squares).

Figure 11a illustrates the reliability diagram of the RF model for the seven wind turbines of the first row, considering all available measurements. In the legend, the number of periods evaluated for each wind turbine (T), along with the average number of DD wind field vectors used to derive the predictive densities ($\overline{N}$) is included. Wind turbines WT3, WT4 and WT5 show reliable forecasts, as their reliability diagrams are close to the diagonal and most of the evaluated quantiles fall within the confidence intervals. In Figure 11b, the reliability diagram of the seven wind turbines when forecasted with the persistence method is depicted. The persistence method shows a poor calibration since nearly 18% of the observations are below the 5% quantile, while only around 88% of the observations are below the 95% quantile. In addition, over half of the evaluated quantiles do not lie within the 95% intervals around the diagonal. Contrary to the RF method, little differences are found among the reliability of the seven wind turbines. Figure 11c,d illustrate the previous reliability diagrams but limiting the analysis to simultaneous periods. In the case of the RF model, WT1, WT2, WT6 and WT7 also highly deviate from the diagonal, while the central wind turbines show a consistently reliable performance. Again, little differences are found among the seven wind turbines for the persistence model, but a generally worse performance than the the RF model is distinguished for wind turbines WT3, WT4 and WT5.

These results allow us to infer that the reliability of the RF model is not directly affected by the reduced sample size of the evaluation set, but by the individual position of each wind turbine within the radar measurement domain. Indeed, the fact that WT3, WT4 and WT5 lie closer to the center of the radar image and have a larger area from which the wind vectors can be advected to the target wind turbine, results in a more skillful forecast for those wind turbines. This hypothesis is further investigated in the following subsection by limiting the spatial availability of the radar measurements.

4.2. Analysis on Limited Radar Availability

Here, we conduct a further analysis of the probabilistic performance of the RF model for individual wind turbines, relating the spatial coverage of the radar scan to the inflow area of each wind turbine, where potential wind vectors can originate from.

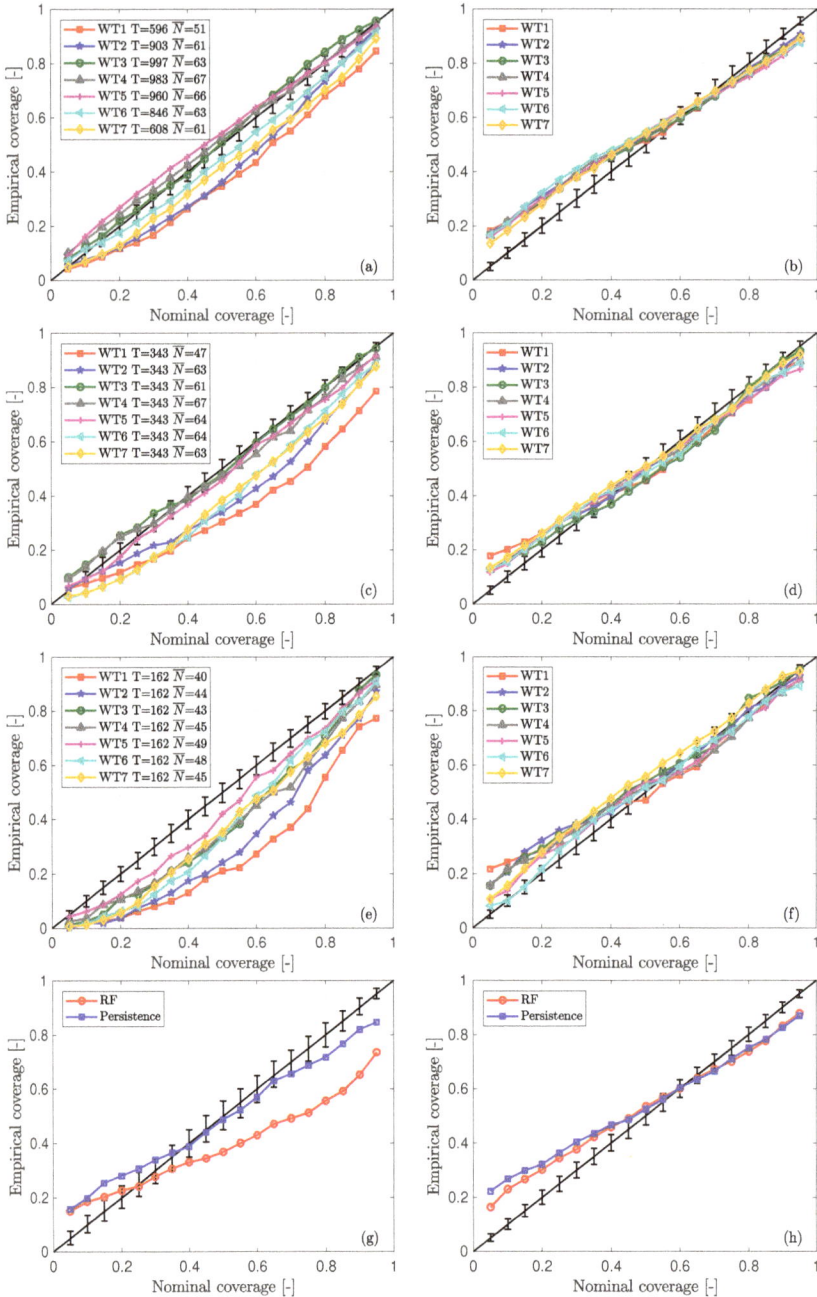

Figure 11. Reliability diagram for all wind turbines (WTs) during available measurements (*T*) for (**a**) the remote sensing-based forecasting model (RF) and (**b**) persistence model, simultaneous periods for (**c**) the RF and (**d**) persistence model, a reduced radar availability case for (**e**) RF and (**f**) persistence model and (**g**) the aggregation of *WT*1 to *WT*7 and (**h**) *WT*3 to *WT*5 for both models. $\overline{N}$ indicates the average number of wind field vectors conforming the wind speed distributions of the RF model. In addition, 95% consistency bars are indicated by the error bars.

Figure 12a depicts the wind speeds that could be observed by each wind turbine for a given south–south westerly direction, setting a forecasting horizon of five minutes. As it can be seen, the turbines in the center have an advantageous positioning within the radar scan, as the range of wind speeds that can be observed is larger than that of the outer wind turbines. To test the hypothesis that the position of the wind turbines with respect to the radar domain influences the reliability of the RF model, we conduct a further experiment. Data measured in the furthest section of the radar domain (dark gray are in Figure 12b) are discarded, as if the radars no longer measure that section.

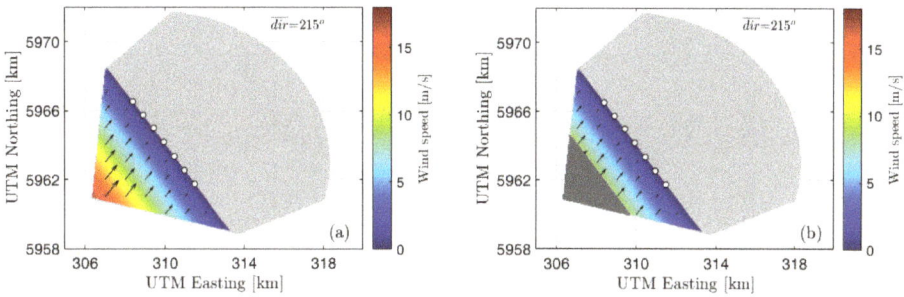

Figure 12. Wind speeds to be forecasted at the wind turbines (white dots) for a horizon of five minutes, a south-southwesterly wind direction and the radar available distance for (**a**) full radar availability; (**b**) limited radar availability.

The results for the probabilistic forecast are presented in Table 2. Regarding the overall skill, the RF performs better than the benchmarks in all WTs except for WT1.

Table 2. Average continous ranked probability score (CRPS) of five-minute ahead forecasts of power for the seven wind turbines evaluated, in the case of a reduced radar availability (Figure 12b). Results (in % of the nominal capacity (P_n)) are shown for the remote sensing-based forecasting (RF) model, persistence and climatology benchmarks.

		WT1	WT2	WT3	WT4	WT5	WT6	WT7
	T	162	162	162	162	162	162	162
CRPS [%]	RF	5.09	3.75	3.45	4.25	3.91	3.49	3.99
	Persistence	4.58	4.56	4.41	4.84	4.10	3.75	4.36
	Climatology	20.73	19.86	19.19	18.83	18.51	17.97	17.57

As for the calibration, Figure 11e shows the reliability diagram for the RF model, in the case of having a reduced radar spatial coverage. It can be seen that the reliability of wind turbines *WT3*, *WT4* and *WT5* strongly decreases when the radar available area is reduced. Furthermore, the reliability of the other wind turbines is almost unchanged. For persistence (Figure 11f), small differences are found among the wind turbines, but here the reduced sample size analysed ($T = 162$) also influences the calibration.

4.2.1. Wind Farm Row Aggregated Power Output

In this section, we evaluate the probabilistic forecast for the average aggregated power produced by the wind turbines. The aggregated power is calculated adding the power distributions of the wind turbines using again a bootstrap resampling technique. Here, we evaluate the aggregated power of the seven wind turbines ($\overline{P}_{17}$) and the aggregation of the central wind turbines *WT3* to *WT5* ($\overline{P}_{35}$). The average aggregated power for both cases is given below:

$$\overline{P}_{17} = \frac{1}{7}\sum_{i=1}^{7} P_i \quad \text{and} \quad \overline{P}_{35} = \frac{1}{3}\sum_{i=3}^{5} P_i. \tag{5}$$

Table 3 summarizes the results for the aggregation of the wind turbines $WT1$ to $WT7$ ($\overline{P}_{17}$) and $WT3$ to $WT5$ ($\overline{P}_{35}$). The RF model has the lowest CRPS when evaluating the aggregation in both cases.

Table 3. Average continous ranked probability score (CRPS), in % of the nominal capacity (P_n), for the five-minute ahead predictive densities of aggregated power of different sets of wind turbines. T indicates the sample size evaluated. Minimum values are shown in bold. Results are shown for the remote sensing-based forecasting (RF) model, persistence and climatology benchmarks.

		$\overline{P}_{17}$	$\overline{P}_{35}$
	T	340	902
	RF	**4.19**	**3.87**
CRPS [%]	Persistence	4.98	4.75
	Climatology	15.51	16.61

Regarding reliability, the quantile–quantile reliability diagram of the RF model for $\overline{P}_{17}$ (Figure 11g) performs worse than the probabilistic extension of persistence. We attribute this results to the fact that there are large differences in the RF model for the seven wind turbines, as mentioned in Section 4.1.1. As $WT3$, $WT4$ and $WT5$ are located at a more favourable position in the radar scanned area, in terms of inflow measurements coverage for the prevailing south–southwesterly direction, we further evaluate the aggregation of power considering only those wind turbines. Figure 11h depicts the reliability diagram of the RF model for the aggregation of wind turbines $WT3$ to $WT5$. In this case, the RF model is better calibrated than that of the $\overline{P}_{17}$ case, and presents even a better performance than persistence. However, in all cases reliability diagrams fall outside of the confidence intervals and it is clear that further work needs to be conducted to increase the reliability of the RF forecasts. However, given that spatio-temporal correlations among wind turbines can not be neglected, a different method to generate confidence intervals should be considered to draw more solid conclusions about the reliability of the evaluated methods for aggregated wind power.

4.3. Evaluation of Single Point Predictions

In this section, we evaluate the performance of single point or deterministic forecasts of wind power for individual and aggregated wind turbines and compare them with the persistence and climatology benchmarks. As stated before, the classical persistence reference forecast is a naive predictor, since it assumes that there is no change in the predicted variable. For climatology, we use the mean of the climatology distributions previously defined. Given that our target evaluation score is the RMSE of the produced power, we use the mean of the predictive density as the single point forecast.

4.3.1. Individual Wind Turbine Power

Table 4 compares the normalised root-mean-square error (NRMSE) of the five-minute ahead power predictions based on the RF model with persistence and climatology for the first row of wind turbines. The metric is normalised with the nominal power of the wind turbines. We show the results considering all periods available for each wind turbine. In four of the seven wind turbines, the NRMSE of the RF model is lower than that of the persistence method. For $WT1$ and $WT7$, persistence is better. For $WT4$, small differences between RF and persistence are found.

Table 4. Normalised root-mean-square-error (NRMSE), in % of the nominal capacity (P_n), of the five-minute ahead forecasts for the seven turbines of the first row. T indicates the number of periods evaluated. Minimum values are shown in bold. Results are shown for the remote sensing-based forecasting (RF) model, persistence and climatology benchmarks.

		WT1	WT2	WT3	WT4	WT5	WT6	WT7
	T	596	903	997	983	960	846	608
NRMSE [%]	RF	10.06	**7.90**	**7.70**	8.93	**8.50**	**8.16**	9.06
	Persistence	**9.40**	8.54	8.68	**8.89**	8.81	8.49	**8.80**
	Climatology	31.34	30.39	30.13	29.84	29.66	29.49	29.19

The way the RF model is able to anticipate strong variations of wind power, given by sudden changes in the flow can be clearly illustrated by exploring some interesting events. Figure 13 shows an episode of nearly 45 min of observed and predicted power for the wind turbines WT1 and WT6 together with the radar images at two time instants. On the left scan, a relatively homogeneous wind field is approaching the wind turbines in the first row, with slightly lower wind speeds in front of WT1 and WT2. Only five minutes later, higher wind speeds are experienced close to wind turbines WT6 and WT7, where two elongated streaks hit those wind turbines. This second period can be clearly identified in the time series at the bottom of Figure 13, where WT6 produces nearly double the power than WT1. The upcoming coherent structures can be observed in the top left image, in front of WT6. This episode highlights the importance of using observations for very short-term forecasting. A naive model such as persistence will predict the arrival of the increase in power for WT6 with a delay of five minutes.

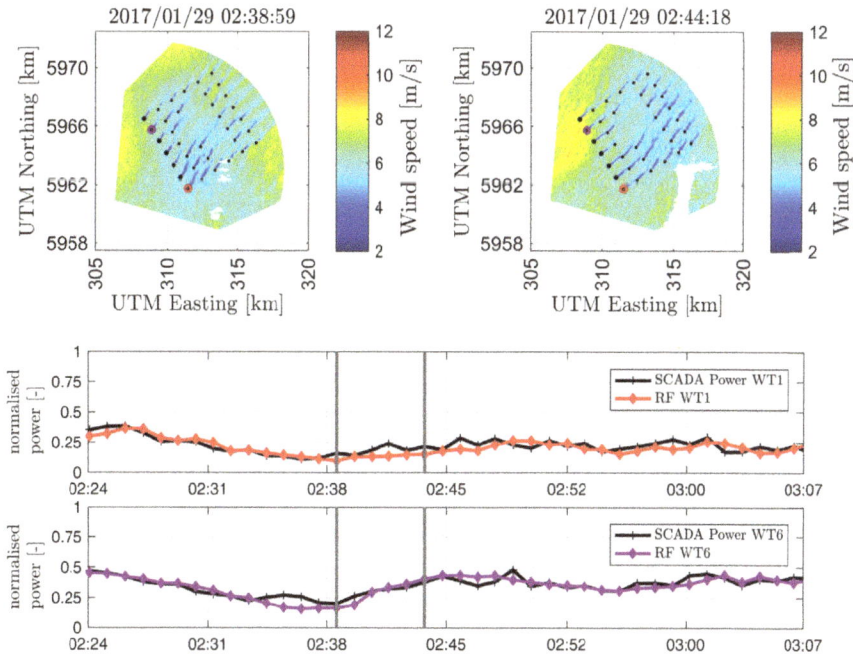

Figure 13. (**Top**) two radar observed wind fields about five minutes apart; (**Bottom**) time series of power produced and predicted with the remote sensing-based forecasting (RF) model for the wind turbines WT1 and WT6. Grey lines indicate the two timestamps above the top panels.

In Table 5, the results for the single point prediction, in the case of having a reduced radar availability are also listed. In this case, the limited radar domain spatial availability results in the RF model performing worse than in the case of having a full spatial radar availability.

Table 5. Normalised root-mean-square-error (NRMSE), in % of the nominal capacity (P_n), of five-minute ahead forecasts for the seven turbines of the first row, for the case of a reduced radar availability (Figure 12b). Results are shown for the remote sensing-based forecasting (RF) model, persistence and climatology benchmarks.

		WT1	*WT2*	*WT3*	*WT4*	*WT5*	*WT6*	*WT7*
	T	162	162	162	162	162	162	162
NRMSE [%]	RF	10.09	**7.35**	**6.83**	8.89	8.01	7.09	**8.01**
	Persistence	**8.02**	7.89	7.48	**8.74**	**7.86**	**6.93**	8.56
	Climatology	35.49	34.15	33.08	32.51	32.04	31.00	30.51

4.3.2. Wind Farm Row Aggregated Power Output

Table 6 comprises the NRMSE for the case of aggregated power. The RF model exhibits the lowest NRMSE compared to climatology and persistence for both aggregated cases, $\overline{P}_{17}$ and $\overline{P}_{35}$. In the case of the aggregation of wind turbines *WT3* to *WT5*, persistence is also a competitive model.

Table 6. Normalised root-mean-square-error (NRMSE), in % of the nominal capacity (P_n), for five-minute ahead forecasts of average aggregated power. Results are shown for the remote sensing-based forecasting (RF) model, persistence and climatology benchmarks.

		$\overline{P}_{17}$	$\overline{P}_{35}$
	T	340	902
NRMSE [%]	RF	**8.02**	**8.19**
	Persistence	8.50	8.22
	Climatology	26.24	28.88

5. Discussion

In the current section, we discuss further uses of Doppler radar observations for wind power very short-term forecasting and address the limitations found in our proposed methodology.

5.1. On Further Use of Doppler Radar Measurements for Wind Power Forecasting

Doppler radars, like the ones used in the BEACon project, can measure up to 32 km, covering an area of more than 100 km^2, which is in line with the size of modern offshore wind farms. Radar measurements covering the whole wind farm, such as the ones presented here, are very valuable to understand the turbine-to-turbine interaction. However, when it comes to forecasting wind power, the radars' spatial coverage strongly influences the predictions. In the present case, an unfavourable scan geometry, with respect to the inflow area upstream of the outer wind turbines in the first row, resulted in degraded power predictions compared to the wind turbines positioned in the centre.

Therefore, an operational wind power forecasting system based on Doppler radar measurements should be configured with the aim of covering a sufficiently wide upstream area. As wake effects will influence the performance of downstream wind turbines, predictions of time series of the accumulated power of a whole wind farm are not straightforward. Both the power of the individual wind turbines, as well as the propagation of transient wind fields inside the wind farm, will strongly depend on the local dynamics and inhomogeneity of the wind farm flow. Further investigations are required to establish a suitable forecasting methodology. One approach might be to limit wind speed forecast to the first row of wind turbines (according to the prevailing wind direction). Additionally, turbine-to-turbine predictions of wind power accounting for wake losses could be used to predict the whole wind farm power generation.

One example time series demonstrated the ability of the radars to predict coherent structures affecting "locally" the power of individual wind turbines, which is beyond the current capabilities of commonly-used statistical methods for very short-term predictions. In a similar manner, it should be possible to probabilistically describe the effect of strong and fast changes in wind speed or direction, which cause wind power ramp events. Those crucial situations require risk-based decisions to be made, and there lies the importance of using a probabilistic approach. Given the current limitations of forecasting ramp events with NWP [43] and statistical [44] methods, radars could serve as a tool for detecting those extreme events. The maximal range of the BEACon radars (32 km) could detect a strong weather front of high wind speed (20 m/s) and give enough time for end-users to react.

As Doppler radars measure with a high temporal and spatial resolution, tracking techniques could also be applied to determine the vorticity and diffusion of identified patterns, extending the available information to determine the future position of the wind field vectors or wind coherent structures. Additionally, vertical information from the radars could be used to better describe gusts and lulls. Following this line, radar measurements at multiple heights could be used to derive a rotor equivalent wind speed which extends to the whole rotor plane and, unlike the hub height wind speed, accounts for the shear.

5.2. On Extension of the Forecasting Horizon

Due to the setup of the experiment and the prevailing south-westerly direction of the analysed data set, the wind field could only be evaluated over a relatively short distance of 5 km limiting the prediction horizon to five minutes. However, the long measurement range and the high spatial and temporal resolution of the used Doppler radars of up to 32 km should enable longer lead times on the order of 10 to 15 min as well. In principle, the proposed methodology should be capable of such horizons.

The proposed methodology is based on the advection of the wind field vectors with their respective motion, i.e., the persistence of their trajectories. Despite the promising results presented in this work, the question remains whether such assumption could be considered for longer horizons, increased turbulence, different types of stratifications, ramp events, strong veering or complex terrains. Based on our understanding, using a probabilistic approach should be able to include some (if not all) related uncertainties.

5.3. On Data Quality

One limitation of remote sensing prediction models is the uncertainty associated with the observations. Forecasting wind power based on wind speed observations introduces a new source of uncertainty. In this work, we overcome this problem by using a power curve based on the local dual-Doppler observations and the observed SCADA data that could, systematically, remove any bias introduced from the different type of measurements used to derive power curves. In a similar way, hybrid models using the radar observations and correcting for time-dependent errors should be explored.

Finally, reduced availability of data is a limitation inherent in remote sensing measurements, as the quality of those measurements highly depends on the meteorological conditions. In this regard, solutions for using Doppler radars in non-optimal meteorological situations need to be explored. Emphasis should be put on using data assimilation techniques, where observations are fed into statistical methods or NWP models, as such methods have shown to be improved when combined with real observations [45,46].

6. Conclusions

This paper investigated the use of DD radar observations to derive deterministic and probabilistic forecasts of wind power in a very short-term horizon of five minutes. An advection Lagrangian persistence technique was introduced to determine the predictive densities of rotor wind speed. In a case study, the proposed methodology was used to forecast the power generated by seven wind turbines in the North Sea, during free-flow conditions. The five-minute ahead predicted mean wind speeds corrected for induction effects showed a high degree of correlation with the observed DD wind speeds $2.5D$ upstream of the wind turbines. The predicted wind speeds densities were transformed into power densities by using a probabilistic power curve. We compared the proposed probabilistic forecast of wind power with the benchmarks persistence and climatology. Our results have shown the superiority of the remote sensing-based forecasting model regarding overall forecasting skill. However, a large spatial radar coverage of the inflow of a wind turbine is necessary to generate reliable density forecasts. The results have also proven that upstream remote sensing observations are especially crucial to detect strong changes in wind power, as shown in an example of a fast and strong predicted increase in power. Based on our results, a DD radar-based forecast might have a positive impact on the integration of offshore wind power into the grid.

Future works should be devoted to forecast the power produced of the whole wind farm, including wake effects, and to detect ramp events where the wind speed and direction changes rapidly. In addition, it is considered promising to further analyse DD radar measurements to enable the extension of the forecasting horizon and to improve the wind power predictions for different meteorological conditions.

Author Contributions: L.V. conducted the research work and wrote the paper. N.G.N. extensively contributed to conceive the methodology, to analyse the data, to structure the paper and to supervise the research work. L.V.-T., L.v.B. and M.K. supervised the research work and contributed to the structure of the paper.

Funding: This project has received funding from the European Union's Horizon 2020 research and innovation programme under the Marie Sklodowska-Curie Grant No. 642108.

Acknowledgments: We acknowledge the reviewers for their comments and suggestions on an earlier version of this manuscript, which enhanced the paper greatly.

Conflicts of Interest: The authors declare no conflict of interest.

Abbreviations

The following abbreviations are used in this manuscript:

DD	Dual-Doppler
R&D	Research and development
D	Wind turbine rotor diameter
SCADA	Supervisory Control and Data Acquisition
P_n	Nominal power
CRPS	Average Continous Ranked Probability Score
IEC	International Electrotechnical Commission
$ecdf$	Empirical cumulative distribution function
RF	Remote sensing-based forecast
P_i^{RF}	Predictive densities of power
ws_i^{RF}	Predictive densities of wind speed
P_{agg}	Average aggregated power
RMSE	Root-mean-square error
NRMSE	Normalised root-mean-square error

References

1. Cutululis, N.; Litong-Palima, M.; Sørensen, P. North Sea Offshore Wind Power Variability in 2020 and 2030. In Proceedings of the 11th International Workshop on Large-Scale Integration of Wind Power into Power Systems, Lisbon, Portugal, 13–15 November 2012.
2. 50Hertz; Amprion; Tennet; TransnetBW. *Leitfaden zur Präqualikation von Windenergieanlagen zur Erbringung von Minutenreserveleistung im Rahmen einer Pilotphase/ Guidelines for Prequalification of Wind Turbines to Provide Minute Reserves during a Pilot Phase*; Technical Report; German Transmission System Operators: Berlin, Germany, 2016.
3. Borggrefe, F.; Neuhoff, K. *Balancing and Intraday Market Design: Options for Wind Integration*; DIW Discussion Papers 1162; German Institute for Economic Research: Berlin, Germany, 2011.
4. EPEXSPOT. Intraday Lead Times. 2017. Available online: https://www.epexspot.com/en/product-info/intradaycontinuous/intraday_lead_time (accessed on 4 February 2018).
5. Giebel, G.; Kariniotakis, G. Wind power forecasting-a review of the state of the art. In *Renewable Energy Forecasting: From Models to Applications*; Woodhead Publishing: Cambridge, UK, 2017; pp. 59–109.
6. Cavalcante, L.; Bessa, R.J.; Reis, M.; Browell, J. LASSO vector autoregression structures for very short—Term wind power forecasting. *Wind. Energy* **2016**, *20*, 657–675. [CrossRef]
7. Erdem, E.; Shi, J. ARMA based approaches for forecasting the tuple of wind speed and direction. *Appl. Energy* **2011**, *88*, 1405–1414. [CrossRef]
8. Pinson, J.W.M.P. Online adaptive lasso estimation in vector autoregressive models for high dimensional wind power forecasting. *Int. J. Forecast.* **2018**, in press. [CrossRef]
9. Blonbou, R. Very short-term wind power forecasting with neural networks and adaptive Bayesian learning. *Renew. Energy* **2011**, *36*, 1118–1124. [CrossRef]
10. Carpinone, A.; Giorgio, M.; Langella, R.; Testa, A. Markov chain modeling for very-short-term wind power forecasting. *Electr. Power Syst. Res.* **2015**, *122*, 152–158. [CrossRef]
11. Pinson, P.; Madsen, H. Adaptive modelling and forecasting of offshore wind power fluctuations with Markov—Switching autoregressive models. *J. Forecast.* **2012**, *31*, 281–313. [CrossRef]
12. Hu, J.; Wang, J.; Ma, K. A hybrid technique for short-term wind speed prediction. *Energy* **2015**, *81*, 563–574. [CrossRef]
13. Pinson, P. Very-short-term probabilistic forecasting of wind power with generalized logit-normal distributions. *J. R. Stat. Soc. Ser. C* **2012**, *61*, 555–576. [CrossRef]
14. Dowell, J.; Pinson, P. Very-short-term probabilistic wind power forecasts by sparse vector autoregression. *IEEE Trans. Smart Grid* **2016**, *7*, 763–770. [CrossRef]
15. Alessandrini, S.; Davò, F.; Sperati, S.; Benini, M.; Delle Monache, L. Comparison of the economic impact of different wind power forecast systems for producers. *Adv. Sci. Res.* **2014**, *11*, 49–53. [CrossRef]
16. Clifton, A.; Clive, P.; Gottschall, J.; Schlipf, D.; Simley, E.; Simmons, L.; Stein, D.; Trabucchi, D.; Vasiljevic, N.; Würth, I. IEA Wind Task 32: Wind Lidar—Identifying and mitigating barriers to the adoption of wind lidar. *Remote Sens.* **2018**, *10*, 406. [CrossRef]
17. Kameyama, S.; Sakimura, T.; Watanabe, Y.; Ando, T.; Asaka, K.; Tanaka, H.; Yanagisawa, T.; Hirano, Y.; Inokuchi, H. Wind sensing demonstration of more than 30 km measurable range with a 1.5 μm coherent Doppler lidar which has the laser amplifier using Er, Yb:glass planar waveguide. In *Lidar Remote Sensing for Environmental Monitoring XIII*; SPIE: Kyoto, Japan, 2012; Volume 8526, p. 85260E.
18. Würth, I.; Brenner, A.; Wigger, M.; Cheng, P. How far do we see? Analysis of the measurement range of long-range lidar data for wind power forecasting. In Proceedings of the German Wind Energy Conference (DEWEK), Bremen, Germany, 17–18 October 2017.
19. Vasiljević, N.; Palma, J.M.L.M.; Angelou, N.; Carlos Matos, J.; Menke, R.; Lea, G.; Mann, J.; Courtney, M.; Frölen Ribeiro, L.; Gomes, V.M.M.G.C. Perdigão 2015: Methodology for atmospheric multi-Doppler lidar experiments. *Atmos. Meas. Tech.* **2017**, *10*, 3463–3483. [CrossRef]
20. Floors, R.; Peña, A.; Lea, G.; Vasiljević, N.; Simon, E.; Courtney, M. The RUNE Experiment—A Database of Remote-Sensing Observations of Near-Shore Winds. *Remote. Sens.* **2016**, *8*, 884. [CrossRef]
21. Van Dooren, M.F.; Trabucchi, D.; Kühn, M. A Methodology for the Reconstruction of 2D Horizontal Wind Fields of Wind Turbine Wakes Based on Dual-Doppler Lidar Measurements. *Remote Sens.* **2016**, *8*, 809. [CrossRef]

22. Valldecabres, L.; Peña, A.; Courtney, M.; von Bremen, L.; Kühn, M. Very short-term forecast of near-coastal flow using scanning lidars. *Wind Energy Sci.* **2018**, *3*, 313–327. [CrossRef]
23. Trombe, P.J.; Pinson, P.; Vincent, C.; Bøvith, T.; Cutululis, N.A.; Draxl, C.; Giebel, G.; Hahmann, A.N.; Jensen, N.E.; Jensen, B.P.; et al. Weather radars—The new eyes for offshore wind farms? *Wind. Energy* **2014**, *17*, 1767–1787. [CrossRef]
24. Meischner, P.; Hagen, M. Weather radars in Europe: Potential for advanced applications. *Phys. Chem. Earth Part B Hydrol. Oceans Atmos.* **2000**, *25*, 813–816. [CrossRef]
25. Hirth, B.D.; Schroeder, J.L.; Gunter, W.S.; Guynes, J.G. Coupling Doppler radar-derived wind maps with operational turbine data to document wind farm complex flows. *Wind. Energy* **2015**, *18*, 529–540. [CrossRef]
26. Nygaard, N.G.; Newcombe, A.C. Wake behind an offshore wind farm observed with dual-Doppler radars. *J. Phys. Conf. Ser.* **2018**, *1037*, 072008. [CrossRef]
27. Marathe, N.; Swift, A.; Hirth, B.; Walker, R.; Schroeder, J. Characterizing power performance and wake of a wind turbine under yaw and blade pitch. *Wind. Energy* **2016**, *19*, 963–978. [CrossRef]
28. Hirth, B.D.; Schroeder, J.L.; Irons, Z.; Walter, K. Dual-Doppler measurements of a wind ramp event at an Oklahoma wind plant. *Wind. Energy* **2016**, *19*, 953–962. [CrossRef]
29. Valldecabres, L.; Nygaard, N.; von Bremen, L.; Kühn, M. Very short-term probabilistic forecasting of wind power based on dual-Doppler radar measurements in the North Sea. *J. Phys. Conf. Ser.* **2018**, *1037*, 052010. [CrossRef]
30. Vignaroli, A.; Svensson, E.; Courtney, M.; Vasiljevic, N.; Lea, G.; Wagner, R.; Nygaard, N. How accurate is the BEACon radar? In Proceedings of the WindEurope Conference, Amsterdam, The Netherlands, 28–30 November 2017.
31. Germann, U.; Zawadzki, I. Scale-Dependence of the Predictability of Precipitation from Continental Radar Images. Part I: Description of the Methodology. *Mon. Weather Rev.* **2002**, *130*, 2859–2873. [CrossRef]
32. Germann, U.; Zawadzki, I. Scale Dependence of the Predictability of Precipitation from Continental Radar Images. Part II: Probability Forecasts. *J. Appl. Meteorol.* **2004**, *43*, 74–89. [CrossRef]
33. International Electrotechnical Commission (IEC). *Wind Energy Generation Systems—Part 12-1: Power Performance Measurements of Electricity Producing Wind Turbines*; IEC: Geneva, Switzerland, 2017.
34. Medici, D.; Ivanell, S.; Dahlberg, J.; Alfredsson, P.H. The upstream flow of a wind turbine: Blockage effect. *Wind. Energy* **2011**, *14*, 691–697. [CrossRef]
35. Gneiting, T.; Balabdaoui, F.; Raftery, A.E. Probabilistic forecasts, calibration and sharpness. *J. R. Stat. Soc. Ser. B* **2007**, *69*, 243–268. [CrossRef]
36. Gneiting, T. Quantiles as optimal point forecasts. *Int. J. Forecast.* **2011**, *27*, 197–207. [CrossRef]
37. Ahsbahs, T.; Badger, M.; Karagali, I.; Larsén, X. Validation of sentinel-1A SAR coastal wind speeds against scanning LiDAR. *Remote. Sens.* **2017**, *9*, 552. [CrossRef]
38. Lange, B.; Højstrup, J. Estimation of offshore wind resources—The influence of the sea fetch. In *Wind Engineering into the 21st Century, Copenhagen, Denmark*; CRC Press/Balkema: Rotterdam, The Netherlands, 1999; Volume 3, pp. 2005–2012.
39. Gonzalez, E.; Stephen, B.; Infield, D.; Melero, J.J. On the use of high-frequency SCADA data for improved wind turbine performance monitoring. *J. Phys. Conf. Ser.* **2017**, *926*, 12009. [CrossRef]
40. Efron, B. Bootstrap Methods: Another Look Jackknife. *Ann. Stat.* **1979**, *7*, 1–26. [CrossRef]
41. Hamill, T.M. Reliability Diagrams for Multicategory Probabilistic Forecasts. *Weather Forecast.* **1997**, *12*, 736–741. [CrossRef]
42. Bröcker, J.; Smith, L.A. Increasing the Reliability of Reliability Diagrams. *Weather Forecast.* **2007**, *22*, 651–661. [CrossRef]
43. Drew, D.R.; Cannon, D.J.; Barlow, J.F.; Coker, P.J.; Frame, T.H. The importance of forecasting regional wind power ramping: A case study for the UK. *Renew. Energy* **2017**, *114*, 1201–1208. [CrossRef]
44. Gallego, C.; Costa, A.; Cuerva, A. Improving short-term forecasting during ramp events by means of Regime-Switching Artificial Neural Networks. *Adv. Sci. Res.* **2011**, *6*, 55–58. [CrossRef]

45. Larson, K.A.; Westrick, K. Short-term wind forecasting using off-site observations. *Wind. Energy* **2006**, *9*, 55–62. [CrossRef]
46. Cheng, W.Y.Y.; Liu, Y.; Bourgeois, A.J.; Wu, Y.; Haupt, S.E. Short-term wind forecast of a data assimilation/weather forecasting system with wind turbine anemometer measurement assimilation. *Renew. Energy* **2017**, *107*, 340–351. [CrossRef]

![remote sensing logo] *remote sensing*

MDPI

Article

Investigation of the Fetch Effect Using Onshore and Offshore Vertical LiDAR Devices

Susumu Shimada [1,*]**, Yuko Takeyama** [2]**, Tetsuya Kogaki** [1]**, Teruo Ohsawa** [3]
and Satoshi Nakamura [4]

[1] National Institute of Advanced Industrial Science and Technology, Koriyama 963-0298, Japan;
    kogaki.t@aist.go.jp
[2] Department of Marine Resources and Energy, Tokyo University of Marine Science and Technology,
    Tokyo 108-8477, Japan; ytakey0@kaiyodai.ac.jp
[3] Graduate School of Maritime Sciences, Kobe University, Kobe 658-0022, Japan; ohsawa@port.kobe-u.ac.jp
[4] National Institute of Maritime, Port and Aviation Technology, Yokosuka 239-0826, Japan;
    nakamura_s@pari.go.jp
*   Correspondence: susumu.shimada@aist.go.jp; Tel.: +81-29-861-3910

Received: 12 August 2018; Accepted: 2 September 2018; Published: 5 September 2018

check for updates

Abstract: An offshore wind measurement campaign using vertical light detection and ranging (LiDAR) devices was conducted at the Hazaki Oceanographic Research Station (HORS) as part of an investigation into determining the optimal distance from the coast for a nearshore wind farm from a meteorological perspective. The research platform was a 427 m long pier located on a rectilinear coastline on the Pacific coast of the central Honshu Island in Japan. The relationship between the ratios of the increase of wind speed near the surface and fetch length within 5 km of the coast was analyzed via LiDAR observations taken at heights from 40 to 200 m. The results showed that the speed of the coastal wind blowing from land to sea gradually increased as the fetch length increased, by approximately 15–20% at 50 m above sea level around a fetch length of 2 km. Moreover, empirical equations were derived by applying the power law to the relationship between the increase of wind speed and fetch lengths at 1–5 km, as obtained from the LiDAR measurements. It was also found that the wind speed increase at a 2 km fetch length was equivalent to the effect of a 50–90 m vertical height increase on the coast in this region.

Keywords: coastal wind measurement; vertical Light Detection and Ranging; NeoWins; fetch effect; Hazaki Oceanographical Research Station; empirical equation

---

1. Introduction

As part of efforts to obtain additional offshore wind observations for the New Energy and Industrial Technology Development Organization, the Offshore Wind Information System (NeoWins) project [1], a coastal wind measurement campaign using light detection and ranging (LiDAR) was conducted from October 2015 to December 2016 at the Hazaki Oceanographic Research Station (HORS) [2] operated by the Port and Airport Research Institute in Japan's Ibaraki Prefecture (see Figure 1). The observations obtained from the measurement campaign were used to evaluate the accuracy of 500 m grid offshore wind simulations using the mesoscale meteorological model Weather Research and Forecasting (WRF) version 3.6.1 [3], which was employed to create a new Japanese offshore wind atlas.

Mesoscale meteorological models are useful for offshore wind resource assessments [4]. However, some previous studies [5–7] have shown that offshore winds simulated by WRF near the coast, where offshore wind farms are likely to be located in Japan in the near future, are less accurate than those simulated over the open ocean. Therefore, the authors have recently focused on the horizontal wind speed gradient near the coast to improve the WRF offshore wind simulation in this location. In addition to the LiDAR over the sea on the HORS research platform, another LiDAR was deployed on land to investigate coastal wind modifications, which is known as the fetch length effect, by comparing the onshore and offshore LiDAR observations. Because the performance of mesoscale models for the fetch length effect has yet to be thoroughly studied, and there were almost no suitable observations for the investigation, the authors believed that focusing on the fetch effect with LiDAR measurements would provide a means for enhancing the accuracy of coastal wind simulations with mesoscale models.

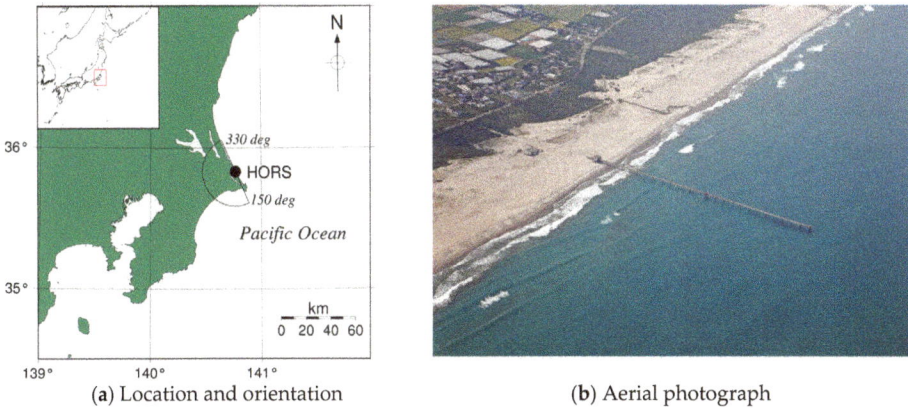

(a) Location and orientation   (b) Aerial photograph

Figure 1. (a) Location and orientation and (b) aerial photograph of the Hazaki Oceanographic Research Station (HORS) research platform.

The fetch length effect in coastal environments has already been extensively studied in the field of coastal engineering. For example, Hasselmann et al. (1973) [8] investigated the relationship between the fetch length and the evolution of the wave spectrum in the Joint North Sea Wave Project, a comprehensive wind wave measurement campaign. They observed the atmospheric and oceanographic parameters for the distance from 2 to 160 km at a site on the Danish coast. From the results of this comprehensive measurement campaign, equations for describing wave characteristics such as wave height and peak spectral period were derived [9].

As well as wind waves for coastal engineering, several studies [10–16] have considered the implications for wind energy applications of the relationship between the wind speed and fetch length over the sea by comparing wind speeds measured onshore and offshore. According to these observational studies, onshore wind speeds increased between 5% and 45% over long marine fetch lengths. However, the increase in wind speed over fetch lengths less than a few kilometers, which are relevant for nearshore wind farm developments, has not been investigated, primarily due to the lack of suitable in situ observations available for such analyses.

The results of previous studies based on in situ observations were reviewed in Barthelmie and Palutikof (1996) [15]. Since the fetch effect depends on local conditions, such as the topography, land use, measurement height, and atmospheric stability, large variations in the relationship between the fetch length and wind speed can be found in these previous studies. For example, Sethuraman and Raynor (1980) [12] showed a wind speed increase of 10–15% at a 5 km fetch length near the sea surface, whereas Lindley et al. (1980) [11] suggested that at a height of 100 m wind speed increases of 5% and 12% are possible for fetch lengths of 7.5 and 20 km, respectively.

Recently, a comprehensive near-shore wind measurement campaign using a total of nine LiDAR systems (three scanning LiDARs, two LiDAR buoys, and four vertical LiDARs) was conducted at a site on the Danish coast as part of the Reducing Uncertainty of Near-shore wind resource Estimates (RUNE) experiment [17]. A RUNE experiment report [18] introduced the horizontal wind speed gradient, which is an index reflecting the fetch effect. During the RUNE experiment, the prevailing winds were strong westerly winds moving from sea to land. The horizontal wind speed gradient that was expected to be dependent on the distance from the shore was largely absent, because an insufficient amount of wind blew from land to sea during the three-month measurement campaign.

In this paper, the authors explain their attempt to explore the increase in wind speed as a function of the fetch length by means of a simple measurement setup using just two vertical LiDARs located on the HORS research platform. From this measurement campaign, they obtained vertical onshore and offshore wind profiles for approximately six months. After analyzing these LiDAR observations, they estimated an optimal coast-to-wind-farm distance from a meteorological perspective, which they believe will be useful for planning nearshore wind farm developments. The authors also believe that the results from these LiDAR observations could provide a useful benchmark dataset for improving the accuracy of numerical models.

2. Experimental Setup

The measurement setup, shown in Figure 2, is a 427 m long pier constructed at a height of 7 m above sea level (ASL) at the HORS research station. Since this research platform was originally established for studies of coastal engineering, it has observed atmospheric and oceanographic parameters for more than 30 years. During the measurement campaign, two vertical profiling LiDAR devices were deployed on the pier. Vertical wind speed profiles at heights of 40 to 200 m from the LiDAR installation level were measured at 20 m intervals with two Windcube V1 units, for which detailed specifications can be found in the literature [19]. One Windcube WLS7-86 (hereafter referred to as LiDAR #1) was attached at the seaside end of the pier, and the other Windcube WLS7-78 (LiDAR #2) was located on the roof of a 3.5 m tall observational hut located at the landside end of the pier. The two LiDAR devices were exactly 400 m apart. Air temperature at 7 m ASL and seawater temperature at a depth of 2 m were observed at the halfway point of the pier, and they were used to evaluate the atmospheric stability during the measurement campaign.

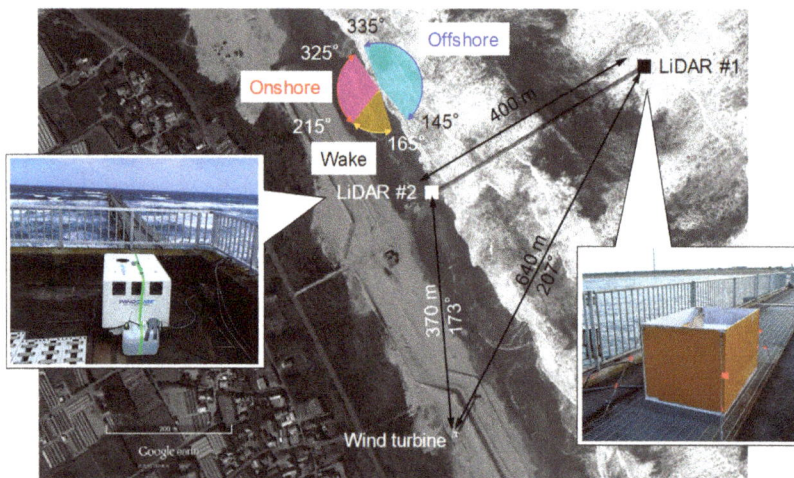

Figure 2. Experimental setup.

Figure 3 describes the topography and land use in the area surrounding of the HORS research platform; the figure covers an area of 20 km × 20 km. The topography distribution is based on the Advanced Spaceborne Thermal Emission and Reflection Radiometer Global Digital Elevation Model [20], which has a spatial resolution of 30 m, and the land use distribution is based on a 100 m resolution land use map provided by the government of Japan [21]. The terrain in this region, which is located outside a sparsely populated industrial area, is flat. Since this section of the coastline is rectilinear, the research platform provides ideal conditions for investigating variations in wind speed related to the fetch length effect. For example, categorizing onshore (land to sea) or offshore (sea to land) winds based on wind direction is frequently applied in coastal wind analysis, but it is difficult to accomplish this in regions with complex coastlines. Due to the straight coastline running from 150° to 330° at the HORS research platform, the authors could clearly distinguish between onshore or offshore winds.

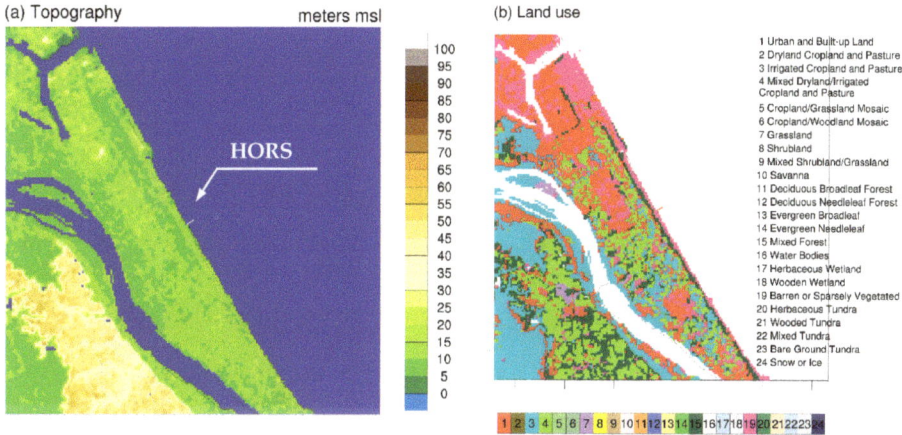

Figure 3. (**a**) Topography and (**b**) land use around the HORS research platform.

It is to be noted that some winds observed by both LiDARs, blowing from between 165° and 215°, were influenced by nearby 1.25 MW wind turbines, which have hub heights of 64.5 m. Since the wind turbines have rotor diameters of 62 m and are built about 5 m above ground level, observations between 40 and 100 m ASL might be significantly influenced by these wind turbines. As a result, the authors categorized the winds with wind directions between 335° and 145° as offshore winds (sea to land), winds with directions between 165° and 215° as turbine wakes, and winds with directions between 215° and 325° as onshore winds (land to sea).

Measurements using the two vertical LiDAR devices began in March 2016. Prior to the actual measurement campaign, LiDAR #2 was located at the seaside end of the pier for two weeks to investigate the instrumental differences between the two LiDAR systems. Comparisons of 10 min mean wind speeds and wind directions at a height of 100 m above the pier for the period of 7–21 March 2016 are shown in Figure 4. The LiDAR devices used in the measurements output both instantaneous values at intervals of a few seconds and 10 min statistics. In this study, 10 min statistics of average and standard deviation were used. Since the LiDAR data acquisition ratio, which is generally associated with data reliability, depended on the atmospheric conditions, the authors used only observations where the ratio exceeded 30% for 10 min.

In Figure 4, the number of samples, average, bias, root mean square difference (RMSD) and coefficient of determination R^2 are also described. Although the data acquisition ratio slightly decreased as the height increased, it was found that the differences between the two instruments were negligibly small. The wind speeds and wind directions had biases of 0.01 m/s and 1.08°, respectively. In addition, they had determination coefficients of more than 0.999. Namely, the authors could confirm that the observed wind speeds contained no unreasonable offsets that might lead to a misinterpretation of the measured data. After this comparison, the measurement campaign started on 23 March 2016, and it concluded on 30 September 2016.

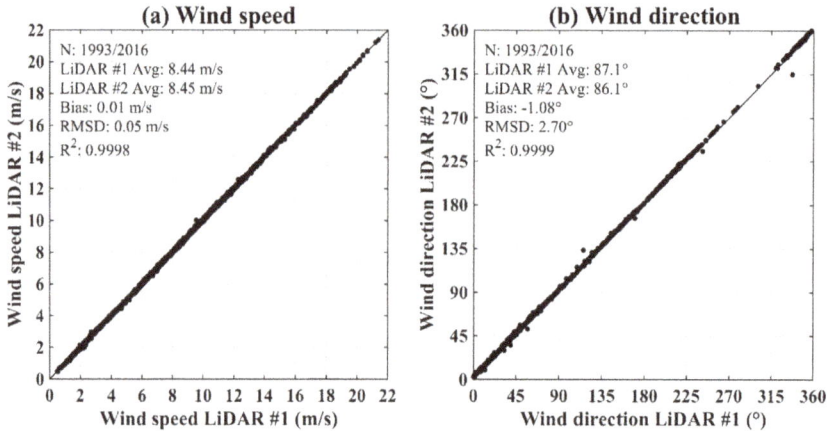

Figure 4. Comparison of (**a**) wind speed and (**b**) wind direction from light detection and ranging (LiDAR) #1 and LiDAR #2 at a height of 100 m above the pier for the period of 7–21 March 2016.

3. Results and Discussion

Figure 5 shows the timeseries of 10 min wind speed observations for LiDAR #1 and LiDAR #2 at 100 m above the LiDAR installation level during the whole measurement campaign. As seen in the figure, LiDAR #1 observations for July 2016 are missing, which is attributable to a problem with an electricity supply failure associated with the high air temperature. Overall, LiDAR #1 data availability during the whole measurement period was 81.6% for 100 m and 70.0% for 200 m. The timeseries of wind directions is also shown in Figure 6. When the observed winds were categorized by the sectors based on the LiDAR #1 wind direction at a height of 100 m, wind results of 59.3%, 25.4%, and 9.0% were recorded for offshore (sea to land), turbine wake, and onshore (land to sea) winds, respectively.

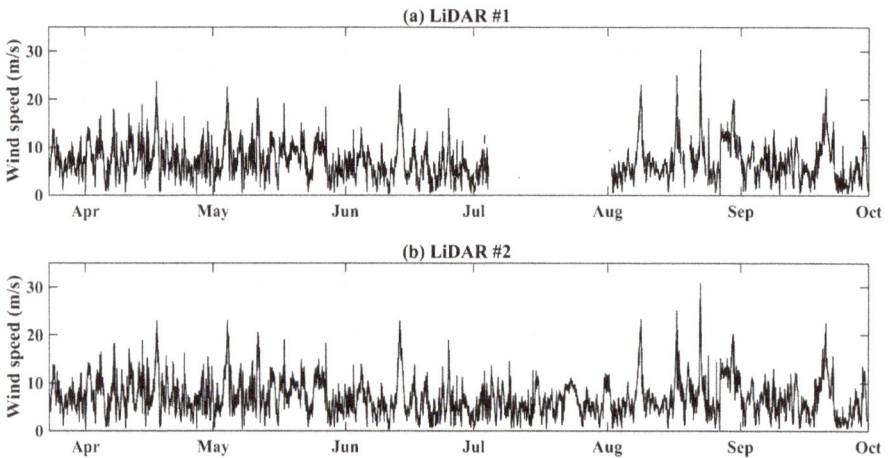

Figure 5. Timeseries of 10 min wind speeds at a height of 100 m for (**a**) LiDAR #1 and (**b**) LiDAR #2 for the period 22 March to 30 September 2016.

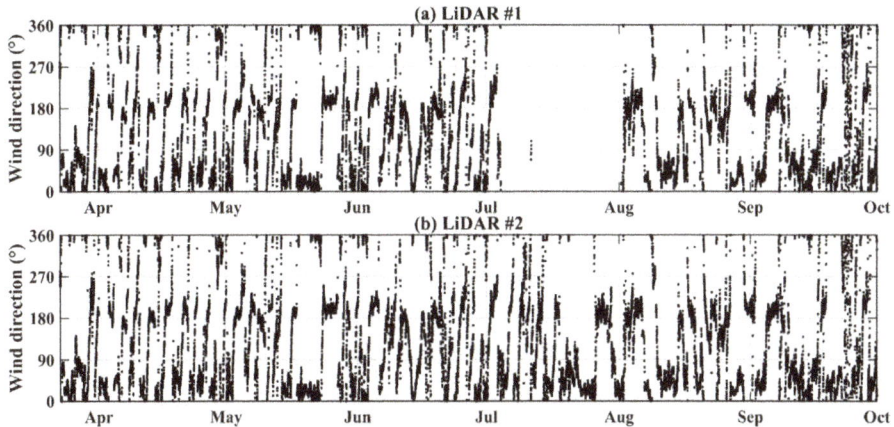

Figure 6. Same as Figure 5, but for wind directions.

The differences between the air and seawater temperature shown in Figure 7 indicate that neutral to stable atmospheric stability conditions prevailed during the measurement period of spring to early autumn. The wintertime seawater temperature is usually higher than air temperature in this region, and vice versa for spring to summer. The fetch effect depends strongly on the atmospheric stability [16], but because the period of the measurement campaign did not cover the winter months with their expected dominance of unstable conditions, an analysis of the dependence of the fetch effect on stability was not part of this study.

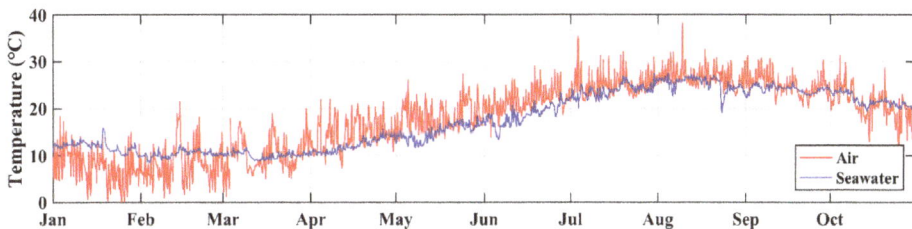

Figure 7. Comparison of 10 min mean air temperature on the pier and seawater temperatures at a depth of 2 m for the period of January to October 2016 at the HORS research platform.

The two LiDAR observation datasets, when visualized, show no obvious significant differences, especially when basic wind climate statistics, such as the average or occurrence frequency, are compared. Figures 8 and 9 show the occurrence frequencies and wind roses at 100 m above the LiDAR installation level for the period of March to September 2016. The Weibull scale parameter *A* and shape parameter *k* are described in Figure 8. In addition, the occurrence frequency at wind speeds greater than 20 m/s is also depicted in the insets in Figure 8. The maximum 10 min wind speeds for LiDAR #1 and LiDAR #2 were 30.4 m/s and 31.1 m/s, respectively, and the occurrence of wind speeds greater than 20 m/s due to extreme events such as typhoons was less than 1% during the measurement period.

Overall, the differences in the occurrence distributions for wind speed and wind direction between the two LiDAR devices, which were only 400 m apart, were small. However, interesting characteristics could be found in the instantaneous values of the wind speed ratios, especially for some wind directions. Accordingly, the authors will now focus on the wind speed ratio between the LiDAR observations.

It is to be noted that, since there was a 3.5 m height difference between the LiDAR installation levels due to the building height, 10 min averages were vertically interpolated onto the same levels before the analysis was carried out using a logarithmic function to calculate the wind speed ratio at the same levels.

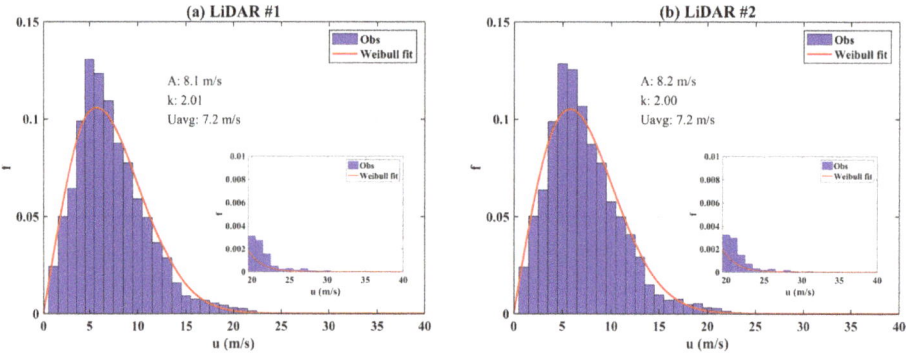

Figure 8. Comparison of occurrence frequency of 10 min wind speed for LiDAR #1 and LiDAR #2 at a height of 100 m from the LiDAR installation levels.

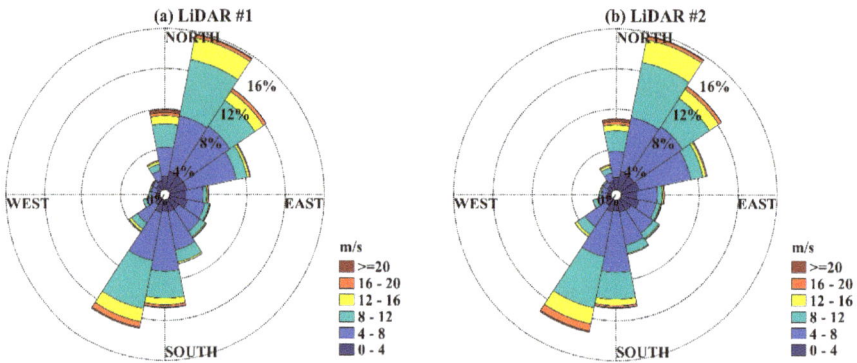

Figure 9. Comparison of wind roses for LiDAR #1 and LiDAR #2 at a height of 100 m from the LiDAR installation levels.

The wind speed ratio for LiDAR #1 to LiDAR #2 at 50 m ASL, as a function of wind direction, is shown in Figure 10. The bin-average and standard deviation were calculated at 5° intervals and then plotted in red for the 10 min values. The 10 min data for calm wind conditions, which corresponded to wind speed less than 2 m/s, were excluded from this analysis. In the figure, two peaks with bin-average ratios of approximately 1.15 to 1.2 can be seen around wind directions of 155° and 325°, which nearly coincide with the coastline. For these directions, winds measured by LiDAR #1 on the pier top had long marine fetch lengths, whereas those measured by LiDAR #2 were entirely onshore winds. This result suggested that the fetch length effect could increase the offshore wind speeds by up to 15–20% when they travelled over water at 50 m ASL. In contrast, the small ratios seen at around 200° might be attributable to the wake from the nearby 1.5 MW wind turbines.

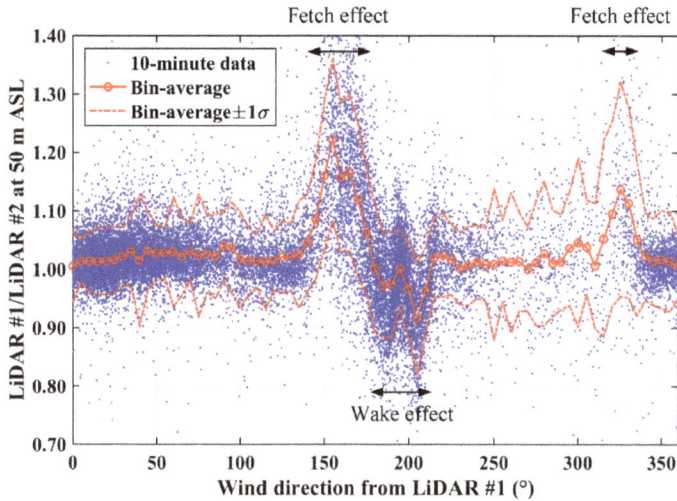

Figure 10. LiDAR #1 to LiDAR #2 wind speed ratios at 50 m above sea level (ASL) as a function of wind direction. Only LiDAR #2 wind speeds greater than 2 m/s were used in the analysis.

Considering the shape of the straight coastline, the fetch length effect should be approximately symmetrical in the figure. However, the wind speed ratios around 155° were slightly higher than those around 325°. This asymmetry might be attributable to the contamination of the wake effect at LiDAR #2 from the nearby wind turbine. The LiDAR #2 observations between 140° and 175° might be impacted by the wind turbine at a direction of 170°, since LiDAR #2 was located only 370 m away from the wind turbine. As a result, the fetch effect around 155° may be enhanced by the wake effect for LiDAR #2.

The strength of the wake and fetch effects was found to depend significantly on height when the same relationships were illustrated with the LiDAR observations at different heights. Figure 11 shows the same relationships as Figure 10, but the relationships for all heights are depicted because, in this figure, the authors considered the height dependency of the wake and fetch effects. Figure 11 shows that the variations of the wind speed ratios dependent on wind directions found at 50 m ASL became ambiguous as the height increased. The wake effect around a direction of 200° disappeared for heights above 130 m. Moreover, the increase in wind speeds around wind directions of 155° and 325°, which might be associated with the fetch length effects, ultimately declined to a few points at 207 m height ASL.

Figures 10 and 11 already suggest the maximum range of an increase of the wind speed due to the fetch effect within the surface layer near the coast, but the results are insufficient for understanding the relationship in detail. Thus, the authors converted the relationships between wind direction and wind speed ratio into fetch length and wind speed by taking into account the characteristics of the research platform. As mentioned above, since the region where the HORS research platform is located on a rectilinear coastline, the fetch length at the top of the pier can be defined simply as a function of wind direction θ as follows:

$$\text{Fetch}(\theta) = L_{pier}/\cos(\theta - \theta_{pier} - 180°), \tag{1}$$

where L_{pier} and θ_{pier} are the pier length of 400 m from the shore and the heading angle of 59°, respectively. Although the pier is only 400 m long, the fetch length effect for a few kilometers can be analyzed when the characteristics of the site's rectilinear coastline are considered.

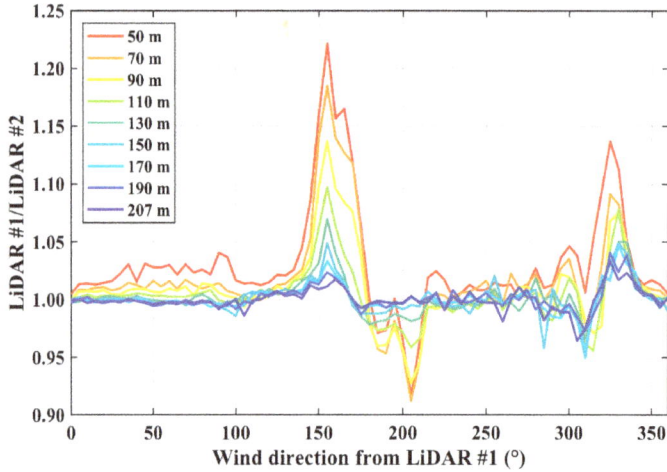

Figure 11. Bin-averaged LiDAR #1 to LiDAR #2 wind speed ratio as a function of wind direction at heights of 50, 70, 90, 110, 130, 150, 170, 190, and 207 m ASL.

The wind speed ratio for LiDAR #1 to LiDAR #2 at 50 m ASL as a function of fetch length is shown in Figure 12. The bin width is 100 m. As can be seen in the figure, the wind speed ratio drops twice at fetch lengths of 500 m and 700 m due to the wind turbine wake effects, and then increases monotonically for fetch lengths of 700 to 1100 m. Moreover, it continues to increase slightly to 1.2 at a fetch length of 1900 m, after which the gradient of the variation appears to be moderate.

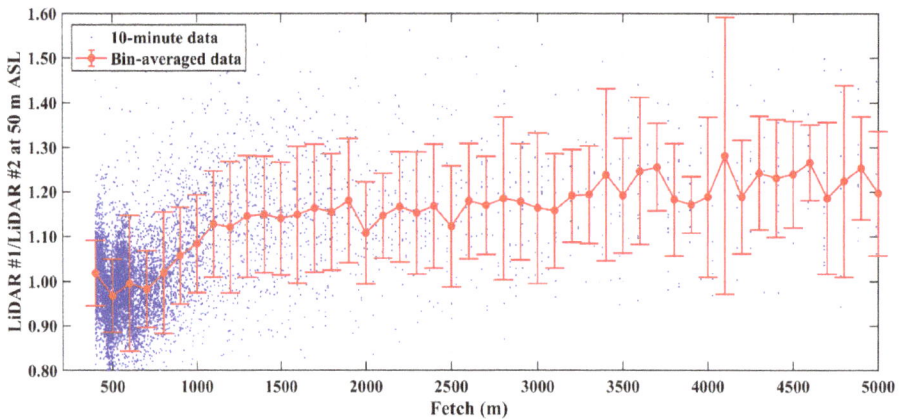

Figure 12. Wind speed ratio as a function of fetch length at 50 m ASL.

From this figure, one can see some interesting features with respect to the fetch length effect. It is difficult to discuss the fetch length effect at 600 m and 800 m, since the wind speeds from LiDAR #1, which was located at the seaside end of the pier, are disturbed by the nearby wind turbines. However, at 400 m, there is no wake effect for both LiDAR #1 and LiDAR #2, since it corresponds to the direction along the pier. Therefore, this might imply that the fetch effect needs some distance to appear, and the distance would be more than 400 m from the shore at a height of 50 m. Moreover, the wind speed at 50 m increased by up to 1.2 times while traveling over 2 km of water, and the fetch length effect appears to become less pronounced at distances exceeding 2 km. These results could provide a limit for the optimal distance for nearshore wind farms.

The relationships for the ratio of turbulence intensity, which is the standard deviation of wind speed for 10 min normalized by the mean value, are shown in Figure 13. Note that because it is difficult to adjust the height difference in the standard deviation, there is a 3 m height difference between the LiDAR #1 and LiDAR #2 values. Generally speaking, offshore winds have less turbulence than onshore winds. As shown in the figure, the turbulence intensity decreases rapidly as the fetch length reaches about 2 km. Additionally, much like the wind speed ratio shown in Figure 12, the ratios of turbulence intensity at fetch lengths of more than 2 km also become almost constant. These results indicated that the transition from onshore to offshore winds might occur within a few kilometers of the coast for flat terrain without forests and settlements.

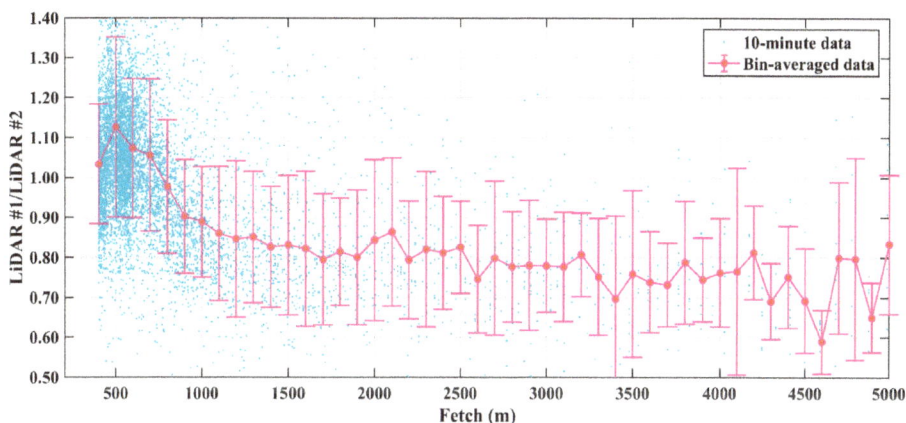

Figure 13. Ratios of turbulence intensity between LiDAR #1 at 47 m and LiDAR #2 at 50 m ASL as a function of fetch length.

Although there have been large variations in the fetch length effect in previous studies, a similar result to Figure 12 can be found in Barthelmie et al. (1996) [13]. They showed that onshore wind speeds at 38 m could increase by up to approximately 20% at a distance of 1630 m by comparing onshore and offshore met mast observations from the Vindeby project [22]. Moreover, they also analyzed the wind speed increase ratio using the Wind Atlas Analysis and Application Program (WAsP) model [23], which is a frequently used application for analyzing wind climates over flat terrain. Specifically, wind speeds observed by a land meteorological mast were input into the WAsP model and wind speed increases, as a function of the distance from the coast, were extracted. As a result, a strong gradient creating a wind speed increase of up to 15% could be found within a fetch length of 3 km, after which the gradient flattened.

In addition, a similar result is found in another numerical study using a simple calculation model based on the internal boundary layer theory by Barthelmie and Palutikof [15]. In their theoretical analysis, the wind speed increase ratio shows its steepest gradient until reaching a 2 km fetch length, after which the ratio stops increasing. The results from the numerical studies are similar to the results obtained from the measurement campaign that are shown in Figure 12. From the results described above, the authors concluded that the results obtained in this study were in line with previous results that were calculated from simple numerical models.

As expected from Figure 11, the increase of the wind speed ratio relevant to the fetch effect becomes unclear as the height increases. The relationships between the fetch length and bin-averaged wind speed ratio at 50 to 207 m ASL are shown in Figure 14. Due to the reduced data availability for levels higher than 50 m, the plots for heights from 70 to 207 m are slightly less stable than those for the lower levels. However, it can unmistakably be seen that the turbine wake and fetch length effects decrease gradually as the height increases, and that almost no clear fetch length-dependent differences can be found above 130 m ASL. Ultimately, the wind speed ratio at a height of 207 m seems to be independent of both the fetch length and turbine wake.

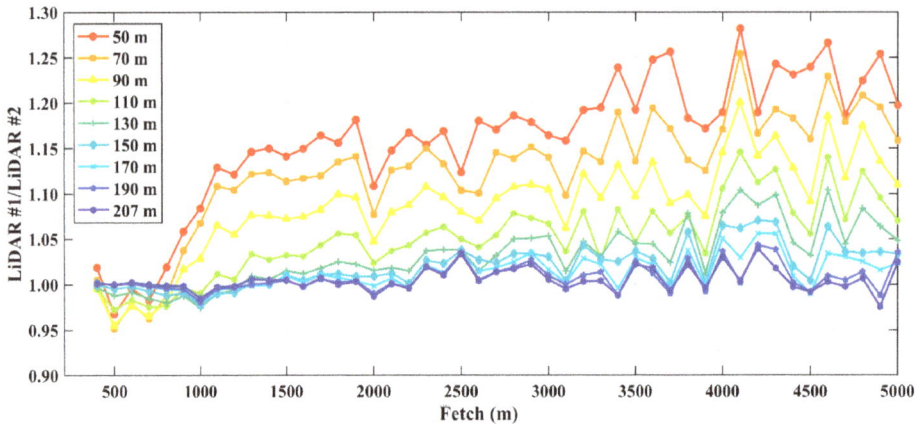

Figure 14. Bin-averaged wind speed ratio as a function of fetch length at heights of 50, 70, 90, 110, 130, 150, 170, 190, and 207 m ASL.

To compare this study's results with those from other sites and numerical models, the relationships of the increase in wind speed and fetch length would be preferable as formulations. Empirical equations derived from the results in Figure 14 are illustrated in Figure 15. The equations are derived by fitting the power law as follows:

$$U_{sea}/U_{land}(x, z) = a(z) \times x^{b(z)}, \tag{2}$$

where U_{sea}/U_{land} is the ratio of wind speed on the land and sea as a function of the fetch length x in meters. The values of coefficients a and b, which are parameters dependent on the height z, are listed in the figure legend. Since the wind speed ratio at a fetch length less than 1000 m seemed to be strongly influenced by wind turbines at this site, the coefficients were derived using only the data obtained with fetch lengths of more than 1000 m. These coefficients might be useful for investigating the fetch length effect at the other sites, as well as for comparing results with numerical models.

Because the authors' analysis of the LiDAR observations revealed that the 50 m wind speed increased up to 15–20% around a fetch length of about 2 km, the last question then becomes, "How attractive or effective is this 20% speedup from a wind resource usage perspective?" Normally, wind speeds can clearly increase independently of fetch length effects as altitude increases due to increased distance from surface roughness. Thus, the authors finally compared the increasing wind speeds resulting from height increases to quantify the advantage of the fetch length effect.

Figure 16 shows mean wind speed profiles and the profiles normalized at the lowest level values for the onshore winds (215° to 325°). Only observations that were taken when all heights were available were used for this analysis. The shear exponents for the LiDAR #1 and LiDAR #2 profiles were 0.267 and 0.277, which corresponded to the values between shrub and forest [24], respectively. The normalized profile for LiDAR #2 shows that the 50 m wind speed increased up to 20% at a 90 m height. This indicates that the fetch length effect over a 2 km distance can be considered equivalent to increasing the height from 50 to 90 m over land. Furthermore, if the height is increased to up to 130 m, the value at 130 m increases to 40%. This is of particular importance because, as the development of taller wind turbines nears completion and they become available, positioning such turbines onshore in coastal areas could provide an acceptable alternative to nearshore wind farm developments in locations where sufficient space is available.

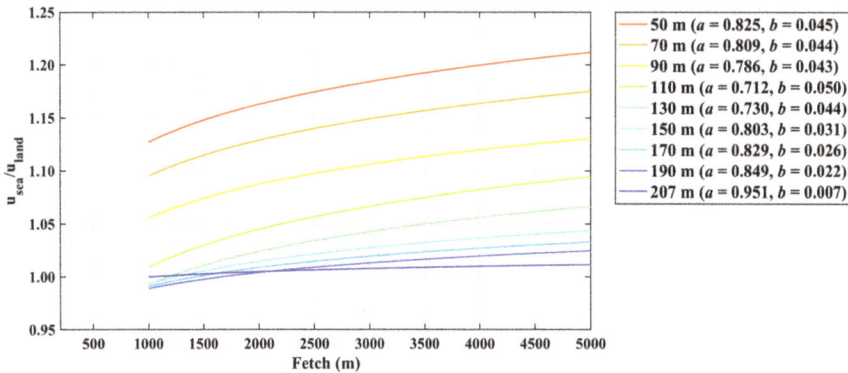

Figure 15. Empirical equations of the ratio of onshore and offshore wind speeds as functions of the fetch length and height, derived from the LiDAR observations for more than a 1000 m fetch length.

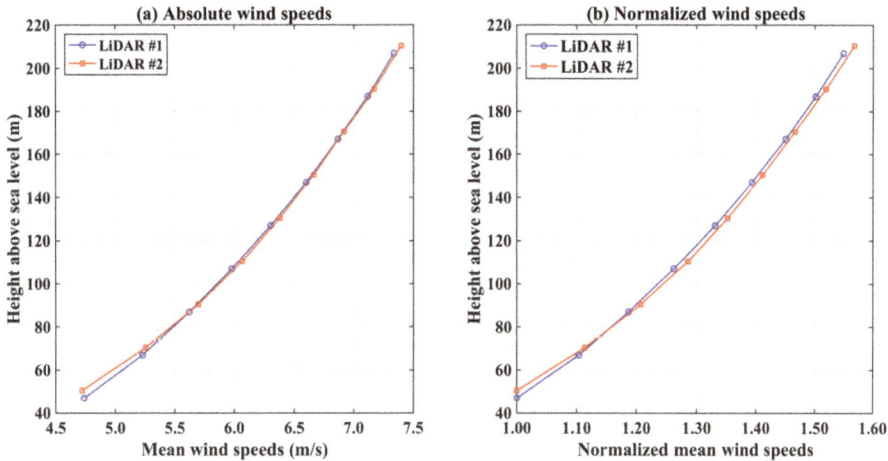

Figure 16. (a) Absolute mean wind speed profiles and (b) mean wind speed profiles normalized by the lowest level values for land sectors (215° to 325°) at the HORS research platform. Only observations that were taken when all heights were available were used for this analysis.

4. Conclusions and Recommendations

In this paper, the authors reported results from an onshore and offshore wind measurement campaign for the period of March to September 2016 using two vertical profiling LiDAR devices at the coastal research platform HORS to investigate increases in wind speed with increasing fetch length from the coast. They began by describing the experimental setup and wind conditions at HORS during the six-month measurement period used in this study, after which the increases in wind speed were examined from comparisons of observations recorded from the two vertical LiDAR devices.

From the wind speed ratios between the two LiDARs, they found that the 50 m wind speed increased by a factor of about 1.15 to 1.2 times when travelling over a long marine fetch length. In addition, observations taken at the top of the pier were found to be significantly influenced by onshore wind turbines located more than several hundred meters away, which is more than 10 times the wind turbine rotor diameter. In addition, the increase in wind speed as a function of fetch length for 400 m to 5 km was demonstrated by applying the characteristics of rectilinear coastline in this region. Furthermore, empirical equations based on the power law to describe the fetch effect dependence, which are useful for comparisons with other sites or numerical models, were introduced.

According to the results obtained, the authors found that the 50 m wind speed on the coast increased monotonically over a 2 km fetch length and approached a ratio of 1.2 at a 1900 m fetch length. The ratio was found to increase slightly after a fetch length of 1900 m. Moreover, the same relationships for the turbulence intensity also showed that it decreased rapidly in the offshore direction, especially up to a 2 km fetch length. These results suggested that the transition from onshore winds to offshore winds that have a higher wind speed and lower turbulence than onshore winds occurred within a few kilometers of the shore. These results also suggested that locating wind turbines a few kilometers away from the coast would be reasonable from a meteorological perspective for efficient near-shore wind energy usage when the coast was not surrounded by complex terrain and vegetation.

The wind speed ratios obtained from the two LiDAR observations had a steep gradient within 2 km of the coast and then gradually flattened further offshore. This strong gradient within a few kilometers was found in previous studies using simple numerical models. These analyses were based on internal boundary layer theory, and a wind analysis application program was run for areas surrounded by flat terrain, similar to the HORS research platform. Although the effect of atmospheric stability was not taken into account in this study due to the measurement period, the relationship between wind speed and fetch length seemed to be applicable for areas with similar geographical conditions, such as flat terrain and little vegetation.

In addition, the dependency of the fetch effect on the height was clearly visualized by means of the LiDAR observations. As a result, the fetch length effect was shown to gradually become less pronounced as the height increased, and it was also found that the fetch length and wake effects from nearby wind turbines were negligible at heights above 130 m. Finally, the impact of the fetch length effect was compared with the impact of increasing height on the wind speed. The result suggested that using taller wind turbines with hub heights more than 100 m on land would be an alternative option to locating turbines offshore when sufficient space was available. The authors finally emphasize that it would be inconceivable to obtain such informative observations without using remote sensing technologies for this study.

In this measurement campaign, the authors obtained informative and valuable offshore wind observations that they believe will provide good benchmark data for the validation of numerical models. In their continuing research, they intend to install a scanning LiDAR system to collect more detailed measurements, strengthen the reproducibility of numerical models, and thus facilitate a more comprehensive understanding of coastal winds.

Author Contributions: Conceptualization, S.S.; data curation, S.S., Y.T., T.K., T.O., and S.N.; funding acquisition, S.S., T.K., and T.O.; methodology, S.S.; project administration, S.S.; resources, S.N.; validation, T.O.; writing—original draft, S.S.

Funding: This work was supported by the Japan Society for the Promotion of Science (JSPS) KAKENHI Grant Number 15K21665. This research was also partially supported by the NeoWins project commissioned by the New Energy and Industrial Technology Development Organization (NEDO).

Conflicts of Interest: The authors declare no conflict of interest.

References

1. NEDO. NEDO Offshore Wind Information System (NeoWins). Available online: http://www.nedo.go.jp/english/news/AA5en_100201.html (accessed on 20 July 2018).
2. PARI. Hazaki Oceanographical Research Station (HORS). Available online: https://www.pari.go.jp/unit/edosy/en/main-facility/2.html (accessed on 26 June 2018).

3.   Skamarock, W.; Klemp, J.; Dudhia, J.; Gill, D.; Barker, D.; Duda, M.; Huang, X.; Wang, W.; Powers, J. *A Description of the Advanced Research WRF Version 3*; NCAR Technical Note NCAR/TN-475+ STR; National Center for Atmospheric Research: Boulder, CO, USA, 2008; 113p.

4.   Sempreviva, A.M.; Barthelmie, R.J.; Pryor, S.C. Review of Methodologies for Offshore Wind Resource Assessment in European Seas. *Surv. Geophys.* **2008**, *29*, 471–497. [CrossRef]

5.   Shimada, S.; Ohsawa, T.; Chikaoka, S.; Kozai, K. Accuracy of the Wind Speed Profile in the Lower PBL as Simulated by the WRF Model. *Sola* **2011**, *7*, 109–112. [CrossRef]

6.   Hahmann, A.N.; Vincent, C.L.; Pena, A.; Lange, J.; Hasager, C.B. Wind climate estimation using WRF model output: Method and model sensitivities over the sea. *Int. J. Climatol.* **2015**, *35*, 3422–3439. [CrossRef]

7.   Floors, R.; Hahmann, A.N.; Pena, A. Evaluating Mesoscale Simulations of the Coastal Flow Using Lidar Measurements. *J. Geophys. Res. Atmos.* **2018**, *123*, 2718–2736. [CrossRef]

8.   Hasselmann, K.P.; Barnett, T.; Bouws, E.; Carlson, H.E.; Cartwright, D.; Enke, K.; Ewing, J.A.; Gienapp, H.E.; Hasselmann, D.; Kruseman, P.; et al. *Measurements of Wind-Wave Growth and Swell Decay during the Joint North Sea Wave Project (JONSWAP)*; Deutches Hydrographisches Institut: Hamburg, Germany, 1973; Volume 8, pp. 1–95.

9.   Hasselmann, K.; Ross, D.B.; Muller, P.; Sell, W. Parametric wave prediction model. *J. Phys. Oceanogr.* **1976**, *6*, 200–228. [CrossRef]

10. Francis, P.E. Effect of Changes of Atmospheric Stability and Surface Roughness on Off-Shore Winds over East Coast of Britain. *Meteorol. Mag.* **1970**, *99*, 130.

11. Lindley, D.; Simpson, P.B.; Hassan, H.; Milborrow, D. An assessment of offshore siting of wind turbine generators. In Proceedings of the 3rd International Symposium on Wind Energy Systems, Cranfield, UK, 26–29 August 1980; BHRA Fluid Engineering: Cranfield, UK, 1980; pp. 17–42.

12. Sethuraman, S.; Raynor, G.S. Comparison of Mean Wind Speeds and Turbulence at a Coastal Site and an Offshore Location. *J. Appl. Meteorol.* **1980**, *19*, 15–21. [CrossRef]

13. Barthelmie, R.J.; Courtney, M.S.; Hojstrup, J.; Larsen, S.E. Meteorological aspects of offshore wind energy: Observations from the Vindeby wind farm. *J. Wind Eng. Ind. Aerodyn.* **1996**, *62*, 191–211. [CrossRef]

14. Barthelmie, R.J.; Grisogono, B.; Pryor, S.C. Observations and simulations of diurnal cycles of near-surface wind speeds over land and sea. *J. Geophys. Res. Atmos.* **1996**, *101*, 21327–21337. [CrossRef]

15. Barthelmie, R.J.; Palutikof, J.P. Coastal wind speed modelling for wind energy applications. *J. Wind Eng. Ind. Aerodyn.* **1996**, *62*, 213–236. [CrossRef]

16. Pryor, S.C.; Barthelmie, R.J. Analysis of the effect of the coastal discontinuity on near-surface flow. *Ann. Geophys.* **1998**, *16*, 882–888. [CrossRef]

17. Floors, R.; Pena, A.; Lea, G.; Vasiljevic, N.; Simon, E.; Courtney, M. The RUNE Experiment—A Database of Remote-Sensing Observations of Near-Shore Winds. *Remote Sens.* **2016**, *8*, 884. [CrossRef]

18. Peña, A. *RUNE Benchmarks*; DTU Wind Energy E, No. 0134(EN); DTU Wind Energy: Roskilde, Denmark, 2017; p. 25.

19. Gottschall, J.; Courtney, M. *Verification Test for three WindCube WLS7 LiDARs at the Høvsøre Test Site*; No. 1732(EN); Danmarks Tekniske Universitet, Risø Nationallaboratoriet for Bæredygtig Energi: Kgs. Lyngby, Denmark, 2010; p. 44.

20. Tachikawa, T.; Hato, M.; Kaku, M.; Iwasaki, A. *Characteristics of ASTER GDEM Version 2*; IEEE: Piscataway, NJ, USA, 2011; pp. 3657–3660.

21. MLIT. National Land Numerical Information Download Service. Available online: http://nlftp.mlit.go.jp/ ksj-e/index.html (accessed on 27 July 2018).

22. Barthelmie, R.J.; Courtney, M.; Højstrup, J.; Sanderhoff, P. *The Vindeby Project: A Description*; Risø National Laboratory: Roskilde, Denmark, 1994; ISBN 87-550-1969-2.

23. Mortensen, N.G.; Landberg, L.; Troen, I.; Lundtang Petersen, E. *Wind Atlas Analysis and Application Program (WAsP)*; Risø National Laboratory: Roskilde, Denmark, 1993.

24. Emeis, S. *Wind Energy Meteorology: Atmospheric Physics for Wind Power Generation*; Springer International Publishing: Berlin, Germany, 2018.

remote sensing

MDPI

Article

The NEWA Ferry Lidar Experiment: Measuring Mesoscale Winds in the Southern Baltic Sea

Julia Gottschall [1,*], Eleonora Catalano [2], Martin Dörenkämper [3] and Björn Witha [4]

[1] Fraunhofer Institute for Wind Energy Systems IWES, 27572 Bremerhaven, Germany
[2] RES Australia, Sydney, NSW 2067, Australia; ele.catalano@gmail.com
[3] Fraunhofer Institute for Wind Energy Systems IWES, 26129 Oldenburg, Germany;
   martin.doerenkaemper@iwes.fraunhofer.de
[4] ForWind, Institute of Physics, University of Oldenburg, 26129 Oldenburg, Germany; bjoern.witha@uol.de
* Correspondence: julia.gottschall@iwes.fraunhofer.de; Tel.: +49-471-14290-354

Received: 31 August 2018; Accepted: 6 October 2018; Published: 12 October 2018

check for updates

Abstract: This article presents the Ferry Lidar Experiment, which is one of the NEWA Experiments, a set of unique flow experiments conducted as part of the New European Wind Atlas (NEWA) project. These experiments have been prepared and conducted to create adequate datasets for mesoscale and microscale model validation. For the Ferry Lidar Experiment a Doppler lidar instrument was placed on a ferry connecting Kiel and Klaipeda in the Southern Baltic Sea from February to June 2017. A comprehensive set of all relevant motions was recorded together with the lidar data and processed in order to obtain and provide corrected wind time series. Due to the existence of the motion effects, the obtained data are essentially different from typical on-site data used for wind resource assessments in the wind industry. First comparisons show that they can be well related to mapped wind trajectories from the output of a numerical weather prediction model showing a reasonable correlation. More detailed validation studies are planned for the future.

Keywords: Doppler lidar; NWP model; mesoscale; Floating Lidar System (FLS), wind resource assessment; wind atlas

1. Introduction

The consortium of the New European Wind Atlas (NEWA [1]) project is currently creating a later publicly available and freely accessible wind atlas covering essentially the countries of the European Union and Turkey. The coverage further includes offshore areas up to 100 km from shore and the complete North and Baltic Sea. The basis of the atlas is a model chain, also developed within the project, comprising mesoscale and microscale flow models that are run to generate the wind time series and statistics making up the atlas. Mesoscale time series data, which represent the largest part of the atlas, will be available with a spatial resolution of 3 km × 3 km and for a length of 30 years. The microscale resolution will be finer than 100 m × 100 m, but here only limited statistics will be provided. The third outcome of the project, besides the wind atlas and model chain, is an experimental database comprising the data from a number of quite unique flow experiments that have been conducted within the project.

The NEWA experiments vary not only in their locations but also in the assessed types of terrains and associated flow phenomena, their durations, and the level and complexity of involved equipment and resulting data volume [2]. Common to all experiments within NEWA is the application of Doppler lidar technology to supplement, and in some cases also completely replace, meteorological towers. This is not by chance, but indicates the breakthrough of a technology that has been continuously developed further for the wind industry within the last 10–15 years [3].

The sites of the NEWA experiments are distributed over the coverage of the later wind atlas (see Figure 1). It is important to emphasize that the experimental data are not integrated in the wind atlas. Instead of supplementing the model data, they are used to test and validate the involved models in dedicated benchmark studies. Each experiment has been defined with a specific focus [2], including a double ridge as in the Perdigao experiment (Portugal) [4], a complex steep terrain with a complex mesoscale flow in Alaiz (Spain), a flow over forested rolling hills in Hornamossen (Sweden), a single forested hill near Kassel (Germany), or a near-shore wind flow in RUNE (Denmark) [5].

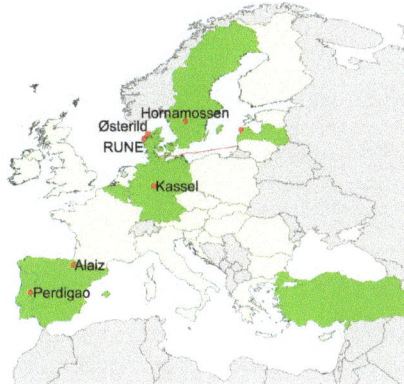

Figure 1. Minimum coverage of the New European Wind Atlas and sites of experiments. The minimum onshore coverage is shown in light green, the NEWA partner countries in a darker lime-green, offshore coverage in light blue, and experimental sites in red. The red line marks the route in the Ferry Lidar Experiment. (Graphic reproduced from Reference [2]).

The Ferry Lidar Experiment is the only offshore experiment within NEWA. Its name originates from its setup: a vertically scanning Doppler lidar is placed on a ferry boat to measure the wind along the ferry route, covering several hundreds of kilometers within one day and travelling back and forth for a period of several months. This rather simple setup is designed for studying mesoscale effects that are, in particular, far-offshore pre-dominant. For distances greater than 30 km from shore and at heights relevant for wind energy exploitation (i.e., greater than about 50 m) microscale effects due to e.g., breaking sea waves do not play an important role [6]. Comparisons with data of meteorological (met) masts show that mesoscale models are capable of resolving the most important features of the marine atmospheric boundary layer and compare well with the in-situ measurements of a mast [7,8]. In this sense, it is common practice to validate and verify the data simulated with a mesoscale model (as, e.g., the known Numerical Weather Prediction (NWP) models) against data from offshore met masts [9]. In most cases, this is a fair comparison, just because of the absence of relevant microscale terrain effects impacting the mast measurements for the height ranges of interest. It has to be kept in mind, however, that an NWP data point is representative for an area of several square kilometers and not just a single spot; this may lead to some smoothing in the simulated data in comparison to spot-like mast measurements. As a second discrepancy, NWP models typically do not give averaged data representative of a certain time interval but instead instantaneous values, e.g., every 10 or 30 min.

In a similar context as in the NEWA project, mesoscale model data were correlated with the measurement data, not just from one offshore met mast but several of those and a number of vertically profiling lidar devices installed on offshore platforms within the NORSEWInD project [10]. The activities within this project proved the usefulness of Doppler lidar technology for assessing offshore wind profiles provided a sufficient data availability is obtained. Like NWP models, the nowadays fully commercially available lidar devices can provide wind velocity data at a number of height levels more or less at the same point in time. This allows a direct assessment of the wind profile for up to several hundreds of meters, with the range depending on the explicit device specifications. With the Ferry Lidar Experiment, the concept of spatially distributed wind profile measurements is further developed. The profiling instrument moves with the ship and covers a distance that is comparable to a typical mesoscale dimension.

The article pursues the following three objectives:

- the ship lidar technology applied in the NEWA Ferry Lidar Experiment is to be introduced;
- the data produced by applying this technology is to be described in detail, such that the generated dataset can be used in future studies;
- and finally, a first comparison of the Ferry Lidar data to mesoscale model data is to be presented and discussed.

Approaching these objectives in the suggested order, the article is structured as follows. Following this introductory section, the ship lidar technology is presented in Section 2. The Ferry Lidar campaign within the NEWA project and the processed dataset from this campaign are described in Sections 3 and 4, respectively. In Section 5, we show an initial comparison between the recorded measurements and simulated data using an NWP model. An outlook in Section 6 and the conclusions in Section 7 complete the article.

2. Ship Lidar Technology

The Ship Lidar System applied for the NEWA Ferry Lidar Experiment has been developed by Fraunhofer IWES since 2009, and is an integrated system comprising the following components:

- a vertically profiling Doppler lidar device of the type Windcube v2 by the manufacturer Leosphere, (Orsay, France) which is the primary measurement sensor, capturing the wind velocity at up to 12 height levels above the instrument;
- a combination of an xSens MTi-G (Enschede, Netherlands) attitude and heading reference sensor (AHRS) and a Trimble SPS361 satellite compass (Sunnyvale, CA, USA), used to record high-resolution motion information that is required to correct the lidar wind data;
- a weather station collecting atmospheric data including air temperature, air pressure, and relative humidity and precipitation, complementing the dataset.

All data are collected and synchronized on a measurement computer. The Windcube v2 lidar device has a sampling resolution of about 0.7 s per line-of-sight (LoS) measurement and measures successively at four azimuthal positions along a cone with a half-opening angle of 28° followed by a vertical beam. From these five measurements the wind velocity vector is reconstructed and updated after each new LoS measurement, resulting in a measurement frequency of about 1.4 Hz. Further information on the ship lidar technology and its components can be found in Reference [11]. Figure 2 shows a typical installation of the system on deck of a medium-size vessel.

Figure 2. Installation of ship lidar System on deck of a medium-size vessel—© Fraunhofer IWES.

An indispensable element of the ship lidar technology is the motion correction algorithm combining recorded lidar and motion data in order to provide motion-corrected wind data. The algorithm applies the basic principles outlined in Reference [12] and referred to in, e.g., Reference [13]; the recorded wind velocity vector is corrected for the translational velocity of the vessel and the involved platform rotations (including the heading of the ship as well as roll and pitch), respectively, according to

$$V_{\text{true}} = TV_{\text{obs}} + \Omega \times TM + V_{\text{CM}} \tag{1}$$

where V_{true} is the desired wind velocity vector in the reference coordinate system, V_{obs} is the measured wind velocity vector in the platform frame of reference, T is the coordinate transformation matrix for a rotation of the platform frame coordinate system to the reference coordinates, Ω is the angular velocity vector of the platform coordinate system, M is the position vector of the platform coordinate system, and finally V_{CM} is the translational velocity vector at the center of motion of the platform with respect to a fixed coordinate system.

For the present configuration of the Fraunhofer IWES Ship Lidar System, a simplified motion correction was applied, where the vessel tilting was essentially ignored due to a negligible impact on the results. This modification was justified in a dedicated verification campaign, in which the system was tested under representative conditions against the fixed reference measurements of the FINO1 met mast in the German North Sea (for further details, see Reference [11]). The results confirm that the ship motions indeed affect the high-frequency data of the lidar but that these effects are averaged out when deriving the 10-min mean values; 10-min averaged lidar and reference data agree within ± 0.5 m/s in a distance of up to about 1.5 km and show a reasonable correlation with a coefficient of determination (R^2) for a linear fit of 0.99. How the motion correction acts on the data of the NEWA Ferry Lidar Experiment is shown in detail in Section 5.

The ship lidar technology should be further seen in the context of the floating lidar technology, a technology that has gained some attention within the offshore wind industry in the past decade (cf. Reference [14]). Floating Lidar Systems (FLSs), referring to a more or less stationary floating platform or buoy equipped with a commercial wind lidar device, particularly show economic benefits in an offshore wind resource assessment when compared to the state-of-the-art met masts, which as a standard, are used for the same application. An FLS can provide wind data of comparable quality as met masts but at considerably lower costs. A ship lidar cannot have the same purpose as a buoy-based measurement system since the movement of the ship has an essential impact on the obtained data, implying a non-stationarity, even if the local motions are compensated quite efficiently. Nonetheless, the development of the one technology can and has benefitted from the other. Also the motion effects on the measurements of an FLS can be described by Equation (1), though with another order of the most relevant motion impacts; while both for FLS and ship lidar heading information is most relevant to correct the wind direction data, the ship lidar wind speed data are essentially affected by the ship's translatory motions, the data of the FLS on the contrary by the tilting at a stationary position.

3. Ferry Lidar Campaign within NEWA

The NEWA Ferry Lidar Experiment started on 7 February 2017, and ended after four months of measurements on 8 June 2017. During this period, the Fraunhofer IWES Ship Lidar System was installed on the vessel Victoria Seaways, which belongs to the DFDS Seaways Group and operates on the route from Kiel, Germany, to Klaipeda, Lithuania, in the Southern Baltic Sea. One trip takes about 20 h, and the vessel spends about 4 h in the harbor each time. Figure 3 shows the average route of the ferry; only small deviations from this route were observed during the period of the campaign.

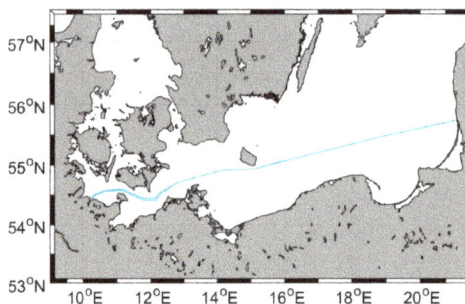

Figure 3. Route of Victoria Seaways, on which the ship lidar system was installed, from Kiel to Klaipeda in the Southern Baltic Sea (reproduced from Reference [15]).

The Victoria Seaways is a so-called ro-ro (for roll-on/roll-off) ship, designed to carry wheeled cargo, and has a passenger capacity of 600 persons. The vessel has an overall length of 199.14 m and a maximum velocity of 23.5 knots. The ship lidar system was installed on Deck 8, which is about 25 m above sea level (asl). Taking this offset into account, the resulting measurement heights of the lidar have been 65 m, 75 m, 90 m, 100 m, 110 m, 130 m, 150 m, 175 m, 200 m, 225 m, 250 m, and 275 m asl. Figure 4 shows a photograph of the vessel while it is in the harbor and of the installation of the ship lidar system.

Figure 4. Ship lidar installation on Victoria Seaways—© Fraunhofer IWES.

Figure 5 shows exemplary trajectories of the recorded and processed wind data (10-min averages) from the ship lidar system in the NEWA Ferry Lidar Experiment. This example covers four days—25–28 February 2017—of the (in total) four months of data. The plot on the left side shows the path of the system within these four days. The ferry was in the harbor four times (twice in Kiel and twice in Klaipeda) during this period, which is clearly visible when comparing uncorrected and corrected wind data (see the right plots in Figure 5), and the periods where the time series of corrected and uncorrected wind speed overlap. For the uncorrected (raw) wind data, the effects of the ship's motions influencing the lidar measurements essentially appeared as offsets in wind speed (negative or positive, depending on the heading of the ship relative to the prevailing wind direction) and direction (depending on the heading of the ship with respect to true North).

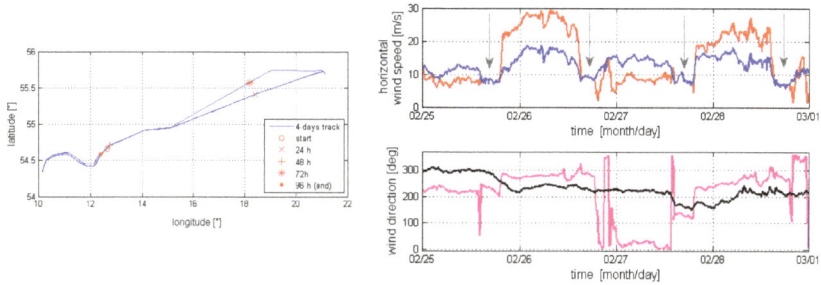

Figure 5. Four days of wind measurements from the ship lidar system: (**left**) path of Victoria Seaways in geo-coordinates, symbols show the position after 1, 2, 3, and 4 days; (**right**) corrected [blue/black] and uncorrected [red/pink] time series of horizontal wind speed [**top**] and direction [**bottom**] for 100 m measurement height. Periods where the ferry was in the harbor and therefore not moving are shown with the gray arrows.

4. Processed Dataset

The processed dataset comprises the corrected time series of horizontal wind speed and direction for the 12 measurement heights together with an availability measure for each data value. This measure represents the percentage of valid data in a 10-min interval where validity was defined on the basis of the respective carrier-to-noise ratio (CNR) value output by the lidar device and the availability of the motion data required to correct the lidar measurements. As a CNR threshold, we have used the value of −23 dB, which was pre-set and recommended by the manufacturer. High-frequency data, defined by one LoS measurement, that have a CNR value below this threshold were ignored for the derivation of the 10-min average.

Additionally, the position data of the ship—in terms of mean, minimum and maximum longitude (lon)/latitude (lat) coordinates, plus the distance between the minimum and maximum—were saved, along with the timestamp of the data vector. Figure 6 gives an impression of the processed data; the plot covers the same time span as in Figure 5 but this time shows the whole (corrected) wind speed and direction profiles instead of only a single time series. Note that the purpose of the plots was not to show and identify the single trajectories at the individual height levels, but rather to underline the profile information provided by the data. The occurrence of different atmospheric stability cases, represented by different wind shear and veer that is defined as wind speed and direction deviations with height, is clearly visible in this presentation. The magnitude of the vertical wind shear is correlated to the atmospheric stratification (e.g., References [16,17]). Periods with little or no wind shear (e.g., during most of February 25) indicate neutral or unstable stratification, while periods with a strong wind shear of up to 7 m/s between 65 and 275 m height (e.g., during most of February 26) indicate stable stratification. Thus, the vertical change of the wind speed with height is an indicator of the presence of different atmospheric stabilities that are mainly caused by air–sea temperature differences.

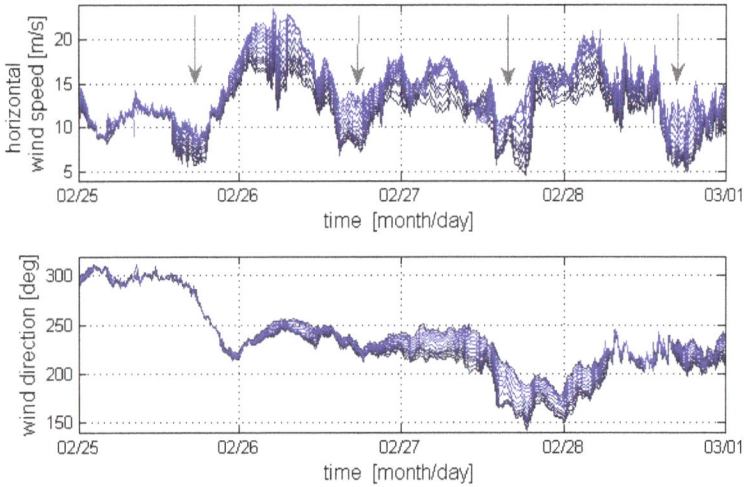

Figure 6. Processed horizontal wind speed and direction for the 12 measurement heights from 65 m to 275 m (darker colors for lower measurement heights). Periods where the ferry was in the harbor are again marked with the gray arrows.

Based on the derived availability measures for the individual 10-min averages, overall and monthly averages for the 12 measurement heights were evaluated (see Table 1). For this, it was additionally assumed that the 10-min availability must be equal to 80% or greater for a measured data point to be valid. This value is a typical threshold for a wind resource assessment. The so assessed availability of the corrected lidar time series was well above 90% for the seven lower measurement heights up to 150 m and only decreased for the upper heights.

Table 1. Data availability of processed wind time series per measurement height and month.

Month	Availability per Measurement Height [%]											
	65 m	75 m	90 m	100 m	110 m	130 m	150 m	175 m	200 m	225 m	250 m	275 m
February 2017	98	98	97	96	95	94	92	88	82	70	59	50
March 2017	97	97	98	97	97	95	93	90	83	75	64	51
April 2017	95	96	96	95	95	94	92	87	71	59	48	39
May 2017	95	95	96	96	96	95	93	88	75	60	44	30
June 2017	92	94	96	96	96	95	93	85	75	60	46	35
Total	96	96	96	96	96	94	92	88	77	66	53	42

4.1. Low-Level Jet Information as Part of the NEWA Ferry Lidar Dataset

The processing of the data from the NEWA Ferry Lidar Experiment was complemented with the derivation of essential low-level jet (LLJ) information. Following the definition of Baas et al. [18], an LLJ is defined at the lowest (local) maximum of the wind speed profile that is at least 2 m/s and 25% faster than the next minimum above. Note that this definition includes the upper and lower edge of the profile defined by the measurement range of the lidar instrument. The LLJ information in the dataset includes the horizontal wind speed and height of the identified profile maximum, the position of the measurement (longitude/latitude coordinates), and the timestamp.

Figure 7 (left plot) shows the wind profiles with LLJs found for the 4-day period already presented in Figure 6. The wind speed maxima appear for measurement heights between 100 m and 175 m, and all on 27 February 2017. The right plots of Figure 7 show all LLJ events found for the month of February (in total 20) with their geo-coordinates in comparison with those found for the month of May (in total 392). This initial comparison indicates that considerably more LLJ events were observed in the spring month than in a comparable period in winter, including a relevant number of wind-speed maxima at the lower limit of the profile range (cf. the color differentiation in Figure 7). This simple observation can be explained by the positive air–sea temperature difference that is typically largest in spring when the water body with its significantly higher heat capacity is still cold while the land at the coastlines typically already heats up. Offshore-oriented winds and the associated step change in the surface roughness can then lead to the development of LLJs. This process is well known (e.g., Reference [19]) and due to the cold waters in late spring very common for the Baltic Sea (e.g., Reference [20]).

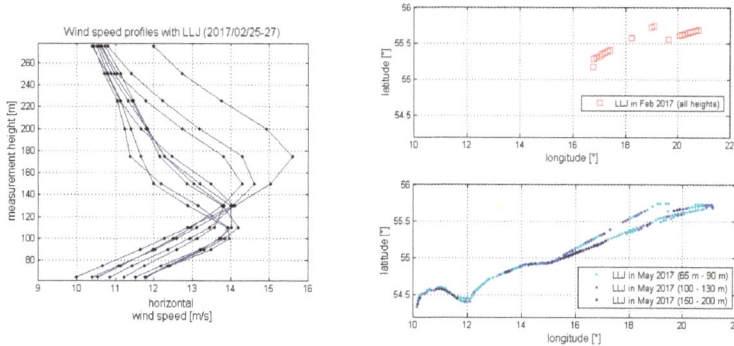

Figure 7. LLJ events observed in the NEWA Ferry Lidar Experiment: (**left**) vertical wind profiles for the period 25–27 February 2017; (**right**) events as markers on the ferry's route with latitude/longitude coordinates, where the top plot is for the month of February and the bottom plot for June in comparison.

4.2. Uncertainty Estimation for Ship Lidar Data

Generally, there are different ways to assess the uncertainty of (wind speed) measurements. In terms of metrology, and to trace back the uncertainty to available standards, a verification or calibration with respect to a suitable reference measurement is required. This principle is also followed in Reference [21] outlining the procedure of how to verify a ground-based vertically profiling Doppler lidar (or more general remote-sensing) instrument for applications in the wind industry. For a ship lidar system, such a verification test can only be realized under high effort, if at all. An attempt was undertaken in Reference [11] with the outcome referred to in Section 2. This result could be used to estimate a verification uncertainty component as one of the most relevant parts of the ship lidar measurement uncertainty. However, in the present case, we believe that, first, the dataset of the verification test was not big enough to derive robust results, and second, the ship lidar configuration in the verification may not have been similar enough to that in the final application; in particular, different types of vessels were used with different motion patterns.

Furthermore, the spatial aspect of the ship lidar measurements needs to be taken into account; the ship lidar measurements are not representative for a single spot in space as, e.g., the measurements of a met mast. The spatial variability of the measurements within the spanned 10-min interval are to be considered just as the temporal variations. This aspect becomes relevant when comparing the ship lidar data with other more "spatial" data (see Section 5). In this sense, we do not think that we could have assigned a single uncertainty figure to the ship lidar data, but rather include uncertainty considerations in the discussion of the data quality and precision in relation to reference datasets.

In this respect, it should also be paid attention to that since the ferry takes one trip per day with recurring departure and arrival times, there was a strong correlation between the location of the ferry, and therefore also the measurements, and time-of-day.

Uncertainty components that could be well estimated are the uncertainties of the input data to the motion correction (i.e., heading and velocity of the ship) and their direct impact on the corrected wind estimates. But since the applied motion measurements were very precise, this contribution was assumed to be negligible in the presented context.

5. Comparison with Mesoscale Model Data

For the measurements in the Ferry Lidar Experiment, the ship lidar system had traversed distances of up to 6 km in one 10-min interval. This distance corresponds to the mesoscale in meteorological observations [22], which suggests relating the ship lidar measurements to the output data of an NWP model having a similar scale. A corresponding initial comparison of the ship lidar measurements with mesoscale model data is presented in this section.

We have used the same model which is used to generate the wind atlas within the NEWA project, namely the Weather Research and Forecasting Model (WRF) [23]. In detail, we used an offshore optimized setup (similar to the setup used in Reference [24]) that was intensively tested in a case study phase for the Wind Atlas generation. Figure 8 shows the model domains centered around the Southern Baltic Sea. WRF time series were generated with a temporal resolution of 30 min and on a 3 km × 3 km grid; further details of the simulations are given in Table 2. Note that the temporal resolution here does not refer to an average but an instantaneous value every 30 min.

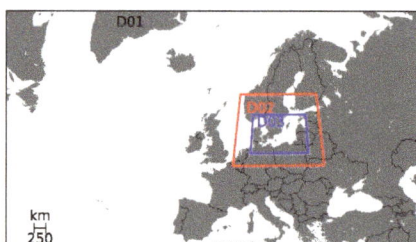

Figure 8. Model domains of WRF runs used for comparison.

Table 2. Details of WRF simulations.

WRF Model Version	3.8.1
Atmospheric Boundary Condition	ERA Interim [25]
Sea Surface Temperatures	OSTIA [26]
Land Use Data	USGS
Horizontal Resolution	D01 (27 km), D02 (9 km), D03 (3 km)
Nesting	1-way
Vertical Resolution	62 levels—with 20 below 1000 m
Microphysics	WSM 5-class scheme
Planetary Boundary Layer Scheme (PBL) Scheme	MYNN (Level 2.5)

In order to compare the ship lidar measurements with the WRF simulations, the measurement data are mapped onto the WRF grid, and for each 10-min interval, the "closest" WRF output value was selected. For a further description of this mapping procedure see Reference [27].

Figure 9 shows the results of the comparison for the period already referred to in the preceding data plots. The mesoscale data were interpolated from the terrain following the hybrid pressure coordinates to 100 m at every time step. Both for the horizontal wind speed and direction, the two time series showed a very similar course. Deviations seem to be in the order of, and not larger than, the scatter, i.e., the deviations between individual data points, in typical WRF-to-met-mast comparisons offshore [9].

Larger deviations are observed for the periods where the ship lies in the harbor (cf. also Figure 5) and could easily be explained by the existence of relevant microscale effects that are not included in the mesoscale model. Further offshore these effects are obviously negligible.

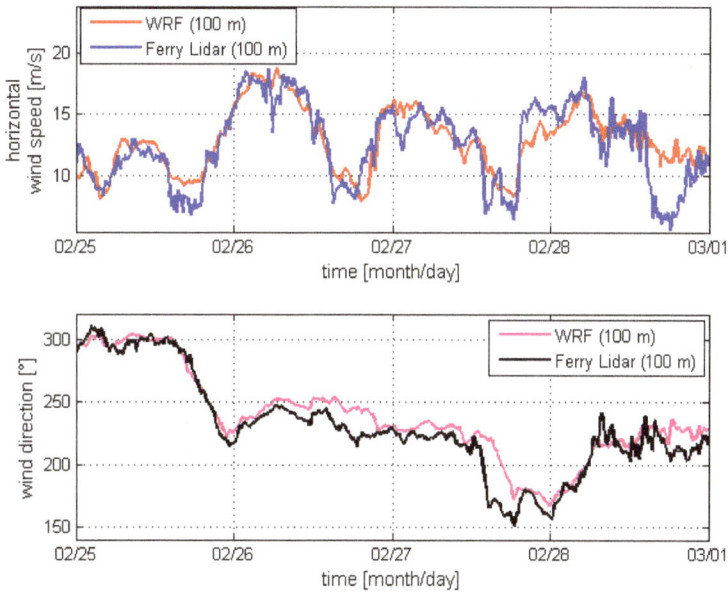

Figure 9. Comparison of the measured and simulated (Ferry Lidar vs. WRF) time series of horizontal wind speed and direction, here for 100 m measurement height only.

To further quantify the observed deviations, we have derived bias and root-mean-square-error (RMSE) values for the deviations in horizontal wind speed and wind direction between the Ferry Lidar and WRF data (see Table 3). For this evaluation, the "harbor effects" were excluded by considering only the data with a longitude coordinate larger than 10.4° and below 20.0°.

The comparison indicates that the 4-day period selected for demonstration indeed shows a better performance than the overall dataset. A deeper analysis may reveal that the agreement varies with the location of the measurement in combination with the wind direction due to the prevailing site effects that were existent at some locations over the track. A corresponding filtering should be applied when further working with the dataset and, in particular, when setting up a benchmark study.

Table 3. Bias and RMSE values for the deviations in horizontal wind speed and wind direction between the Ferry Lidar and WRF data.

	Horizontal Wind Speed [m/s]		Wind Direction [°]	
	4-Day Period (25–28 February 2017)	All Data	4-Day Period (25–28 February 2017)	All Data
Bias	0.09	0.35	−5.2	−6.3
RMSE	1.31	1.90	8.3	29.2

6. Outlook

First comparisons show a good correlation of the measurement data from the NEWA Ferry Lidar Experiment recorded using a ship-mounted lidar device with mesoscale model data simulated using the WRF model and mapped onto the ship's route in time and space. This opens up promising possibilities, overcoming the difficulty of verifying mesoscale model output data against in-situ data of, e.g., met masts that represent a very small and fixed volume only. Data of a ship lidar system could instead be representative of a line covering a certain space that is similar to the resolution of a mesoscale model simulation for a respective time scale. Comparing the time series of an NWP model with the data trajectories of the measurement system moving in space, and finding good correlations, may increase the trust in the model data and further promote their use as primary data for a wind resource assessment where on-site measurements are otherwise challenging. Studying the correlations between the two data sources along well-defined tracks may help to assess and understand where NWP models show larger and possibly systematic discrepancies.

The comparisons presented in this contribution represent only a first step. Within the NEWA project, the data of the Ferry Lidar Experiment are provided as a basis for more detailed benchmark studies that are planned and currently prepared for the near future. Furthermore, apart from that, the measurement data are part of the open accessibly NEWA Experimental Database, which will be available to the broader public for manifold investigations after the end of project, i.e., from May 2019.

7. Conclusions

The article presents the Ferry Lidar Experiment conducted within the NEWA project from February to June 2017. It introduces the ship lidar technology, which was applied on a ferry route from Kiel to Klaipeda in the Southern Baltic Sea. The obtained dataset of measurements is described in detail and made comprehensible. It is discussed how the data differ from met mast measurements as typical on-site data used for wind resource assessments in the wind industry. Ship lidar data stand out due to their covering of the detailed wind profile with a good data availability of more than 80% for measurement heights up to 175 m and still a reasonable availability (above 50%) for heights up to 250 m. From this data basis, e.g., low level jet information can be derived.

Measured wind time series are further compared to output data of an NWP model. The results of this initial comparison demonstrate that mesoscale simulations with their domain sizes in the order of the ship track, i.e., the distance the ship covers in the applied averaging period of 10 min, compare well to the ship lidar measurements.

To our knowledge, the NEWA Ferry Lidar Experiment provides the first dataset of wind profile measurements using lidar technology covering wind-energy relevant scales along a track of that length and distance (i.e., several hundreds of kilometers).

Author Contributions: J.G. prepared the original draft and managed the measurement campaign. E.C. developed and performed the data analysis and produced most of the results. M.D. provided the simulated data and contributed in numerous discussions. B.W. contributed in numerous discussions and supported the interpretation of the results. All authors reviewed and edited the manuscript until it reached the final stage.

Funding: This research: as part of the NEWA project, was funded by the German Federal Ministry for Economic Affairs and Energy (ref. no. 0325832A/B) on the basis of a decision by the German Bundestag with further financial support from NEWA ERA-NET Plus, topic FP7-ENERGY.2013.10.1.2.

Acknowledgments: The authors would like to acknowledge the contribution of DFDS, for giving us access to the Victoria Seaways and supporting the measurement campaign. Special thanks go to Gerrit Wolken-Möhlmann for his comments to the manuscript and his great efforts in developing and demonstrating the Ship Lidar technology in earlier studies as the ones referred to in the article. The presented simulations were performed at the HPC Cluster EDDY, located at the University of Oldenburg (Germany) and funded by the BMWi (ref. no. 0324005).

Conflicts of Interest: The authors declare no conflict of interest.

References

1. Petersen, E.L.; Troen, I.; Jørgensen, H.E.; Mann, J. Are local wind power resources well estimated? *Environ. Res. Lett.* **2013**, *8*, 011005. [CrossRef]
2. Mann, J.; Angelou, N.; Arnqvist, J.; Callies, D.; Cantero, E.; Chávez Arroyo, R.; Courtney, M.; Cuxart, J.; Dellwik, E.; Gottschall, J.; et al. Complex terrain experiments in the New European Wind Atlas. *Philos. Trans. A Math. Phys. Eng. Sci.* **2017**, *375*, 20160101. [CrossRef] [PubMed]

3. Peña, A.; Hasager, C.B.; Badger, M.; Barthelmie, R.J.; Bingöl, F.; Cariou, J.-P.; Emeis, S.; Frandsen, S.T.; Harris, M.; Karagali, I.; et al. Remote Sensing for Wind Energy, DTU Wind Energy-E-Report-0084 (EN). 2015. Available online: http://orbit.dtu.dk/files/111814239/DTU_Wind_Energy_Report_E_0084.pdf (accessed on 31 August 2018).
4. Witze, A. World's largest wind-mapping project spins up in Portugal. *Nature* **2017**, *542*, 282–283. [CrossRef] [PubMed]
5. Floors, R.; Pena, A.; Lea, G.; Vasiljevic, N.; Simon, E.; Courtney, M. The RUNE Experiment—A Database of Remote-Sensing Observations of Near-Shore Winds. *Remote Sens.* **2016**, *8*, 884. [CrossRef]
6. Lange, B.; Larsen, S.; Hojstrup, J.; Barthelmie, R. Importance of thermal effects and sea surface roughness for offshore wind resource assessment. *J. Wind Eng. Ind. Aerod.* **2004**, *92*, 959–988. [CrossRef]
7. Hahmann, A.; Witha, B.; Sile, T.; Dörenkämper, M.; Söderberg, S.; Navarro, J.; Leroy, G.; Folch, A.; Garcia Bustamante, E.; Gonzalez-Rouco, F. WRF sensitivity experiments for the mesoscale NEWA wind atlas production run. In Proceedings of the European Geosciences Union (EGU) General Assembly, Vienna, Austria, 8–13 April 2018.
8. Jimenez, B.; Durante, F.; Lange, B.; Kreutzer, T.; Tambke, J. Offshore Wind Resource Assessment with WAsP and MM5: Comparative Study for the German Bight. *Wind Energy* **2007**, *10*, 121–134. [CrossRef]
9. Olsen, B.T.; Hahmann, A.N.; Sempreviva, A.M.; Badger, J.; Jørgensen, H.E. An intercomparison of mesoscale models at simple sites for wind energy applications. *Wind Energy Sci.* **2017**, *2*, 211–228. [CrossRef]
10. Information on the NORSEWInD Project. Available online: http://www.norsewind.eu/norse/index.php (accessed on 29 August 2018).
11. Wolken-Möhlmann, G.; Gottschall, J.; Lange, B. First verification test and wake measurement results using a Ship-LIDAR System. *Energy Procedia* **2014**, *53*, 146–155. [CrossRef]
12. Edson, J.B.; Hinton, A.A.; Prada, K.E.; Hare, J.E.; Fairall, C.W. Direct covariance flux estimates from mobile platforms at sea. *J. Atmos. Ocean. Technol.* **1998**, *15*, 547–562. [CrossRef]
13. Strobach, E.J. The Impact of Coastal Terrain on Offshore Wind and Implications for Offshore Wind Energy. Ph.D. Thesis, University of Maryland, Baltimore, MD, USA, 2017.
14. Gottschall, J.; Gribben, B.; Stein, D.; Würth, I. Floating lidar as an advanced offshore wind speed measurement technique: Current technology status and gap analysis in regard to full maturity. *WIREs Energy Environ.* **2017**, *6*, e250. [CrossRef]
15. Karagali, I.; Hasager, C.; Badger, M.; Hahmann, A.; Volker, P.; Pena, A.; Gottschall, J.; Catalano, E.; Mann, J. Mapping offshore winds in the New European Wind Atlas. In Proceedings of the Offshore Wind Energy Conference, London, UK, 6–8 June 2017.
16. Peña, A.; Gryning, S.E.; Hasager, C.B. Measurements and Modelling of the Wind Speed Profile in the Marine Atmospheric Boundary Layer. *Bound.-Layer Meteorol.* **2008**, *129*, 479–495. [CrossRef]
17. Tambke, J.; Claveri, L.; Bye, J.A.; Poppinga, C.; Lange, B.; Bremen, L.V.; Durante, F.; Wolff, J.O. Offshore Meteorology for Multi-Mega-Watt Turbines. In Proceedings of the European Wind Energy Conference (EWEC), Athens, Greece, 27 February–2 March 2006.
18. Baas, P.; Bosveld, F.C.; Klein Baltink, H.; Holtslag, A.A.M. A Climatology of Nocturnal Low-Level Jets at Cabauw. *J. Appl. Meteorol. Climatol.* **2009**, *48*, 1627. [CrossRef]
19. Emeis, S. *Wind Energy Meteorology: Atmospheric Physics for Wind Power Generation*, 2nd ed.; Springer International Publishing: Cham, Switzerland, 2018; 255p.
20. Smedman, A.-S.; Högström, U.; Bergström, H. Low Level Jets—A Decisive Factor for Off-Shore Wind Energy Siting in the Baltic Sea. *Wind Eng.* **1996**, *20*, 137–147.
21. IEC 61400-12-1:2017 (International Standard). *Wind Energy Generation Systems—Part 12-1: Power Performance Measurements of Electricity Producing Wind Turbines*; IEC: Geneva, Switzerland, 2017.
22. WMO (World Meteorological Organization). Guide to Meteorological Instruments and Methods of Observation (CIMO guide) No. 8 (2014 edition, updated in 2017). Available online: http://www.wmo.int/pages/prog/www/IMOP/CIMO-Guide.html (accessed on 11 October 2018).
23. Skamarock, W.C.; Klemp, J.B.; Dudhia, J.; Gill, D.O.; Barker, D.M.; Duda, M.G.; Huang, X.Y.; Wang, W.; Powers, J.G. *A Description of the Advanced Research WRF Version 3*; Technical Report NCAR/TN-475+STR; NCAR—National Center for Atmospheric Research: Boulder, CO, USA, 2008.

24. Dörenkämper, M.; Optis, M.; Monahan, A.; Steinfeld, G. On the offshore advection of boundary-layer structures and the influence on offshore wind conditions. *Bound.-Layer Meteorol.* **2015**, *155*, 459–482. [CrossRef]

25. Dee, D.P.; Uppala, S.M.; Simmons, A.J.; Berrisford, P.; Poli, P.; Kobayashi, S.; Andrae, U.; Balmaseda, M.A.; Balsamo, G.; Bauer, D.P.; et al. The ERA-Interim reanalysis: Configuration and performance of the data assimilation system. *Q. J. R. Meteorol. Soc.* **2011**, *137*, 553–597. [CrossRef]

26. Donlon, C.J.; Martin, M.; Stark, J.; Roberts-Jones, J.; Fiedler, E.; Wimmer, W. The Operational Sea Surface Temperature and Sea Ice Analysis (OSTIA) system. *Remote Sens. Environ.* **2012**, *116*, 140–158. [CrossRef]

27. Catalano, E. Assessment of Offshore Wind Resources through Measurements from a Ship-Based LiDAR System. Master's Thesis, Genova University, Genova, Italy, 2017.

remote sensing

MDPI

Article

vEstimation of the Motion-Induced Horizontal-Wind-Speed Standard Deviation in an Offshore Doppler Lidar

Miguel A. Gutiérrez-Antuñano [1], Jordi Tiana-Alsina [2], Andreu Salcedo [1] and Francesc Rocadenbosch [1,3,*]

[1] CommSensLab, Unidad de Excelencia María de Maeztu, Department of Signal Theory and Communications, Universitat Politècnica de Catalunya (UPC), E-08034 Barcelona, Spain; miguel.angel.gutierrez@upc.edu (M.A.G.-A.); andreusalbos@gmail.com (A.S.)
[2] Nonlinear Dynamics, Nonlinear Optics and Lasers (DONLL), Department of Physics (DFIS), Universitat Politècnica de Catalunya (UPC), E-08222 Terrassa, Spain; jordi.tiana@upc.edu
[3] Institut d'Estudis Espacials de Catalunya (IEEC), Universitat Politècnica de Catalunya, E-08034 Barcelona, Spain
* Correspondence: roca@tsc.upc.edu

Received: 30 October 2018; Accepted: 11 December 2018; Published: 14 December 2018

check for
updates

Abstract: This work presents a new methodology to estimate the motion-induced standard deviation and related turbulence intensity on the retrieved horizontal wind speed by means of the velocity-azimuth-display algorithm applied to the conical scanning pattern of a floating Doppler lidar. The method considers a ZephIR™300 continuous-wave focusable Doppler lidar and does not require access to individual line-of-sight radial-wind information along the scanning pattern. The method combines a software-based velocity-azimuth-display and motion simulator and a statistical recursive procedure to estimate the horizontal wind speed standard deviation—as well as the turbulence intensity—due to floating lidar buoy motion. The motion-induced error is estimated from the simulator's side by using basic motional parameters, namely, roll/pitch angular amplitude and period of the floating lidar buoy, as well as reference wind speed and direction measurements at the study height. The impact of buoy motion on the retrieved wind speed and related standard deviation is compared against a reference sonic anemometer and a reference fixed lidar over a 60-day period at the IJmuiden test site (the Netherlands). Individual case examples and an analysis of the overall campaign are presented. After the correction, the mean deviation in the horizontal wind speed standard deviation between the reference and the floating lidar was improved by about 70%, from 0.14 m/s (uncorrected) to −0.04 m/s (corrected), which makes evident the goodness of the method. Equivalently, the error on the estimated turbulence intensity (3–20 m/s range) reduced from 38% (uncorrected) to 4% (corrected).

Keywords: wind energy; remote sensing; Doppler wind lidar; velocity-azimuth-display algorithm; resource assessment; offshore; turbulence intensity

1. Introduction

In recent years, offshore wind energy has become a trustable and mature technology for electricity generation [1]. At the end of 2016, 14 GW cumulative offshore wind capacity proved the importance of this technology in the energy mix, with Europe being the main area of development but also with a significant contribution from China [2]. Although most of the commercial developments of floating lidars are being carried out in shallow waters (0–30 m), their benefits are not limited to these depths and there is a tendency to go further off-coast to higher depths [3], where the advantages of floating lidar technology versus conventional anemometry are more significant.

Different remote sensing technologies have been used in wind energy, including satellite measurements in offshore environments [4,5], radar [6], sodar [7–9], and combined techniques [10,11]. Nevertheless, due to the high requirements of the industry regarding resolution and accuracy, lidar has been the most used technology for different applications in the wind energy sector since the appearance of the first commercial units. These applications include turbine control [12], resource assessment [13–15], wakes [16–18], and power curve measurements in flat terrain [19], among others.

In the resource assessment phase of offshore wind farms, floating lidars have become an alternative to conventional fixed metmasts because lidar allows more flexibility in the deployment in a cost-effective way [13,20–23]. In 2015, the Carbon Trust published a roadmap for the commercial acceptance of this technology in the wind industry [24]. A state-of-the-art report and recommended practices developed by the IEA Wind Task 32 [25] can be found in [26], and several validation tests and commercial developments in [27–33].

The increasing use of floating lidar systems in the offshore wind energy sector motivates the need to assess and compensate the effect of motion on floating lidar measurements [34,35]. It has been shown that both the static [36] and dynamic tilt [37–40] of the lidar instrument induce errors in the retrieved horizontal wind speed (HWS). Different approaches can be considered to reduce the impact of sea-waves-induced motion on the wind speed measured by floating lidar devices: mechanical [41,42] and numerical compensation methods [43,44].

Turbulence intensity (TI), which is defined as the ratio between the standard deviation of the HWS to the mean HWS, has a critical impact on wind turbine production, loads and design. The IEC61400-1 Normal Turbulence Model describes the TI threshold a wind turbine is designed for, and defines the wind turbine class of the machine that describes the external conditions that must be considered [45–47]. The lidar-observed TI is not identical to the "true" TI that can be measured by point-like measurements from cup anemometers. The lidar-observed TI is affected by the spatial (i.e., probe length) and temporal averaging (i.e., scanning time) of the Doppler lidar instrument and by the motion effects of sea waves on the lidar buoy. While spatial/temporal averaging effects on the measured TI can be found elsewhere [48–51], here we aim at *studying the effects of lidar motion on the measured TI and their statistical correction*. To simplify the mathematical framework to be presented next, we numerically assessed the motion-corrected HWS standard deviation under simple harmonic motion conditions of the lidar buoy for a given HWS and wind direction (WD). Towards this end, we considered a software motion simulator to emulate the motion of sea waves under these simplified motion conditions and the velocity-azimuth-display (VAD) algorithm [52] to retrieve the motion-corrupted HWS. Furthermore, simulation results were validated against experimental results as part of the IJmuiden test campaign.

This paper is organized as follows: Section 2 begins with a short description of the measurement instrumentation at IJmuiden and follows with a description of the methods used: it revisits the velocity-azimuth-display simulator, presents simulation examples of dynamic tilting of the lidar buoy under different initial conditions, and describes the proposed methodology to compute the standard deviation of the HWS error induced by lidar motion. Section 3 discusses the simulator's results from the IJmuiden data. Three study cases for different sea and atmospheric conditions are analysed. The overall performance of the proposed methodology for the whole 60-day measurement campaign is also presented. Finally, Section 4 gives concluding remarks.

2. Materials and Methods

2.1. Materials

The ZephIR™300 lidar used in this work is a continuous-wave focusable Doppler lidar adapted for offshore measurements that uses a conical scanning pattern combined with the velocity-azimuth-display algorithm to retrieve the wind velocity. The scan period is 1 s and each scan is composed of 50 lines of sight. The lidar can retrieve the wind vector between 10 and 200 m in height in user-defined steps of 1 m, although not simultaneously. The latter is a consequence of the focusing principle of the instrument, which also yields a height-dependent spatial resolution (e.g., 15 m at 100 m in height).

As described in [53], a validation campaign of the floating lidar was performed at the IJmuiden test site [54,55]. The aim of this campaign was to assess the accuracy of the EOLOS lidar buoy against metmast IJmuiden [24]. The main instruments used were: (i) a moving ZephIR™300 lidar in a buoy, measuring at 27, 58, and 85 m above the Lowest Astronomical Tide (LAT); (ii) a reference ZephIR™300 lidar placed on the metmast platform and measuring at 90, 115, 140, 165, 190, 215, 240, 265, 290, and 315 m above LAT, both measuring sequentially at each height; and (iii) sonic anemometers at 27, 58, and 85 m above LAT. The ZephIR™300 lidar has shown to be a reliable device for on- and offshore wind-energy applications. More detailed information about these sensors and additional sensors in the metmast can be found in [54]. Additionally, data from inertial measurement units were used to characterise the motion of the lidar buoy. In the present work, data from 1 April to 1 June 2015 were used.

2.2. The Velocity–Azimuth-Display Motion Simulator

The velocity-azimuth-display (VAD) algorithm enables the retrieval of the three components of the wind-speed vector from a vertically-pointing, conically-scanning Doppler lidar, as is the case of the ZephIR™300. Under the assumption of a constant wind vector, it can be shown that the radial wind speed component along the lidar line of sight as a function of the scan time follows a sinusoidal pattern (the so-called VAD pattern). The wind speed components can be retrieved from the amplitude and offset and this sinusoidal pattern by using geometrical considerations and a simple least-squares fitting procedure [52,56].

In previous works [40], we have presented the formulation of the VAD motion simulator under the assumption of a time-invariant and horizontally-homogeneous wind field. The simulator uses Euler's angles to compute the rotated line-of-sight vector at a given time in response to simultaneous pitch, roll, and yaw tilting angles (three degrees of freedom). Because the rotation matrix of the buoy can numerically be computed as a function of discrete time in response to harmonic excitations in these three angles, it is possible to compute the rotated set of lines of sight of the lidar for each conical scan in response to buoy motion. When the VAD retrieval algorithm is applied to the radial wind speed onto the rotated set of lines of sight, the motion-induced HWS is retrieved with a temporal resolution of 1 s (scan period of the ZephIR™300). In principle, the simulation process is complicated by the existence of three degrees of freedom, each one being described by three variables (i.e., amplitude, phase, and frequency) representing the sinusoidal excitation. In practice, dependence on the yaw is not considered because yaw motion can be considered static as compared to the lidar scan period. Therefore, wind direction errors caused by yaw motion are corrected by means of the buoy compass. The fact that the scan phase of the lidar scanning pattern (i.e., the starting line of sight of the scanning pattern at time zero) is completely uncorrelated with buoy roll/pitch movement forced us to carry out the study by defining different constraints on these variables (this is further discussed in Sections 2.3 and 2.4). Thus, two simple cases were considered in the publication above: static and dynamic buoy tilting. The latter was limited to specific constraints: (i) only one degree-of-freedom (either roll or pitch); (ii) zero initial phase of the angular movement; and (iii) zero scan phase of the VAD scanning pattern.

In the present paper, we overcome these constraints by considering: (i) the combined contributions from both roll and pitch degrees of freedom; (ii) all possible phases in roll and pitch motion; and (iii) all possible phases in the VAD scan. To illustrate the importance of these parameters, Figure 1 plots the simulated error on the VAD-retrieved HWS (Equation (1) next) under roll-only lidar motion (one degree of freedom) as a function of the scan phase (*x*-axis), motional angular period (*y*-axis), and motional phase (Figure 1a–d). The error on the retrieved *HWS* is defined as

$$Z = \overline{HWS} - HWS, \tag{1}$$

where *HWS* is the real wind speed and $\overline{HWS}$ is the VAD-retrieved HWS.

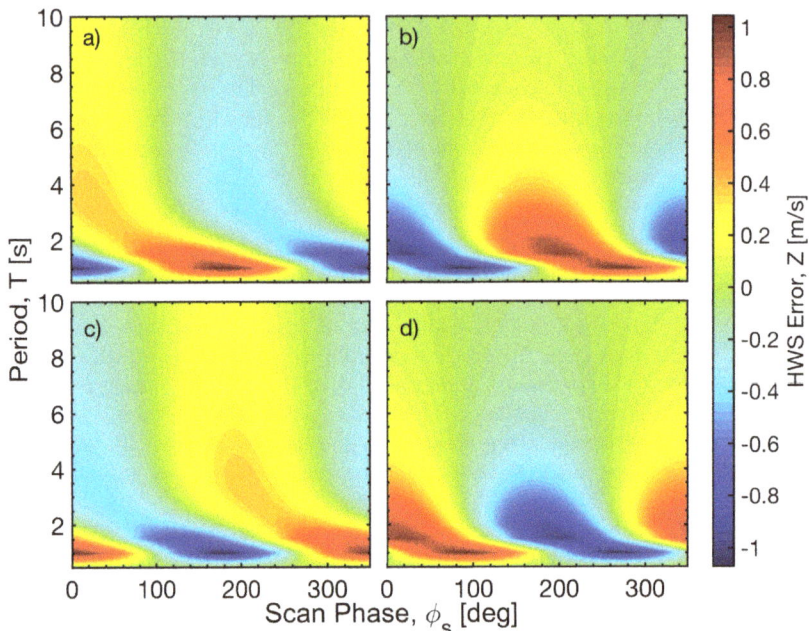

Figure 1. Simulated horizontal wind speed (HWS) error, Z (Equation (1)), under roll-only motion as a function of velocity-azimuth-display (VAD) scan phase (*x*-axis) and motional period, *T*, (*y*-axis). Roll phase (ϕ_r) is equal to 0 deg (**a**), 90 deg (**b**), 180 deg (**c**), and 270 deg (**d**). Roll amplitude is 3.5 deg, wind vector is $(0, 10, 0)$ m/s, and measurement height is 100 m in all panels [40].

This plot shows that, for 10 m/s HWS and 3.5 deg tilt, the HWS error increases to ±10% depending on the lidar scan phase. When comparing top and bottom panels in Figure 1, which account for 180-degree difference in roll phase, positive HWS errors in the top panels translate into negative ones in the bottom panels and vice-versa. Therefore, both the initial phase of movement and that of the VAD scan should be taken into account to evaluate the impact of lidar motion on the HWS error.

2.3. Motion-Induced HWS Error Variance

In this section, we introduce the methodology used to estimate the HWS error variance induced by lidar motion. This assumes no "a priori" information about the radial wind component measured by each line of sight of the scanning pattern.

As mentioned in Section 2.2, the VAD simulator retrieves the motion-corrupted HWS (1 s resolution) in response to roll and pitch harmonic motion, lidar scan phase, and HWS and WD at a given measurement height. In turn, each degree of freedom (roll/pitch) is characterised by three variables—namely, amplitude, period, and phase. Therefore, the HWS retrieved by the VAD motion simulator can be expressed as

$$\overline{HWS} = h(HWS, WD, H, A_r, \phi_r, T_r, A_p, \phi_p, T_p, \phi_s), \tag{2}$$

where h is the nonlinear function modelling the VAD-fitting algorithm, H is the measurement height, and A, ϕ, and T are the amplitude, phase, and period associated to sinusoidal roll/pitch motional excitation, $A \cdot sin(2\pi f t + \phi)$, with $f = \frac{1}{T}$ (subscripts r and p stand for roll and pitch angles, respectively), and ϕ_s is the conical scan phase of the lidar.

Horizontal wind speed (HWS), wind direction (WD), and roll/pitch amplitudes and periods ($A_{r/p}$, $T_{r/p}$, respectively) are deterministic variables because they can be measured experimentally (e.g., HWS and WD from metmast anemometers or a reference fixed lidar, and roll/pitch amplitudes and periods from inertial measurement units on the buoy). In contrast, roll/pitch motional phases, $\phi_{r/p}$, and VAD scan phase, ϕ_s, become random variables because buoy initial motion conditions ($\phi_{r/p}$) cannot be recovered from inertial measurement unit measurements, nor is the scan phase (ϕ_s) available from the lidar.

For convenience, we define HWS-error function g as Equation (2) above, constrained to the set of deterministic conditions $\vec{S} = (HWS, WD, A_p, T_p, A_r, T_r)$ (i.e., given HWS, WD, and buoy attitude) minus the true HWS,

$$Z = g(\phi_r, \phi_p, \phi_s) = h|_{\vec{S}} - HWS. \tag{3}$$

The motion-induced HWS error variance can be estimated for the first and second raw moments of Z as

$$Var(Z) = E(Z^2) - E(Z)^2. \tag{4}$$

By using the expectation theorem [57], the first two raw moments of Z can be computed as

$$E(Z^n) = \int_{-\infty}^{\infty}\int_{-\infty}^{\infty}\int_{-\infty}^{\infty} g(\phi_r, \phi_p, \phi_s)^n f_{\Phi_r \Phi_p \Phi_s}(\phi_r, \phi_p, \phi_s) d\phi_r d\phi_p d\phi_s, \tag{5}$$

where $f_{\Phi_r \Phi_p \Phi_s}(\phi_r, \phi_p, \phi_s)$ is the joint probability distribution function for the random-variable set of phases, Φ_r, Φ_p, and Φ_s; and $n = 1, 2$. At this point, and following standard notation in probability theory [58], we use uppercase Greek letters to denote random variables and lowercase letters to denote the values for these variables.

Formulation of the multivariate distribution function $f_{\Phi_r \Phi_p \Phi_s}(\phi_r, \phi_p, \phi_s)$ can largely be simplified by introducing different properties describing the statistics of random variables Φ_r, Φ_p and Φ_s. We hypothesise that information about any one of these three variables gives no information about the other two, which is equivalent to saying that phases Φ_r, Φ_p and Φ_s are independent random variables. This will be further discussed in Section 2.4. As a result, joint density function $f_{\Phi_r \Phi_p \Phi_s}$ factors out as the product of univariate functions f_{Φ_r}, f_{Φ_p} and f_{Φ_s}, as $f_{\Phi_r \Phi_p \Phi_s} = f_{\Phi_r} f_{\Phi_p} f_{\Phi_s}$. This enables us to rewrite Equation (5) as

$$E(Z^n) = \int_0^{2\pi}\int_0^{2\pi} f_{\Phi_r}(\phi_r) f_{\Phi_p}(\phi_p) \left[\int_0^{2\pi} g(\phi_r, \phi_p, \phi_s)^n f_{\Phi_s}(\phi_s) d\phi_s\right] d\phi_r d\phi_p, \tag{6}$$

where it has been used that random variables Φ_r, Φ_p, and Φ_s are *uniformly* distributed in $[0, 2\pi)$ so that

$$f_v(v) = \frac{1}{2\pi}, v \in [0, 2\pi) \; with \; v = \phi_r, \phi_p, \phi_s. \tag{7}$$

The hypothesis of uniform distribution in $[0, 2\pi)$ for scan phase Φ_s is well-justified on account of the fact that, despite the 1 s temporal resolution of the lidar, measurements are not exactly delivered every second due to lidar refocusing and internal checkings.

We define

$$g'_n(\phi_r, \phi_p) = \int_0^{2\pi} g(\phi_s)^n \Big|_{\Phi_r=\phi_r, \Phi_p=\phi_p} f_{\Phi_s}(\phi_s)d\phi_s, \tag{8}$$

which can physically be understood as the n-th raw moment of the HWS error due to random variable scan phase, Φ_s, for a given pair of roll and pitch phases, $\Phi_r = \phi_r$ and $\Phi_p = \phi_p$. Equivalently, Equation (8) can be written as

$$g'_n(\phi_r, \phi_p) = E(g(\phi_s)^n \Big|_{\Phi_r=\phi_r, \Phi_p=\phi_p}, \tag{9}$$

which is the expected value of $g(\phi_s)^n$ for a particular pair of motional phases $\Phi_r = \phi_r$ and $\Phi_p = \phi_p$. Because f_{Φ_s} is a uniform probability density function, the expected value is just the arithmetic mean of $g(\phi_s)^n$ along the Φ_s dimension.

By substituting Equation (8) into Equation (6), Equation (6) takes the form

$$E(Z^n) = \int_0^{2\pi} f_{\Phi_r}(\phi_r) \left[\int_0^{2\pi} g'_n(\phi_r, \phi_p) f_{\Phi_p}(\phi_p)d\phi_p \right] d\phi_r. \tag{10}$$

By comparing Equation (10) to Equation (6) above, it emerges that we reduced the calculus from the tri-dimensional domain $[\Phi_r, \Phi_p, \Phi_s]$ in Equation (6) to the bi-dimensional domain $[\Phi_r, \Phi_p]$ in Equation (10). The same procedure above can be repeated recursively to reduce Equation (10) from the bi-dimensional domain $[\Phi_r, \Phi_p]$ to the one-dimensional domain, $[\Phi_r]$. Thus, in similar fashion to Equation (8), we define

$$g''_n(\phi_r) = \int_0^{2\pi} g'_n(\phi_p) \Big|_{\Phi_r=\phi_r} f_{\Phi_p}(\phi_p)d\phi_p, \tag{11}$$

which can also be written as (counterpart of Equation (9))

$$g''_n(\phi_r) = E(g'_n(\phi_p) \Big|_{\Phi_r=\phi_r}. \tag{12}$$

Substitution of Equation (11) into Equation (6) yields

$$E(Z^n) = \int_0^{2\pi} g''_n(\phi_r) f_{\Phi_r}(\phi_r)d\phi_r, \tag{13}$$

or, equivalently,

$$E(Z^n) = E(g''_n(\phi_r)), \tag{14}$$

which is to say that the raw moments of the HWS error function Z can be calculated by using a three-step procedure given by Equations (9), (12) and (14), where the contribution from each random variable (i.e., roll phase, Φ_r, pitch phase, Φ_p, and scan phase, Φ_s) are successively averaged out.

The practical computational procedure of Equations (9), (12) and (14) is as follows: for a given set of simulation parameters $\vec{S} = (HWS, WD, H, A_p, T_p, A_r, T_r)$, the HWS error (Equation (1)) is calculated by the motion simulator of Section 2.2 in the $[0 - 2\pi] \times [0 - 2\pi] \times [0 - 2\pi]$ domain of random phases Φ_r, Φ_p, and Φ_s by using a grid of $24 \times 24 \times 24$ evenly spaced points between 0 and 2π. This gives a 3D matrix of HWS error values similar to the 2D matrix represented in Figure 1, but in three dimensions. Then, the HWS error is averaged along the Φ_s (scan phase) dimension of the matrix for every pair of roll/pitch phase values (ϕ_r, ϕ_p) to obtain g'_1 (1st raw moment, Equation (9)). Next, this procedure is repeated recursively over the Φ_p dimension of g'_1 (now a 2D instead of a 3D matrix) to yield g''_1 (a 1D matrix or vector, Equation (12)), and finally, over the Φ_r dimension of g''_1, which yields the scalar $E(Z)$ (Equation (6)). This three-step procedure is repeated twice to compute $E(Z)$ and $E(Z^2)$. Finally, the sought-after HWS error variance, $Var(Z)$, is obtained from Equation (4). The standard deviation of the motion-induced HWS error, σ_Z, is computed as the square root of the variance.

2.4. Roll/Pitch Correlation Hypothesis

As described by vector $\vec{S}$ (Equation (3)), besides the input parameters directly related to the wind (i.e., HWS and WD), the simulator requires roll and pitch angular amplitude and period information to describe buoy attitude. This information is derived from 5 Hz inertial measurement unit data on the buoy [53]. We hypothesise that, if significant correlation between roll and pitch periods and between roll and pitch amplitudes is found, these two angular variables can be considered equivalent and, therefore, a single amplitude and period can meaningfully be used to describe motion in both axes. Thus, for each 10 min timestamp, we computed the motional amplitude as the average roll and pitch angular amplitude, and the motional period as the average roll and pitch period. This is to say that buoy attitude can be given by significant wave height and wave period, which is a state-of-the-art practice in oceanography and wind energy to model the sea state. To evaluate this hypothesis, Figure 2 shows roll–pitch scatter plots for both amplitude and period variables as measured by inertial measurement units during the study period. The pitch-to-roll correlation coefficients in angular amplitude and period were 0.88 and 0.54, respectively, demonstrating the validity of the correlation hypothesis for the amplitude and a comparatively weaker correlation for the period. The correlation coefficient is equivalent to the cross-covariance at zero time lag (see inset). Further experimental analysis showed that this comparatively lower correlation is due to the bi-modality behaviour of the angular period, which means that two dominant motional periods (or frequencies) coexist in many measurement records. In this case, the single-frequency harmonic motion model becomes an oversimplification of reality, this being the main limitation of the method.

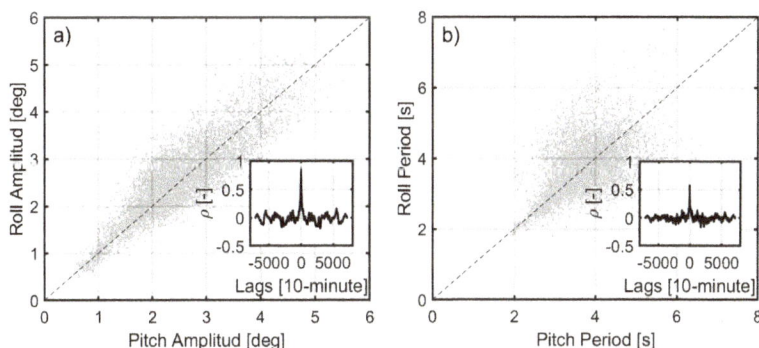

Figure 2. Scatter plots for 10-min-averaged roll and pitch angles. (**a**) angular amplitude; (**b**) angular period. Dashed lines correspond to the 1:1 reference line. Insets show the roll–pitch cross-covariance for different time lags.

2.5. Wind Direction Exclusion

In previous works [40] limited to one degree of freedom in angular motion (i.e., roll or pitch only), the authors have shown that wind direction has a relevant impact on the HWS error. In addition, under one-degree-of-freedom harmonic motion, it has been shown that the HWS error exhibits sinusoidal dependence with wind direction.

Under the two-degrees-of-freedom model and the approximation of nearly correlated roll and pitch motion (Section 2.4), the HWS error was simulated for different wind directions (0, 30, 60, ..., 330 deg) and periods (1, 1.5, 2, 2.5, ..., 10 s) for a particular pair of values, HWS (10 m/s) and angular amplitude (3.5 deg). Figure 3 shows the increase of the motion-induced HWS error standard deviation for low angular periods and that the error standard deviation does not depend on wind direction.

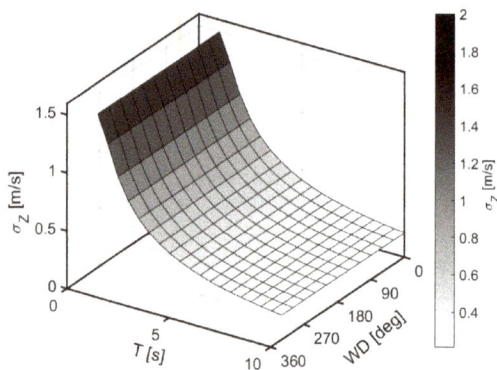

Figure 3. Simulator results of motion-induced HWS error standard deviation, σ_Z as a function of motional period, T (x-axis) and wind direction (y-axis). HWS is 10 m/s HWS, roll-and-pitch amplitude is 3.5 deg.

A plausible explanation is as follows: the fact that roll and pitch are approximately linearly correlated in amplitude and period enables an equivalent one-degree-of-freedom treatment of buoy motion (buoy tilt "amplitude" and buoy tilt "period"). Because the HWS error standard deviation follows a sinusoidal variation with wind direction [40] and roll and pitch axes are orthogonal ($\pi/2$ phase shift between roll and pitch sinusoidal variation with wind direction), the error standard deviation, which is the quadratic sum of roll and pitch error standard deviations, remains constant with wind direction. Similar simulations were carried out for other HWSs and angular amplitude conditions, showing analogous behaviour with wind direction. Therefore, under the approximation of correlated roll and pitch motion, wind direction was excluded from the analysis.

3. Results and Discussion

To validate the simulator's performance in Section 2.2 when estimating the motion-induced HWS error standard deviation on the floating lidar (in the buoy), data from metmast IJmuiden (Section 2.1) was used. Two sensors were chosen as reference: (i) the ZephIR™300 lidar and (ii) the sonic anemometers in the metmast. The intercomparison was carried out at 10 min temporal resolution.

On one hand, the advantage of using the fixed lidar as reference is that we were comparing two identical lidars although configured to sequentially measure at a different number of heights (the lidar in the metmast measured at 10 heights while the lidar in the buoy at only 3). On the other hand, the advantage of using sonic anemometers is that this technology is more accepted by the wind industry and more similar to the cup anemometer, the official sensor reference in the state-of-the-art. This is because both sonic and cup anemometers perform point-like measurements as opposed to the volume scanning technique of the lidar.

There is only one measurement height in common for the three collocated devices: 85 m. Therefore, this height was the one used in for the comparison.

3.1. Binning

As discussed in Section 2.3, an underlying requirement of the proposed methodology is the assumption of uncorrelated- and uniformly-distributed phases ϕ_r, ϕ_p, and ϕ_s in the floating lidar for each HWS and buoy motional condition under study. To better fulfill this requirement, a binning procedure was applied to the whole campaign dataset (6985 10-min records). As a result, each bin contained measurement records with similar HWSs and motional conditions but not necessarily (and usually not) having correlative timestamps. As a result of this timestamp "mixing" into a bin (also called time "scrambling"), the requirement of uncorrelated and uniformly distributed phases (Section 2.3) into a bin was reinforced. The chosen binning variables were: HWS, angular amplitude, and period in equally spaced bins of width 1 unit ((m/s), (deg), and (s), respectively) centred on integer values (bin edges at [0.5 1.5], [1.5 2.5] units, etc.).

Table 1 shows the 25 most frequent cases in the IJmuiden campaign. The most common HWSs were between 3 and 12 m/s, amplitudes were between 2 and 4 degrees, and motional periods were between 3 and 4 s. The total set of measurement cases is considered in Figure 6 and Section 3.4. The conditions of the site during the study period included HWS between 2 and 21 m/s, angular amplitudes between 1 and 5 deg, and periods between 2 and 5 s.

Table 1. The 25 most frequent HWS and motional cases in the IJmuiden campaign. "Case no." is the bin number sorted by decreasing frequency of event occurrence ("1" indicating the most frequent case); HWS (m/s) stands for 10-min mean horizontal wind speed; AA (deg) stands for motion angular amplitude; T (s) stands for period; Count no. is the bin count number; and σ_Z (m/s) is the motion-induced HWS error standard deviation estimated by the simulator after Equation (4).

Case No.	HWS (m/s)	AA (deg)	T (s)	Count No.	σ_Z (m/s)
1	8	3	4	288	0.18
2	5	2	4	247	0.07
3	9	3	4	237	0.20
4	7	2	4	208	0.10
5	6	2	4	198	0.09
6	7	3	4	196	0.16
7	6	3	4	182	0.13
8	6	2	3	180	0.12
9	3	2	4	175	0.04
10	7	2	3	174	0.14
11	10	3	4	169	0.22
12	5	2	3	166	0.10
13	4	2	4	164	0.06
14	8	2	4	157	0.12
15	8	2	3	133	0.16
16	11	3	4	130	0.25
17	5	3	4	130	0.11
18	9	3	3	112	0.27
19	8	3	3	108	0.24
20	7	3	3	106	0.21
21	12	3	4	100	0.27
22	11	4	4	95	0.33
23	2	1	3	91	0.02
24	4	2	3	86	0.08
25	3	1	3	80	0.03

3.2. Variance of the Sum of Partially Correlated Variables

Next, we discuss how to combine the motion-induced HWS error standard deviation, σ_Z, estimated by the simulator (Section 2.3), with the reference HWS standard deviation, σ_{ref}, which is measured from either the lidar on the metmast, $\sigma_{ref(lidar)}$, or the sonic anemometer, $\sigma_{ref(sonic)}$, in order to estimate the *motion-corrected HWS standard deviation*, σ_{corr}. The latter is the key output of our study to be compared with the HWS standard deviation measured by the floating lidar, σ_{moving}.

According to the law of propagation of errors, the corrected variance, σ_{corr}^2, of the sum of two variables (the real wind speed (or reference), HWS, and the motion-induced HWS error, Z; Equation (1)) is written as [57]

$$\sigma_{corr}^2 = \sigma_{ref}^2 + \sigma_Z^2 + 2\,cov(ref, Z), \tag{15}$$

where σ^2 stands for variance (i.e., the square of the standard deviation) and $cov(ref, Z)$ is the covariance between the reference HWS and the motion-induced HWS error.

Equation (15) above states that the standard deviation of the HWS measured by the moving lidar not only depends on the variance from both the wind (intrinsic turbulence) and the motion-induced error, but also on the covariance between these two variables. In the limit cases of: (i) uncorrelated variables (U), $cov(ref, Z) = 0$, and (ii) linearly correlated variables (C), $cov(ref, Z) = \sigma_{ref} \cdot \sigma_Z$, Equation (15) reduces to

$$\sigma_{corr}^{U} = \sqrt{\sigma_{ref}^2 + \sigma_Z^2}, \tag{16}$$

$$\sigma_{corr}^{C} = \sigma_{ref} + \sigma_Z. \tag{17}$$

In what follows, and unless otherwise stated, the motion-corrected HWS standard deviation σ_{corr} is calculated assuming partial correlation between these variables (i.e., by using Equation (15)). The term $cov(ref, Z)$ is computed from the correlation coefficient between the reference HWS, ref, and the expected value of the motion-induced HWS error, $E(Z)$. Here, we use the mathematical definition $cov(ref, Z) = \rho_{ref,Z} \cdot \sigma_{ref} \cdot \sigma_Z$, where $\rho_{ref,Z}$ is the correlation coefficient, and σ_{ref} and σ_Z are the standard deviations of the 10-min reference HWS and 10-min motion-induced HWS error, respectively. In practice, and considering that the binning process ensures similar motional characteristics in each bin (Section 3.1), we computed a single ordered pair (reference HWS, $E(Z)$) per bin (109 simulations) and a single correlation coefficient given these 109 bins ($\rho = 0.78$), which is representative of the motional conditions of the overall sample under study.

3.3. Analysis of Particular Cases

In order to discuss the goodness of the proposed methodology to estimate the motion-induced HWS standard deviation, this section tackles three representative cases (or bins) from Table 1: cases no. 2, 18, and 25. The first case gave good estimation of the motion-induced HWS standard deviation; the second one, overestimation; and the third one, underestimation.

Figure 4 plots the standard deviation of the HWS with and without correction (Equation (15)), using the lidar on the metmast as reference. The sample size associated with each of these three cases is listed in the "Count no." column of Table 1.

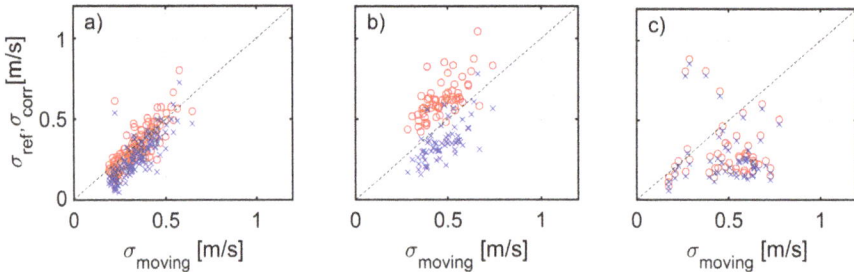

Figure 4. Selected discussion case examples from Table 1. (**a**) case no. 2, HWS = 5 m/s; angular amplitude (AA) = 2 deg; period (T) = 4 s; (**b**) case no. 18, HWS = 9 m/s; AA = 3 deg; T = 3 s; (**c**) case no. 25, HWS = 3 m/s; AA = 1 deg; T = 3 s). All panels: the x-axis represents the 10-min HWS standard deviation of the floating lidar, denoted σ_{moving}. The y-axis represents (in blue crosses) the standard deviation of the reference-lidar HWS (denoted σ_{ref}) and (in red circles) the standard deviation of the motion-corrected HWS (denoted σ_{corr}). The dashed black line represents the 1:1 reference line.

Figure 4a (case no. 2) shows 247 10-min measurements for which the proposed methodology accurately estimated the standard deviation of the motion-induced HWS error. Before applying Equation (15) correction, uncorrected values fell below the 1:1 line, which indicates that the moving lidar "saw" a higher standard deviation. After Equation (15) correction, most of the measurements laid on the 1:1 reference line.

Figure 4b,c, which are representative of case nos. 18 and 25, respectively, show two opposite situations: on one hand, for case no. 18 (Figure 4b), the simulator overestimated the influence of motion and the corrected values laid above the 1:1 line. Further investigation showed that this can be caused by the lack of consistency of the roll/pitch correlation hypothesis (Section 2.4) due to most measurements undergoing bi-modal motion behaviour. On the other hand, case no. 25 (Figure 4c) showed corrected values falling nearly always below the 1:1 line, which means that the estimated correction given by the motion simulator was too low. Further inspection indicated that this underestimation was caused by untrustworthy retrieval of the HWS by the VAD algorithm, as made evident by too-high spatial variation (SV) values from the ZephIR™300 lidar (Figure 5, to be discussed in Section 3.4). The spatial variation is a lidar internal parameter related to the goodness of fit that reveals whether the measurement data is consistent or not with the sinusoidal model assumed by the VAD algorithm. Thus, high SV values are related to a poor VAD fitting, and they are usually found in low HWS, where Taylor's frozen-eddies hypothesis is no longer true and the lidar does not measure a homogeneous wind along the VAD scanning area.

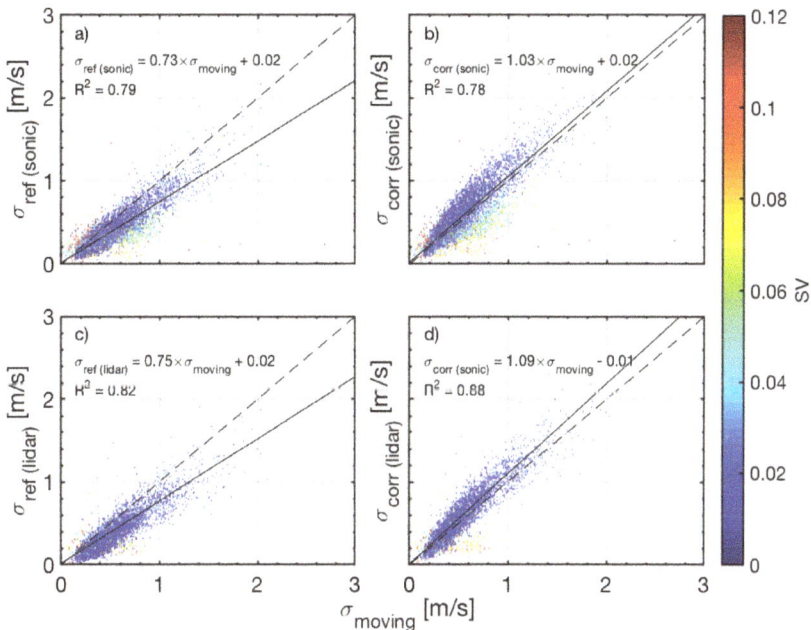

Figure 5. Analysis of the whole campaign (109 cases, 6985 10-min measurement records) by using as reference the sonic anemometer (**a**) and the lidar on metmast (**b**), without motion correction (**c**); $\sigma_{ref(sonic/lidar)}$ in the *y*-axis) and with motion correction (**d**); $\sigma_{corr(sonic/lidar}$ in the *y*-axis). The *x*-axis represents the HWS standard deviation of the floating lidar, denoted σ_{moving}. Each point is a 10-min record. Dashed lines represent the 1:1 line. Solid lines plot the regression lines. Colour bar indicates spatial variation.

Table 2 gives mean difference (MD) and root mean square error (RMSE) indicators for case nos. 2, 18, and 25 in Figure 4 without and with motion correction:

Table 2. Statistical indicators with and without motion correction for the selected discussion case examples from Table 1. MD stands for mean deviation and RMSE stands for root mean square error (see text and Equations (18) and (19)). MD and RMSE units are (m/s).

Case No.	Count No.	Reference Sonic				Reference Lidar			
		Corrected		Uncorrected		Corrected		Uncorrected	
		MD	RMSE	MD	RMSE	MD	RMSE	MD	RMSE
2	247	0.08	0.15	0.14	0.19	0.02	0.08	0.08	0.11
18	112	−0.10	0.18	0.14	0.21	−0.12	0.15	0.12	0.15
25	80	0.20	0.26	0.22	0.28	0.20	0.31	0.22	0.33

The motion-corrected mean deviation is defined as

$$MD_{corr} = \frac{\sum_i (\sigma_{moving,i} - \sigma_{corr(x),i})}{N},$$ (18)

where N is the case "count no." (Table 1), σ_{moving} is the HWS standard deviation measured by the floating lidar (already introduced in Section 3.2), and $\sigma_{corr(x)}$ is the motion-corrected HWS standard deviation (Equation (15)) of the reference instrument, where $x = lidar$ denotes the reference fixed lidar and $x = sonic$ denotes the sonic anemometer. Subscript i is the count-number index, that is, i went from $i = 1$ to $i = 247$ for case no. 2.

The motion-corrected root mean-square error is defined as

$$RMSE_{corr} = \sqrt{\frac{\sum_i (\sigma_{moving,i} - \sigma_{corr(x),i})^2}{N}}.$$ (19)

Similarly, uncorrected MD and RMSE indicators are computed by substituting $\sigma_{corr(x),i}$ with $\sigma_{ref(x),i}$, the reference HWS standard deviation, in Equations (18) and (19) above. These indicators are denoted MD_{ref} and $RMSE_{ref}$, respectively.

The mean deviation gives an estimation of the systematic error, equivalently, the amount of bias, while the RMSE is the quadratic mean of differences, with an ideal value of 0 indicating a perfect fit.

As shown in Table 2, the mean deviation for case no. 2 improved from 0.08 (uncorrected) to 0.02 m/s after motion correction. The RMSE also improved from 0.11 to 0.08 m/s. For overestimation case no. 18, the mean deviation changed sign from 0.12 to −0.12 m/s and, for underestimation case no. 25, the mean deviation virtually did not change (from 0.22 to 0.20 m/s). In over/underestimated case nos. 18 and 25, the RMSE did not improve after motion correction by Equation (15). All things considered, these indicators were consistent with the discussion carried out for Figure 4a–c, and they were therefore used to quantitatively analyse the overall campaign in the following.

3.4. Analysis of the Whole Campaign

In this section, we discuss *overall performance* of the motion-corrected HWS standard deviation, σ_{corr}, calculated via Equation (15) and, for comparison, via Equations (16) and (17), for the whole measurement campaign at IJmuiden (6985 10-min records clustered into 109 cases).

In similar fashion to Figure 4 but for the whole campaign, Figure 5 compares the HWS standard deviation of the moving lidar, σ_{moving}, to the motion-corrected standard deviation (Equation (15)) of the sonic and fixed-lidar reference devices ($\sigma_{corr(sonic)}$ and $\sigma_{corr(lidar)}$, respectively; right panels) and to the uncorrected ones (left panels; labelled $\sigma_{ref(sonic)}$ and $\sigma_{ref(lidar)}$), respectively). Linear regression parameters and correlation coefficients, superimposed on Figure 5, clearly improved after applying the correction methodology for both the sonic and the fixed-lidar references. Therefore, better agreement between the floating lidar and the instrumental references was obtained. Despite the improvement, there was a tendency to slightly overestimate the motion-corrected standard deviation, $\sigma_{corr,(x)}$, $x =$ *sonic, lidar*, for both the sonic and lidar references.

To further investigate this issue, each point in the scatter plots was colour-coded according to the spatial variation given by the floating lidar. Blue dots, which are associated to low spatial variation, exhibited good correlation while poorly correlated points were associated to spatial-variation figures above 0.06. These high figures were usually due to errors in the VAD-retrieved HWS caused by inhomogeneity of the wind. This means that regression-line results could better approach the ideal 1:1 line by filtering out these outliers on a spatial variation criterion, which is out of the scope of the present work.

To quantitatively discuss the whole campaign via mean difference and root mean square error indicators (Equations (18) and (19), Table 3 presents the results for all 109 cases in the campaign, for both the fixed lidar and sonic references. Results are graphically depicted in the histogram of Figure 6 for the lidar reference only. Figure 6 shows that the motion-uncorrected mean difference, MD_{ref}, had a positive bias of 0.13 m/s when using the fixed lidar as reference. This bias accounts for the systematic error in the measured HWS standard deviation caused by floating lidar motion as previously reported in [27,59]. After motion correction, the mean difference MD_{corr} reduced to the virtually unbiased figure of -0.03 m/s when using the fixed lidar as reference. The negative sign indicates the tendency to overestimate, as mentioned previously. This accounts for an 80% reduction in absolute value. Using the sonic anemometer as reference, the MD reduced from 0.12 to -0.03 m/s (histogram not shown). The RMSE reduced from $RMSE_{ref} = 0.17$ (uncorrected) to $RMSE_{corr} = 0.12$ m/s (motion corrected) when using the lidar reference (this accounts for a 29% reduction) and from 0.18 to 0.16 m/s when using the sonic reference. This is considered evidence of the accuracy of the proposed methodology in estimating the motion-induced standard deviation.

Table 3. Performance of the variance-combination laws of Section 3.2. (C) stands for linearly correlated variables, (PC) for partially correlated, and (U) for uncorrelated.

	\multicolumn{6}{c}{Variance-Combination Law for σ_{corr}}	Uncorrected, σ_{ref}	
	(C) Equation (17)		(PC) Equation (15)		(U) Equation (16)			
	Sonic	Lidar	Sonic	Lidar	Sonic	Lidar	Sonic	Lidar
MD	−0.06	−0.05	−0.03	−0.03	0.08	0.08	0.12	0.13
RMSE	0.17	0.13	0.16	0.12	0.15	0.13	0.18	0.17

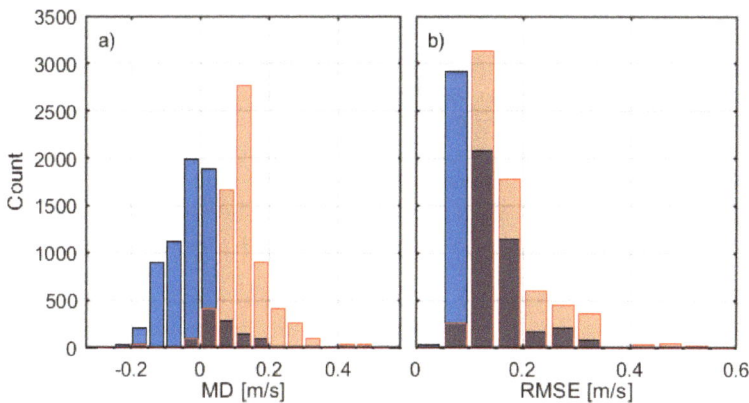

Figure 6. Histogram of the main statistical parameters. (**a**) mean difference; (**b**) root mean square error using the fixed lidar as reference; for all panels: blue = motion corrected, red = uncorrected.

As a remark, Figure 7 shows similar HWS motion-corrected results to Figure 5d, but under the limit hypotheses of uncorrelation (Equation (16)) and linear correlation (Equation (17)) between the reference horizontal wind speed, HWS, and the motion-induced HWS error, Z. Figure 7 shows that the uncorrelated case and the linear-correlated case can respectively be understood—in a statistical sense over the whole population—as lower (Equation (16)) and upper (Equation (17)) bounds of the proposed motion correction. According to the definition of correlation coefficient, $0 <= |\rho| <= 1$, Equation (15) lies in between these two limit cases ($\rho = 0$, $\rho = 1$). This is corroborated in Table 3, which shows MD and RMSE indicators when the lidar and the sonic anemometer are used as references, for the three combination hypotheses discussed in Section 3.2: partially correlated (PC), uncorrelated (U) and correlated (C) variables. It emerges that the approximation of partial correlation yielded the best results, as shown by the lowest MD and RMSE figures in Table 3.

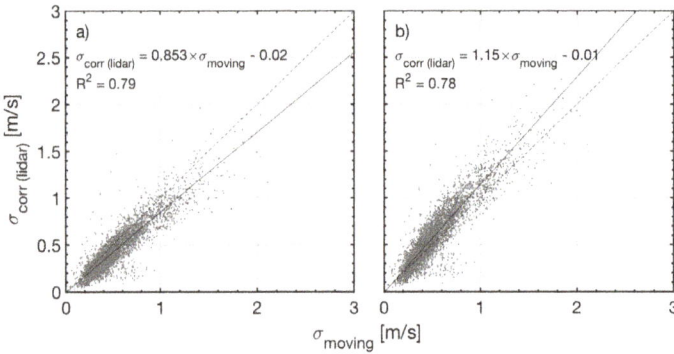

Figure 7. Comparison between 10-min floating-lidar HWS standard deviation measurements and motion-corrected ones by using Equation (16) versus Equation (17). (**a**) uncorrelation hypothesis (Equation (16)); (**b**) linear-correlation hypothesis (Equation (17)). The dashed line indicates the 1:1 line and the solid line shows the linear regression.

3.5. Turbulence Intensity

Analogously to Figure 5d, Figure 8a compares the TI of the floating lidar, TI_{moving}, to the motion-corrected TI of the fixed-lidar reference, $TI_{corr(lidar)}$. Dots are colour-coded according to their spatial variation parameter. HWSs below 3 m/s, which are usually out of the production regime of commercial wind turbines and tend to numerically distort the TI, were filtered out to enhance the readability of the graph. Although some scattering is present in the pattern of dots, the regression line (slope = 0.86, intercept = 0.01) shows a similar tendency to that of Figure 5d of approaching the 1:1 line after motion correction. Quantitatively, by defining similar MD and RMSE indicators for the TI (counterpart of Equations (18) and (19) by changing standard deviation, σ, into TI), the MD for the moving lidar reduced from 0.016 (uncorrected) to 0.003 (motion corrected). In terms of RMSE, the reduction was from 0.018 to 0.012, which despite being not very important implies an approximate 30% reduction in the dispersion of data. In addition, most of the points falling far from the 1:1 line had high spatial variation figures, typically SV > 0.06, which is characteristic of low HWS.

Figure 8b illustrates the successful application of the motion-correction algorithm by superimposing: (i) the TI measured by the uncorrected fixed-lidar reference ($TI_{ref(lidar)}$, red); (ii) the TI derived from the motion-corrected lidar reference ($TI_{corr(lidar)}$, grey); and (iii) the TI measured by the moving floating lidar (TI_{moving}, black) as a function of the 10-min HWS. To aid visual interpretation, average TIs using a 1.0 m/s binwidth were also plotted in red, white, and black traces, respectively. As expected, the *apparent* TI measured by the floating lidar (black trace) was higher than the *true one* measured by the reference lidar (red trace). After application of motion correction to the reference TI, $TI_{ref(lidar)}$ (red dots/red trace), the motion-corrected TI, $TI_{corr(lidar)}$ (grey dots/white trace), approximately followed the floating lidar TI, TI_{moving} (black dots/black trace). At this point, it must be said that, in practice, the correction is to be applied to the TI measured by the floating lidar so as to shift it down. However, this does not change the line of reasoning. Quantitatively, the mean value of the TI measured by the fixed lidar in the 3–20 m/s HWS range was $TI_{ref(lidar)} = 0.047$ and the TI measured by the floating lidar was $TI_{moving} = 0.065$. After motion correction, the mean value of the reference-corrected TI was $TI_{corr(lidar)} = 0.067$, which was only -0.002 apart from TI_{moving} and drastically reduced the initial difference between floating lidar and the reference lidar TI from 0.018 to -0.002. These differences account for an error reduction from 38.3% to 4.3%.

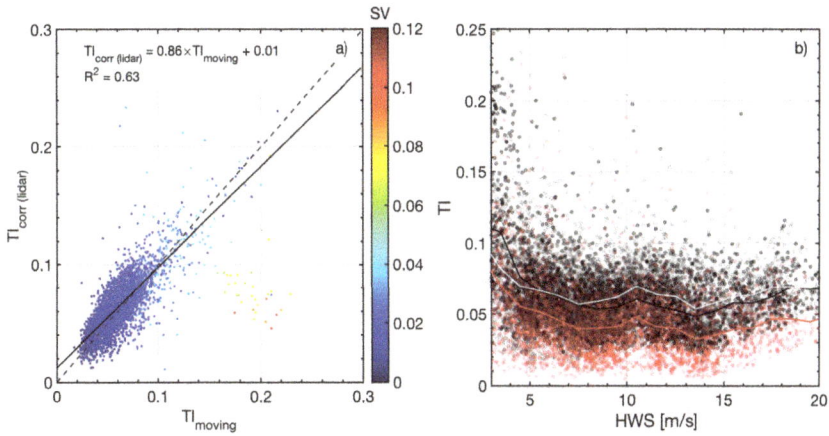

Figure 8. Turbulence intensity (TI) results for the whole campaign. (**a**) comparison between the motion-corrected TI of the fixed-lidar reference, $TI_{corr(lidar)}$ and the floating-lidar TI, TI_{moving}. The colour bar indicates spatial variation. The dashed line indicates the 1:1 line; (**b**) plots of TI versus HWS (see text): Red dots = uncorrected fixed-lidar reference, $TI_{ref(lidar)}$. Grey dots = motion-corrected lidar reference, $TI_{corr(lidar)}$. Black dots = floating lidar, TI_{moving}. Traces with the same colours plot average TIs using a 1.0-m/s binwidth.

As discussed in Section 1, in order to improve the design layout of offshore wind farms and selection of the appropriate wind turbine model, turbulence intensity measurements of a floating lidar are needed. Performance results from Sections 3.4 and 3.5 showed that, in the environmental conditions considered, the proposed methodology has the potential to estimate the influence of motion on TI measurements with the ZephIR™300 lidar.

4. Conclusions

We presented a methodology to estimate the 10-min motion-induced standard deviation and turbulence intensity on the retrieved HWS for a ZephIR™300 lidar at a given measurement height without accessing individual line-of-sight information of the lidar scanning pattern or individual 1-s data.

The proposed methodology includes a software-based motion simulator that reproduces the VAD algorithm used to retrieve the HWS under simple-harmonic motional conditions and a moment-computation recursive procedure to estimate the motion-induced HWS error standard deviation, σ_Z, as well as the motion-induced TI.

The motion simulator input parameters are the 10-min average HWS and 10-min motional amplitude and period of the floating lidar buoy as well as initial roll/pitch motional phases and lidar scan phase (ϕ_r, ϕ_p, and ϕ_s, respectively). A binning procedure is used to group measurement records into bins with similar HWS and motional conditions. The procedure is aimed at computing the 10-min

HWS error standard deviation in each bin by internally sweeping these phases in the $[0, 2\pi)$ range, which therefore become blind inputs to the user.

The method relies on the approximation that roll/pitch amplitudes and periods are linearly correlated on a 10-min basis and that, consequently, only one motional amplitude and period is needed. This one-degree-of-freedom approximation combined with that of simple harmonic motion are the main limitations of the method. Under these hypotheses, the motion-induced HWS standard deviation was proven to be independent of wind direction, which allows this variable to be neglected in the computations (wind direction errors caused by yaw motion are always corrected by means of the buoy compass).

According to error-propagation laws, the motion-corrected HWS standard deviation (Equation (15)), which combines the motion-induced HWS error and the reference HWS, was shown to depend on the correlation between these two variables and the degree of approximation by which it is estimated. Uncorrelated ($\rho = 0$) and linearly-correlated ($|\rho| = 1$) sub-cases were interpreted as upper and lower bounds of the motion-corrected HWS standard deviation, respectively.

The performance of the proposed methodology was tested as part of a 60-day study period at offshore metmast IJmuiden by using a sonic anemometer and a fixed lidar as reference instruments. The motion-corrected HWS standard deviation and that of the reference HWS (from either the fixed lidar or the sonic anemometer) were compared to the measured floating-lidar HWS standard deviation for the 109 most frequent cases of the campaign. This indicated an overall improvement in the average MD from 0.13 (uncorrected) to −0.03 m/s (motion corrected) and an average RMSE reduction from 0.17 to 0.12 m/s, which essentially means that the floating-lidar and the motion-corrected HWS standard deviation laid on the ideal 1:1 line with a dispersion equal to the RMSE.

When analysing the whole campaign as a function of the spatial variation, the most poorly correlated points were associated with mid-to-high spatial variations (SV > 0.06). Wider dispersion arose when using the sonic anemometer as reference, which was caused by the inherently different wind measurement principle of the sonic as compared to the lidar. Analysis in terms of TI showed similar improvement, made evident by a reduction in the difference between the reference-lidar and the floating-lidar TI from 0.018 (uncorrected) to −0.002 (motion corrected).

Despite these good results, they must be interpreted with caution because performance is based on MD and RMSE criteria over the whole statistical sample and not on an individual measurement basis. Overall, in the environmental conditions considered, the proposed methodology holds promise for use in the estimation of the influence of motion on TI measurements with the ZephIR™300 lidar. These results should be extended to other conditions and set-ups, which, if proven effective, could eventually be used to correct TI measurements of floating lidars as standalone devices.

Author Contributions: This work was developed as part of M.A.G.-A.'s doctoral thesis supervised by F.R. and J.T.-A. Development of the simulator (Section 2.2), analysis, and figures were done by M.A.G.-A. and J.T.-A. Database creation and binning was contributed by M.Sc. student A.S. Paper conceptualisation, mathematical framework, and scientific text editing was completed by F.R.

Funding: This work was funded by the Spanish Ministry of Economy and Competitiveness (MEC)—European Regional Development (FEDER) funds under TEC2015-63832-P project and by European Union H2020, ACTRIS-2 (GA-654109). CommSensLab is a "María de Maeztu" Unit of Excellence (MDM-2016-0600) funded by the Agencia Estatal de Investigación, Spain.

Acknowledgments: The authors gratefully acknowledge LIM-UPC and EOLOS for the endless tests carried out at their premises and during the commissioning phase at IJmuiden (the Netherlands).

Conflicts of Interest: The authors declare no conflict of interest.

Remote Sens. **2018**, *10*, 2037

Abbreviations

The following abbreviations are used in this manuscript:

HWS   Horizontal Wind Speed
LAT   Lowest Astronomical Tide
MD   Mean Difference
SV   Spatial Variation
RMSE   Root Mean Square Error
TI   Turbulence Intensity
VAD   Velocity–Azimuth Display
WD   Wind Direction

References

1. Global Wind Energy Council. *Global Wind Energy Outlook 2016*; Technical Report; Global Wind Energy Council: Brussels, Belgium, 2016.
2. Global Wind Energy Council. *Global Wind Energy Rerport 2016: Annual Market Update*; Technical Report; Global Wind Energy Council: Brussels, Belgium, 2017.
3. Roland Berger. *Offshore Wind toward 2020: On the Pathway to Cost Competitiveness*; Technical Report; Roland Berger: Munich, Germany, 2013.
4. Barthelmie, R.; Pryor, S. Can Satellite Sampling of Offshore Wind Speeds Realistically Represent Wind Speed Distributions? *J. Appl. Meteorol.* **2003**, *42*, 83–94. [CrossRef]
5. Chang, R.; Zhu, R.; Badger, M.; Hasager, C.B.; Zhou, R.; Ye, D.; Zhang, X. Applicability of Synthetic Aperture Radar wind retrievals on offshore wind resources assessment in Hangzhou Bay, China. *Energies* **2014**, *7*, 3339–3354. [CrossRef]
6. Hirth, B.D.; Schroeder, J.L.; Gunter, W.S.; Guynes, J.G. Measuring a utility-scale turbine wake using the TTUKa mobile research radars. *J. Atmos. Ocean. Technol.* **2012**, *29*, 765–771. [CrossRef]
7. Barthelmie, R.; Folkerts, L.; Ormel, F.; Sanderhoff, P.; Eecen, P.; Stobbe, O.; Nielsen, N. Offshore wind turbine wakes measured by SODAR. *J. Atmos. Ocean. Technol.* **2003**, *20*, 466–477. [CrossRef]
8. Vogt, S.; Thomas, P. SODAR—A useful remote sounder to measure wind and turbulence. *J. Wind Eng. Ind. Aerodyn.* **1995**, *54*, 163–172. [CrossRef]
9. Lang, S.; McKeogh, E. LIDAR and SODAR Measurements of Wind Speed and Direction in Upland Terrain for Wind Energy Purposes. *Remote Sens.* **2011**, *3*, 1871–1901. [CrossRef]
10. International Energy Association. *State of the Art of Remote Wind Speed Sensing Techniques Using Sodar, Lidar and Satellites*; Technical Report; International Energy Association: Paris, France, 2007.
11. Sempreviva, A.M.; Barthelmie, R.J.; Pryor, S.C. Review of Methodologies for Offshore Wind Resource Assessment in European Seas. *Surv. Geophys.* **2008**, *29*, 471–497. [CrossRef]
12. Scholbrock, A.; Fleming, P.; Schlipf, D.; Wright, A.; Johnson, K.; Wang, N. Lidar-Enhanced Wind Turbine Control: Past, Present, and Future. In Proceedings of the 2016 American Control Conference (ACC), Boston, MA, USA, 6–8 July 2016; pp. 1399–1406.
13. Rodrigo, J.S. State-of-the-Art of Wind Resource Assessment. Deliverable D7, CENER, 2010. Available online: https://cordis.europa.eu/project/rcn/93290_en.html (accessed on 14 December 2012).
14. Clifton, A.; Courtney, M. *15. Ground-Based Vertically Profiling Remote Sensing for Wind Reource Assessment*; Technical Report; IEA Wind Expert Group Study on Recommended Practices; IEA: Paris, France, 2013.
15. Li, J.; Yu, X.B. LiDAR technology for wind energy potential assessment: Demonstration and validation at a site around Lake Erie. *Energy Convers. Manag.* **2017**, *144*, 252–261. [CrossRef]
16. Trabucchi, D.; Trujillo, J.J.; Kühn, M. Nacelle-based Lidar Measurements for the Calibration of a Wake Model at Different Offshore Operating Conditions. *Energy Procedia* **2017**, *137*, 77–88. [CrossRef]
17. Krishnamurthy, R.; Reuder, J.; Svardal, B.; Fernando, H.; Jakobsen, J. Offshore Wind Turbine Wake characteristics using Scanning Doppler Lidar. *Energy Procedia* **2017**, *137*, 428–442. [CrossRef]
18. van Dooren, M.; Trabucchi, D.; Kühn, M. A Methodology for the Reconstruction of 2D Horizontal Wind Fields of Wind Turbine Wakes Based on Dual-Doppler Lidar Measurements. *Remote Sens.* **2016**, *8*, 809. [CrossRef]

19. International Electrotechnical Commission. *IEC 61400-12 Wind Turbine Power Performance Testing*; Technical Report; International Electrotechnical Commission: Geneva, Switzerland, 1998.

20. Williams, B.M. New Applications of Remote Sensing Technology for Offshore Wind Powert. Master's Thesis, University Delawre, Newark, DE, USA, 2013.

21. Pichugina, Y.L.; Banta, R.M.; Brewer, W.A.; Sandberg, S.P.; Hardesty, R.M. Doppler Lidar–Based Wind-Profile Measurement System for Offshore Wind-Energy and Other Marine Boundary Layer Applications. *J. Appl. Meteorol. Climatol.* **2011**, *51*, 327–349. [CrossRef]

22. Courtney, M.S.; Hasager, C.B. Remote sensing technologies for measuring offshore wind. In *Offshore Wind Farms*; Elsevier: Amsterdam, The Netherlands, 2016; Chapter 4, pp. 59–82.

23. Antoniou, I.; Jorgensen, H.E.; Mikkelsen, T.; Frandsen, S.; Barthelmie, R.; Perstrup, C.; Hurtig, M. Offshore wind profile measurements from remote sensing instruments. In Proceedings of the European Wind Energy Association Conference & Exhibition, Athens, Greece, 27 February–2 March 2006.

24. Carbon Trust. *Carbon Trust Offshore Wind Accelerator Roadmap for the Commercial Acceptance of Floating LIDAR Technology*; Technical Report; Carbon Trust: London, UK, 2013.

25. Clifton, A.; Clive, P.; Gottschall, J.; Schlipf, D.; Simley, E.; Simmons, L.; Stein, D.; Trabucchi, D.; Vasiljevic, N.; Würth, I. IEA Wind Task 32: Wind Lidar Identifying and Mitigating Barriers to the Adoption of Wind Lidar. *Remote Sens.* **2018**, *10*, 406. [CrossRef]

26. Bischoff, O.; Wurth, I.; Gottschall, J.; Gribben, B.; Hughes, J.; Stein, D.; Verhoef, H. *Recommended Practices for Floating Lidar Systems*; Technical Report; IEA Wind Task 32; IEA: Paris, France, 2016.

27. Gottschall, J.; Wolken-Möhlmann, G.; Viergutz, T.; Lange, B. Results and conclusions of a floating-lidar offshore test. *Energy Procedia* **2014**, *53*, 156–161. [CrossRef]

28. Schuon, F.; González, D.; Rocadenbosch, F.; Bischoff, O.; Jané, R. KIC InnoEnergy Project Neptune: Development of a Floating LiDAR Buoy for Wind, Wave and Current Measurements. In Proceedings of the DEWEK 2012 German Wind Energy Conference, Bremen, Germany, 7–8 November 2012.

29. Sospedra, J.; Cateura, J.; Puigdefàbregas, J. Novel multipurpose buoy for offshore wind profile measurements EOLOS platform faces validation at ijmuiden offshore metmast. *Sea Technol.* **2015**, *56*, 25–28.

30. Mathisen, J.P. Measurement of wind profile with a buoy mounted lidar. *Energy Procedia* **2013**, *30*, 12.

31. Kyriazis, T. Low cost and flexible offshore wind measurements using a floating lidar solution (FLIDAR™). In Proceedings of the EWEA Conference, Vienna, Austria, 4–7 February 2013.

32. Hung, J.B.; Hsu, W.Y.; Chang, P.C.; Yang, R.Y.; Lin, T.H. The performance validation and operation of nearshore wind measurements using the floating lidar. *Coast. Eng. Proc.* **2014**, *1*, 11. [CrossRef]

33. Hsuan, C.Y.; Tasi, Y.S.; Ke, J.H.; Prahmana, R.A.; Chen, K.J.; Lin, T.H. Validation and Measurements of Floating LiDAR for Nearshore Wind Resource Assessment Application. *Energy Procedia* **2014**, *61*, 1699–1702. [CrossRef]

34. Gottschall, J.; Gribben, B.; Stein, D.; Würth, I. Floating lidar as an advanced offshore wind speed measurement technique: Current technology status and gap analysis in regard to full maturity. *Wiley Interdiscip. Rev. Energy Environ.* **2017**, *6*. [CrossRef]

35. Gottschall, J.; Wolken-Möhlmann, G.; Lange, B. About offshore resource assessment with floating lidars with special respect to turbulence and extreme events. *J. Phys. Conf. Ser.* **2014**, *555*, 012043. [CrossRef]

36. Mangat, M.; des Roziers, E.B.; Medley, J.; Pitter, M.; Barker, W.; Harris, M. The impact of tilt and inflow angle on ground based lidar wind measurements. In Proceedings of the EWEA 2014, Barcelona, Spain, 10–13 March 2014.

37. Pitter, M.; Burin des Roziers, E.; Medley, J.; Mangat, M.; Slinger, C.; Harris, M. Performance Stability of Zephir in High Motion Enviroments: Floating and Turbine Mounted. Available online: https://bit.ly/2EuDY5i (accessed on 14 December 2018).

38. Wolken-Möhlmann, G.; Lilov, H.; Lange, B. Simulation of motion induced measurement errors for wind measurements using LIDAR on floating platforms. *Fraunhofer IWES Am Seedeich* **2011**, *45*, 27572.

39. Bischoff, O.; Würth, I.; Cheng, P.; Tiana-Alsina, J.; Gutiérrez, M. Motion effects on lidar wind measurement data of the EOLOS buoy. In Proceedings of the First International Conference on Renewable Energies Offshore, Lisbon, Portugal, 24–26 November 2014.

40. Tiana-Alsina, J.; Rocadenbosch, F.; Gutierrez-Antunano, M.A. Vertical Azimuth Display simulator for wind-Doppler lidar error assessment. In Proceedings of the 2017 IEEE International Geoscience and Remote Sensing Symposium (IGARSS), Fort Worth, TX, USA, 23–28 July 2017. [CrossRef]

41. Nicholls-Lee, R. A low motion floating platform for offshore wind resource assessment using Lidars. In Proceedings of the ASME 2013 32nd International Conference on Ocean, Offshore and Arctic Engineering, Nantes, France, 9–14 June 2013.

42. Tiana-Alsina, J.; Gutiérrez, M.A.; Würth, I.; Puigdefàbregas, J.; Rocadenbosch, F. Motion compensation study for a floating Doppler wind lidar. In Proceedings of the Geoscience and Remote Sensing Symposium, Milan, Italy, 26–31 July 2015.

43. Bischoff, O.; Schlipf, D.; Würth, I.; Cheng, P. Dynamic Motion Effects and Compensation Methods of a Floating Lidar Buoy. In Proceedings of the EERA DeepWind 2015 Deep Sea Offshore Wind Conference, Trondheim, Norway, 4–6 February 2015.

44. Gottschall, J.; Lilov, H.; Wolken-Möhlmann, G.; Lange, B. Lidars on floating offshore platforms; about the correction of motion-induced lidar measurement errors. In Proceedings of the EWEA 2012, Copenhagen, Denmark, 16–19 April 2012.

45. International Electrotechnical Commission. *IEC 61400-1 2005 Wind Turbine Power Performance Testing*; Technical Report; International Electrotechnical Commission: Geneva, Switzerland, 2005.

46. Manwell, J.F.; McGowan, J.G.; Rogers, A.L. *Wind Energy Explained: Theory, Design and Application*; Number Book, Whole; Wiley: Chichester, UK, 2009.

47. Hansen, K.S.; Barthelmie, R.J.; Jensen, L.E.; Sommer, A. The impact of turbulence intensity and atmospheric stability on power deficits due to wind turbine wakes at Horns Rev wind farm. *Wind Energy* **2012**, *15*, 183–196. [CrossRef]

48. Sathe, A.; Mann, J.; Gottschall, J.; Courtney, M. Can wind lidars measure turbulence? *J. Atmos. Ocean. Technol.* **2011**, *28*, 853–868. [CrossRef]

49. Sathe, A. Influence of Wind Conditions on Wind Turbine Loads and Measurements of Turbulence Using Lidars. Ph.D. Thesis, Delft University, Delft, The Netherlands, 2012.

50. Sathe, A.; Banta, R.; Pauscher, L.; Vogstad, K.; Schilpf, D.; Wylie, S. *Estimating Turbulence Statistics and Parameters from Ground- and Nacelle-Based Lidar Measurements*; Technical Report; Technical University of Denmark: Lyngby, Denmark, 2015.

51. Wagner, R.; Mikkelsen, T.; Courtney, M. *Investigation of Turbulence Measurements With a Continuous Wave, Conically Scanning LiDAR*; Technical Report; DTU: Lyngby, Denmark, 2009.

52. Henderson, S.W.; Gatt, P.; Rees, D.; Huffaker, M. Wind LIDAR. *Laser Remote Sensing*; Optical Science and Engineering; CRC Press: Boca Raton, FL, USA, 2005; Chapter 7.

53. Gutierrez-Antunano, M.A.; Tiana-Alsina, J.; Rocadenbosch, F.; Sospedra, J.; Aghabi, R.; Gonzalez-Marco, D. A wind-lidar buoy for offshore wind measurements: First commissioning test-phase results. In Proceedings of the 2017 IEEE International Geoscience and Remote Sensing Symposium (IGARSS), Fort Worth, TX, USA, 23–28 July 2017; doi:10.1109/igarss.2017.8127280.

54. Werkhoven, E.J.; Verhoef, J.P. Offshore Meteorological Mast Ijmuiden Abstract of Instrumentation Report. Available online: https://bit.ly/2PCjjxg (accessed on 14 December 2018).

55. Poveda, J.; Wouters, D.; Nederland, S.E.C. Wind Measurements at Meteorological Mast Ijmuiden. Available online: https://bit.ly/2QYbqHj (accessed on 14 December 2018).

56. Clifford, S.F.; Kaimal, J.C.; Lataitis, R.J.; Strauch, R.G. Ground-based remote profiling in atmospheric studies: An overview. *Proc. IEEE* **1994**, *82*, 313–355. [CrossRef]

57. Barlow, R.J. *Statistics: A Guide to the Use of Statistical Methods in the Physical Sciences*; Manchester Physics Series; Wiley: Chichester, UK, 1989.

58. Papoulis, A. *Probability, Random Variables, and Stochastic Processes*; McGraw-Hill: New York, NY, USA, 1965.

59. Gutiérrez-Antuñano, M.A.; Tiana-Alsina, J.; Rocadenbosch, F. Performance evaluation of a floating lidar buoy in nearshore conditions. *Wind Energy* **2017**, *20*, 1711–1726. [CrossRef]

![remote sensing logo](remote sensing)

Article

Assessing Global Ocean Wind Energy Resources Using Multiple Satellite Data

Qiaoying Guo [1,2,3], Xiazhen Xu [4], Kangyu Zhang [1,2,3], Zhengquan Li [5], Weijiao Huang [6], Lamin R. Mansaray [1,2,7], Weiwei Liu [1,2], Xiuzhen Wang [8], Jian Gao [8] and Jingfeng Huang [1,2,3,*]

1   Institute of Applied Remote Sensing and Information Technology, Zhejiang University, Hangzhou 310058, China; qiaoyingguo@zju.edu.cn (Q.G.); kangyuzhang@zju.edu.cn (K.Z.); lmansaray@zju.edu.cn (L.R.M.); weiweiliu@zju.edu.cn (W.L.)
2   Key Laboratory of Agricultural Remote Sensing and Information Systems, Hangzhou 310058, China
3   State Key Laboratory of Satellite Ocean Environment Dynamics, Second Institute of Oceanography, State Oceanic Administration, Hangzhou 310012, China
4   Jiangsu Climate Centre, Jiangsu Meteorological Bureau, Nanjing 210009, China; Xuxz0119@126.com
5   Zhejiang Climate Centre, Zhejiang Meteorological Bureau, Hangzhou 310007, China; lzq110119@163.com
6   Department of Land Management, Zhejiang University, Hangzhou 310058, China; huangweijiao@zju.edu.cn
7   Department of Agro-meteorology and Geo-informatics, Magbosi Land, Water and Environment Research Centre (MLWERC), Sierra Leone Agricultural Research Institute (SLARI), Freetown PMB 1313, Sierra Leone
8   Institute of Remote Sensing and Earth Sciences, Hangzhou Normal University, Hangzhou 311121, China; wxz05160516@126.com (X.W.); gaojiannj@126.com (J.G.)
*   Correspondence: hjf@zju.edu.cn; Tel.: +86-571-8898-2830

Received: 29 October 2017; Accepted: 9 January 2018; Published: 12 January 2018

Abstract: Wind energy, as a vital renewable energy source, also plays a significant role in reducing carbon emissions and mitigating climate change. It is therefore of utmost necessity to evaluate ocean wind energy resources for electricity generation and environmental management. Ocean wind distribution around the globe can be obtained from satellite observations to compensate for limited in situ measurements. However, previous studies have largely ignored uncertainties in ocean wind energy resources assessment with multiple satellite data. It is against this background that the current study compares mean wind speeds (MWS) and wind power densities (WPD) retrieved from scatterometers (QuikSCAT, ASCAT) and radiometers (WindSAT) and their different combinations with National Data Buoy Center (NDBC) buoy measurements at heights of 10 m and 100 m (wind turbine hub height) above sea level. Our results show an improvement in the accuracy of wind resources estimation with the use of multiple satellite observations. This has implications for the acquisition of reliable data on ocean wind energy in support of management policies.

Keywords: wind energy resources; QuikSCAT; WindSAT; ASCAT; global ocean

1. Introduction

Climate change is a global issue that impacts on all human beings: an associated rising of sea level, extreme hydrologic events (such as floods and droughts) and urban heat island effects are projected to occur with climate change. Such changes have already affected human health due to extreme heat, cold, drought, storms and crop failures [1]. Scientists have demonstrated that global warming over millennial time scales is due to greenhouse gas emissions produced by human activities [2,3]. Energy use efficiency and renewable energy generation can benefit public health and the global climate system by displacing emissions from fossil-fuelled electricity generation units [4]. Expanding renewable energy, especially wind power, is a central strategy for reducing carbon emissions and mitigating climate change [5,6]. Therefore, the evaluation of wind resources plays a significant role in the selection of appropriate sites for the establishment of wind farms, wind energy development, and national energy policy formulation. In most countries, coastal areas have become heavily urbanized and industrialized as a consequence of agglomeration. In effect, more energy will be required in these areas. As environmental sustainability continues to occupy the center stage of the global development agenda, offshore wind has been regarded as a potential renewable energy source that can be generated through a network of wind turbines. Currently, more than 91% of all the offshore wind installations worldwide are in European waters, particularly in the North Sea [7].

Although offshore wind energy has huge potential in powering the global economy, there are spatial and temporal variabilities in the distribution of wind power, dynamics which are worth investigating for the development of efficient and sustainable offshore wind energy resources. Offshore wind energy resources are mainly estimated from in situ wind measurements [8], satellite data, numerical simulation results [9], and reanalysis data [10–12]. With progress in microwave remote sensing, a great deal of satellite-derived data have been obtained and applied in the study of wind energy resources, including sea surface wind distribution data derived from Synthetic Aperture Radars (SAR) and scatterometers, such as the Earth Resources Satellite ERS-2 SAR (1995–2011) [13,14], Environment Satellite (ENVISAT) Advanced Synthetic Aperture Radar (ASAR) (2002–2012) [14–19], RADARSAT-1 SAR (1995–2013) [20], SeaWinds onboard QuikSCAT (1999–2009) [17,19–27], ASCAT onboard METOP-A (2007–present) [17,18,27,28] and OceanSat-2 scatterometer (OSCAT, 2009–present) [28,29]. Wind fields retrieved from SAR imagery have a high spatial resolution (<100 m). However, from previous research, there are less than 1500 overlapping SAR samples [14,15,17], and SAR cannot obtain observations of the whole ocean. Scatterometers and passive microwave radiometers can provide global sea surface wind fields at a relatively coarse resolution of approximately 12.5–50 km with two observations per day from single satellite data.

A number of studies have been conducted on ocean wind energy assessment at a variety of spatial scales (from local through regional to global). Early studies evaluated offshore or global ocean wind power resources mainly derived from single satellite data. It should be emphasized however that, generally, only a maximum of two observations per day (at descending and ascending passes) are obtainable from a single satellite. As diurnal ocean wind variations are apparent, statistics derived from a single satellite are limited both synoptically and spatially and therefore using multiple sources of satellite data are imperative to gain a more comprehensive recording and analysis of ocean wind energy in space and time [14,17]. A growing number of studies are now focusing on wind resources assessment based on multiple satellite data [17–20,27,28], albeit with the reduction of uncertainties largely ignored.

The purpose of the current study is to estimate the uncertainty associated with the number of satellite observations and its impact on the accuracy of ocean wind resources assessment derived from multiple satellite data. The spatial variability of global ocean wind energy resources is assessed at heights of 10 m and 100 m to provide relevant data on the selection of wind energy sites.

2. Data

2.1. QuikSCAT, WindSAT and ASCAT Data

In this study, two types of time series satellite data—which can provide sea surface wind fields at a spatial resolution of 0.25° × 0.25° and at a height of 10 m above sea level—have been utilized, including scatterometer (QuikSCAT and ASCAT) and radiometer (WindSAT) wind data. The scatterometer operates by transmitting microwave pulses to the ocean surface and then measuring the microwave pulses returned to the satellite sensor. This backscattered signal is physically related to surface roughness. For water surfaces, the surface roughness is highly correlated with the near-surface wind speed and direction at a height of 10 m above sea level. The GMF is the Geophysical Model Function which relates the observed backscatter ratio to surface wind speed and direction at a height of 10 m above sea level. However, the passive microwave radiometer wind vector data are retrieved from the microwave brightness temperatures measurements using Radiative Transfer Model (RTM).

SeaWinds scatterometer is the main instrument on the QuikSCAT satellite and operates at Ku band (13.4 GHz) which is sensitive to rain. Ku-2011 GMF was used. However, the advanced scatterometer (ASCAT) onboard Metop-A satellite is working at C band (5.3 GHz), and C band GMF (C-2015) had been used. The impact of rain on wind retrieval is less severe for ASCAT C-band data than for QuikSCAT Ku-band data. The wind products contain rain flags and researchers can remove rain effects from the datasets by discarding these data. This has been done in the current analysis. The WindSAT fully polarimetric radiometer, as a passive microwave sensor, operates in five discrete channels: 6.8, 10.7, 18.7, 23.8, and 37.0 GHz. All are fully polarimetric except the 6.8 and 23.8 GHz channels that have only dual polarization.

Wind products used herein are the daily gridded maps from Remote Sensing Systems [30]. Table 1 summarizes the information from multiple satellite data including the maximum numbers of different satellite data in the two over pass times and per month. There are two observations based on a single satellite sensor per day. The total number of satellite data is about 9000–12,000 from 20°S to 18°N, and is about 12,000–14,000 among all satellite data at 20–35°S and 18–35°N. The total number of satellite data is about 14,000–16,000 at 35–45°S and 35–42°N, and is about 16,000–21,609 among all satellite data at 45–60°S and 42–80°N. However there are 2000–10,000 at latitude 60–90°S and some areas of the North Pole (the north of Asia and North America). Furthermore, rain flags may be more prevalent in the equatorial East Pacific and the eastern part of the Indian Ocean because these regions have fewer numbers than adjacent areas (9000–10,000).

Table 1. Total numbers of QuikSCAT, WindSAT and ASCAT wind data in descending/ascending (des./asc.) mode and per month in local solar time.

Satellite Platform	QuikSCAT	Coriolis	MetOp-A
Instrument	SeaWinds	WindSAT	ASCAT
Band of operation	Ku (13.4 GHz)	5 discrete channels: 6.8, 10.7, 18.7, 23.8, and 37.0 GHz	C (5.3 GHz)
Total	6944	8372	6379
Ascending	3514	4197	3194
Descending	3430	4175	3185
January	549	665	486
February	512	609	445
March	567	701	549
April	560	683	521
May	562	690	555
June	580	618	537
July	610	696	554
August	656	736	557
September	620	739	532
October	612	751	552
November	539	735	535
December	577	749	556
Time period	1999–2009	2003-current	2007-current
Time of datasets	1999.07–2009.11	2003.02–2015.12	2007.03–2015.12
Descending node time	06:00	06:00	9:30
Ascending node time	18:00	18:00	21:30
Spatial resolution	0.25° × 0.25°	0.25° × 0.25°	0.25° × 0.25°
Produce version	V4	V7.0.1	V2.1

2.2. NDBC Buoys Data

The National Data Buoy Center (NDBC) [31] provides average hourly wind vector measurements recorded at 39 buoys around North America (shown in Figure 1). These NDBC buoys are commonly operated and maintained by American government organizations to report on winds, waves and other ocean conditions at strategic locations for the purposes of ocean navigation, search and rescue operations, and scientific research. The 39 buoy measurements were selected as validation data and they provide a longer time series of meteorological observations relative to QuikSCAT, WindSAT and ASCAT data from 1999 to 2015. These NDBC buoys provided hourly wind vector measurements at a height of 4 or 5 m above sea level. The wind profile method (in Section 3.1) has been used in this study to extrapolate wind speed to heights of 10 m and 100 m above sea level in order to compare with satellite data at the same height. The distances from buoys to coastline are greater than 30 km, and the water depths of the buoys' positions range from 16 m to 5230 m. There are 32 buoys at water depths greater than 50 m so a large amount of the buoys may not be influenced by coastal effects.

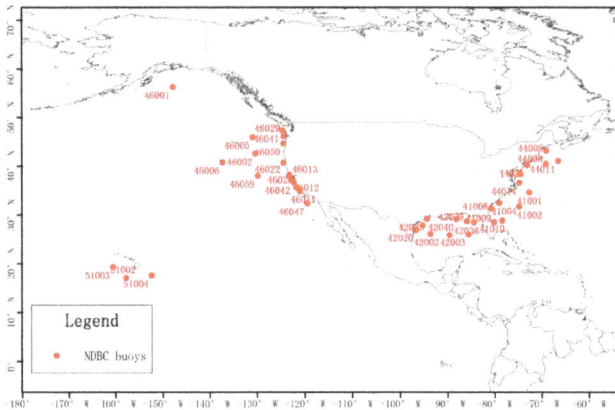

Figure 1. Location of National Data Buoy Center (NDBC) buoys providing hourly wind vector measurements used for comparison with satellite data.

3. Methodology

3.1. Wind Profile Method

In order to compare the wind energy resources derived from buoys with those from satellite data at the same height, the extrapolated wind speed V at heights of 10 m or 100 m (z) is calculated from Equation(1) [32]:

$$V(z) = \frac{u_*}{\kappa} \left[\ln\left(\frac{z}{z_0}\right) - \psi_m \right] \tag{1}$$

where V is the wind speed at height z. κ is the von Karman constant (~0.4). The parameter ψ_m is a correction for atmospheric stability effects (here ψ_m is set to zero to estimate neutral winds, following Badger et al. [32]). In this paper, the effect of atmospheric stability is ignored due to the lack of relevant data. The long-term stability correction (ψ_m) is usually positive under stable conditions and negative under unstable conditions, and this may lead to an underestimation of mean wind speeds when assuming neutral atmospheric conditions. Knowing the spatial distribution of ψ_m is beneficial to improving the accuracy of wind speed extrapolation. The parameter ψ_m ranges from -2.5 to 0 at a height of 100 m in the south Baltic Sea based on SAR and WRF model data, and there is a wind speed difference of 0.5 m/s with and without the long-term atmospheric stability correction at a height of 100 m based on meteorological mast observations (Fino-2) [32]. Takeyama et al. [33] pointed out that atmospheric stability can cause an error of about -1 to $+1$ m/s without stability correction at a height of 10 m based on SAR images over the Japanese coastal waters. The sea surface roughness length z_0 can be estimated from Equation (2):

$$z_0 = \alpha_c \frac{u_*^2}{g} \tag{2}$$

where α_c is the Charnock's parameter (here set to 0.0144, following Badger et al. [32]), and g is the gravitational acceleration of the Earth. The sea surface friction velocity u_* can be estimated by combining and solving iteratively from Equations (1) and (2) when the wind speed at a single level is known.

3.2. Wind Resource Assessment Method

The mean wind speed (MWS) is the wind speed averaged over a given time period and is given as:

$$\overline{V} = \frac{1}{N} \sum_{i=1}^{N} V_i \tag{3}$$

where $\overline{V}$ is the mean wind speed (m/s), V_i is the wind speed (m/s) at measurement i at a given height, N is the total number of measurements.

The wind power density (WPD) may be estimated by statistical method (Equation (4)) [34,35] and the Weibull probability distribution function of two parameters (Equation (5)) [17,19,21,27,32,34,35].

$$E_{statistic} = \frac{1}{2} \rho \overline{V_i^3} \tag{4}$$

$$E_{weibull} = \frac{1}{2} \rho C^3 \Gamma (1 + \frac{3}{k}) \tag{5}$$

where E is the wind power density (W/m^2), and ρ is the standard sea-level air density (1.225 kg/m^3) [11,12,19,27,28]. C is the scale parameter (m/s), and k is the dimensionless shape parameter. Several methods have been applied to calculate Weibull parameters C and k, such as the method using mean and standard deviation of wind speed samples. In this study, we have used the formulae [21,27,34,35] given as follows:

$$k = (\sigma / \overline{V})^{-1.086} \tag{6}$$

$$C = \overline{V} / \Gamma (1 + \frac{1}{k}) \tag{7}$$

where σ is the standard deviation of wind speed. Γ is the gamma function.

In this study, we assume the probability density function of the wind speeds to follow the Weibull distribution, and based on a comparison of the WPD derived by the statistical method and that by the Weibull distribution function from 39 buoys (in Section 2.2), the RMSE is 4.8 W/m^2 at a height of 10 m and 9.9 W/m^2 at a height of 100 m.

4. Results

4.1. Evaluation of MWS and WPD Derived from Multiple Satellite Data Compared with Buoy Measurement Data

The mean wind speeds (MWS) and wind power densities (WPD) calculated from 39 NDBC buoys during 1999–2015 have been compared with those derived from QuikSCAT, WindSAT and ASCAT, and their different combinations at heights of 10 m and 100 m above sea level as presented in Tables 2–5. In this section, the RMSE, Bias, correlation coefficient (Corr.), R^2 and Slope are used to compare satellite-derived MWS/WPD with 39 buoy-derived MWS/WPD at heights of 10 m and 100 m. Tables 2–5 reveal that MWS/WPD derived from ASCAT have better accuracies than those from WindSAT and QuikSCAT in terms of RMSE. QuikSCAT and ASCAT overestimated the MWS and WPD derived from buoys (positive biases and slopes greater than 1.00), while WindSAT shows a tendency to underestimate the MWS and WPD in terms of negative biases and slopes. From the Remote Sensing Systems, QuikSCAT shows similar wind speeds to those from ASCAT, and wind speeds from QuikSCAT/ASCAT are slightly higher than those from WindSAT (the differences between wind speeds are within 0.1 m/s), wind speeds from QuikSCAT have lower errors than those from WindSAT in terms of bias compared with aircraft measurements. This may be due to the different sensor configurations and wind retrieval algorithms.

QuikSCAT + WindSAT-derived MWS/WPD at heights of 10 m and 100 m show lower errors in terms of RMSE and higher correlations than QuikSCAT-derived MWS/WPD and WindSAT-derived MWS/WPD. WindSAT + ASCAT-derived MWS at heights of 10 m and 100 m show lower errors (lower RMSE, biases and slopes are equal to 0.00 and 1.00 respectively) and higher correlations than WindSAT-derived MWS and ASCAT-derived MWS. WindSAT + ASCAT-derived WPD at heights of 10 m and 100 m show lower errors in terms of RMSE and higher correlations than WindSAT-derived WPD and ASCAT-derived WPD. QuikSCAT + WindSAT + ASCAT-derived WPD at heights of 10 m and 100 m show the lowest errors in terms of RMSE and highest correlations. The result of this comparison shows that a better accuracy of MWS/WPD may be derived from multiple satellite data than from single satellite data.

Table 2. Statistics of the comparison between buoy-derived mean wind speeds (MWS) and satellite-derived MWS (m/s) at a height of 10 m above sea level.

Different Combinations of Satellite Data	RMSE	Bias	Corr.	R^2	Slope	N
QuikSCAT	0.39	0.23	0.91	0.78	1.03	4134–7063
WindSAT	0.36	−0.07	0.90	0.81	0.99	3474–7643
ASCAT	0.33	0.09	0.90	0.77	1.01	2717–4017
QuikSCAT + WindSAT	0.27	0.10	0.94	0.87	1.01	8307–14,586
QuikSCAT + ASCAT	0.35	0.18	0.91	0.78	1.02	7209–11,080
WindSAT + ASCAT	0.28	0.00	0.93	0.86	1.00	6272–11,086
QuikSCAT + WindSAT + ASCAT	0.27	0.10	0.94	0.86	1.01	11,229–18,029

Here, the intercept of linear regression is set to zero, and the number of satellite observations at different buoy positions is indicated by N (min-max).

Table 3. Statistics of the comparison between buoy-derived wind power densities (WPD) and satellite-derived WPD (W/m^2) at a height of 10 m above sea level.

Different Combinations of Satellite Data	RMSE	Bias	Corr.	R^2	Slope	N
QuikSCAT	55.4	34.6	0.91	0.82	1.09	4134 7063
WindSAT	53.5	−19.7	0.88	0.78	0.95	3474–7643
ASCAT	42.2	4.7	0.91	0.81	1.01	2717–4017
QuikSCAT + WindSAT	37.0	9.9	0.93	0.87	1.02	8307–14,586
QuikSCAT + ASCAT	47.5	23.3	0.91	0.83	1.06	7209–11,080
WindSAT + ASCAT	40.5	−9.2	0.92	0.84	0.97	6272–11,086
QuikSCAT + WindSAT + ASCAT	36.9	8.7	0.93	0.87	1.02	11,229–18,029

Table 4. Statistics of the comparison between buoy-derived MWS and satellite-derived MWS (m/s) at a height of 100 m above sea level.

Different Combinations of Satellite Data	RMSE	Bias	Corr.	R^2	Slope	N
QuikSCAT	0.48	0.29	0.91	0.78	1.03	4134–7063
WindSAT	0.45	−0.09	0.90	0.81	0.99	3474–7643
ASCAT	0.40	0.11	0.90	0.77	1.01	2717–4017
QuikSCAT + WindSAT	0.33	0.12	0.94	0.87	1.01	8307–14,586
QuikSCAT + ASCAT	0.43	0.22	0.91	0.79	1.03	7209–11,080
WindSAT + ASCAT	0.34	0.00	0.93	0.86	1.00	6272–11,086
QuikSCAT + WindSAT + ASCAT	0.34	0.12	0.94	0.86	1.01	11,229–18,029

Table 5. Statistics of the comparison between buoy-derived WPD and satellite-derived WPD (W/m^2) at a height of 100 m above sea level.

Different Combinations of Satellite Data	RMSE	Bias	Corr.	R^2	Slope	N
QuikSCAT	105.0	65.6	0.91	0.82	1.09	4134–7063
WindSAT	101.1	−37.0	0.88	0.77	0.95	3474–7643
ASCAT	79.2	8.3	0.91	0.81	1.01	2717–4017
QuikSCAT + WindSAT	70.2	18.8	0.93	0.87	1.03	8307–14,586
QuikSCAT + ASCAT	89.6	44.0	0.91	0.83	1.06	7209–11,080
WindSAT + ASCAT	76.8	−17.5	0.92	0.84	0.97	6272–11,086
QuikSCAT + WindSAT + ASCAT	69.8	16.4	0.93	0.87	1.02	11,229–18,029

In order to verify the impact of data sampling density (random sampling numbers of satellite observations) on the accuracy of satellite-derived MWS and WPD at 10 m and 100 m above sea level, we acquired 100 MWS/WPD random samples (500, 1000, 1500, ... , 11,000) from all satellite observations (QuikSCAT + WindSAT + ASCAT) by sampling 100 times repeatedly. The 100 satellite-derived MWS/WPD samples were compared with the buoy-derived MWS/WPD at each buoy position (39 NDBC buoy in total) and at different sampling densities using RMSE. Figures 2 and 3 illustrate the variation of mean RMSE as a function of sampling density at 10 m and 100 m above sea level. The result shows that more satellite observations may reduce the uncertainty in MWS and WPD estimation at 10 m and 100 m above sea level. This result is consistent with the research results of Barthelmie, Pryor and Hasager [17,36].

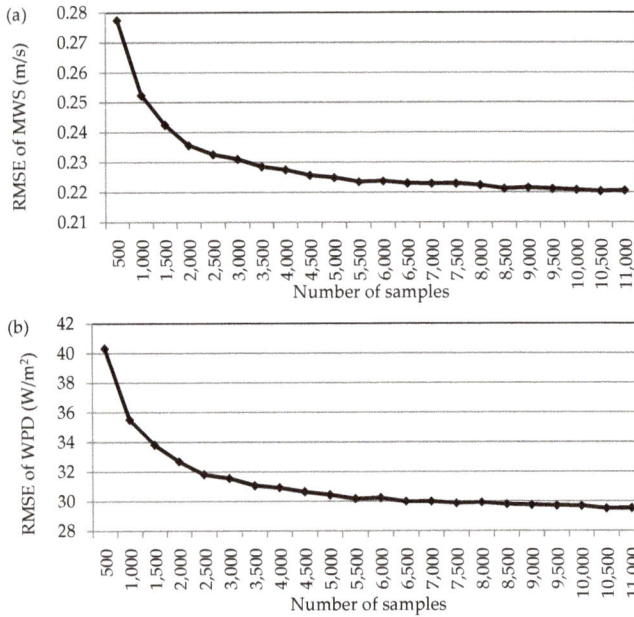

Figure 2. The mean RMSE of MWS and WPD from random sampling of all satellite observations (QuikSCAT + WindSAT + ASCAT) with different numbers of samples compared to the buoy-derived MWS and WPD at a height of 10 m above sea level. (**a**) MWS, and (**b**) WPD.

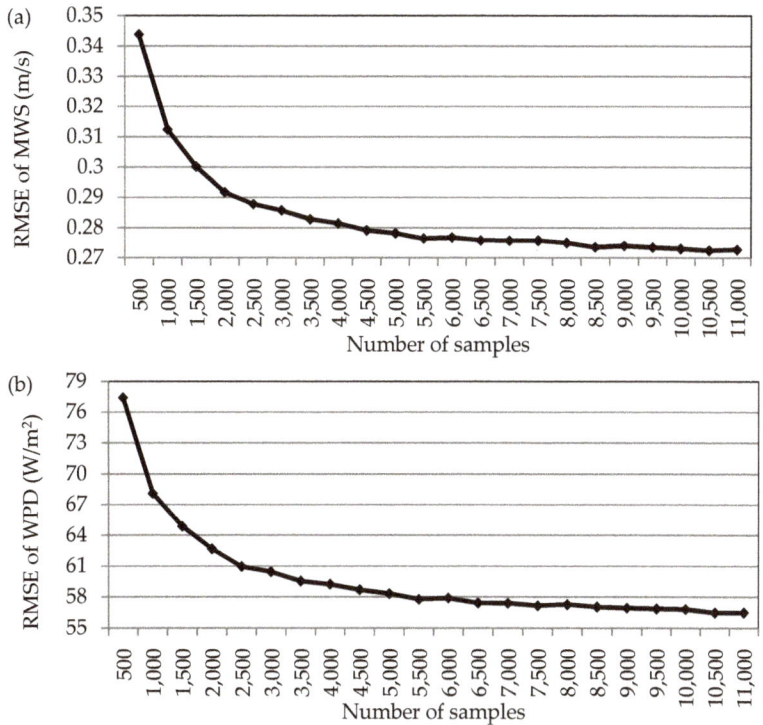

Figure 3. The mean RMSE of MWS and WPD from random sampling of all satellite observations (QuikSCAT + WindSAT + ASCAT) with different numbers of samples compared to the buoy-derived MWS and WPD at a height of 100 m above sea level. (**a**) MWS, and (**b**) WPD.

4.2. Spatial Variability of Global Ocean Wind Energy Resources Using Multiple Satellite Data

The geographic distribution of global MWS and WPD at 10 m above sea level using multiple satellite data (QuikSCAT + WindSAT + ASCAT) is shown in Figure 4. It can be observed that the distribution of wind energy exhibits significant regional differences across the global ocean. The MWS at most global ocean regions are higher than 3 m/s, with the areas of highest MWS being primarily distributed around the Southern Hemisphere westerlies (8–13 m/s). The Northern Hemisphere westerlies are also areas of relatively high MWS (8–11 m/s), whereas MWS in the low latitudes are in the range of 3–10 m/s. The MWS at the Northern European Seas is 6.5–10 m/s, a result similar to that of Hasager et al. [17] who used ASAR, ASCAT and QuikSCAT and Badger et al. [37] based on ASAR. High wind areas in the middle of the Indian Ocean (7–10 m/s) and the East China Sea (7–10 m/s) may mainly be influenced by the winter monsoon, and the strong wind in the Arabian Sea (6–9 m/s) and the South China Sea (6.5–9 m/s) may mainly be caused by the summer monsoon as discussed by Liu et al. [21] based on QuikSCAT. The highest wind speeds in China are found in the southeastern region, especially along the coastline of Fujian Province and the Strait of Taiwan. This finding is also consistent with Jiang et al. [26] who used QuikSCAT. The MWS of Southeastern Brazil at a height of 10 m above sea level is 6–9 m/s which is similar to the result of Pimenta et al. [22] based on QuikSCAT. The global oceanic MWS values at a height of 10 m above sea level reported by our study are slightly higher than those by Atlas et al. [38], which were based on SSM/I (Special Sensor Microwave Imager) data acquired from 1987 to 1994.

The WPD at most global ocean regions are higher than 200 W/m^2. The areas of higher WPD at 10 m above sea level are mainly distributed around the Southern (400–1600 W/m^2) and Northern (400–1300 W/m^2) Hemisphere westerlies. This observed phenomenon may be influenced by the westerly winds. However, WPD in the low latitudes are about 50–600 W/m^2. The coast of Somalia and Southeast China show relatively higher WPD (>400 W/m^2) which is largely attributable to tropical monsoon weather conditions typical in these areas. On the other hand, WPD in the equatorial regions are generally less than 200 W/m^2. Equatorial regions are generally characterized by low atmospheric pressure conditions with a muted seasonal cycle. The high WPD at the North Atlantic Ocean may be due to the ocean–atmosphere interaction around the warm current of the Mexico Gulf and the cold eddy of the Labrador Sea in winter [21]. The distribution of global oceanic WPD at 10 m above sea level presented in this paper is largely consistent with the results of Zheng and Pan [11].

The geographic distribution of global MWS and WPD at 100 m above sea level using multiple satellite data (QuikSCAT + WindSAT + ASCAT) is shown in Figure 5. The MWS at most global ocean regions are higher than 4 m/s, with the areas of highest MWS being primarily distributed around the Southern Hemisphere westerlies (9–16 m/s). The areas of Northern Hemisphere westerlies also have relatively high MWS (9–14 m/s), whereas MWS in the low latitudes are in the range of 4–12 m/s. The MWS at the North Sea and South China Sea are 8–12 m/s and 6.5–10 m/s respectively, a result similar to that of Hasager et al. [39] who used SSM/I data from 1988 to 2013 at a height of 100 m above sea level.

The areas of higher WPD at 100 m above sea level are mainly distributed around the Southern (800–3200 W/m^2) and Northern (800–2800 W/m^2) Hemisphere westerlies. This observed phenomenon may be influenced by the westerly winds. However, WPD in the low latitudes are about 150–1200 W/m^2.

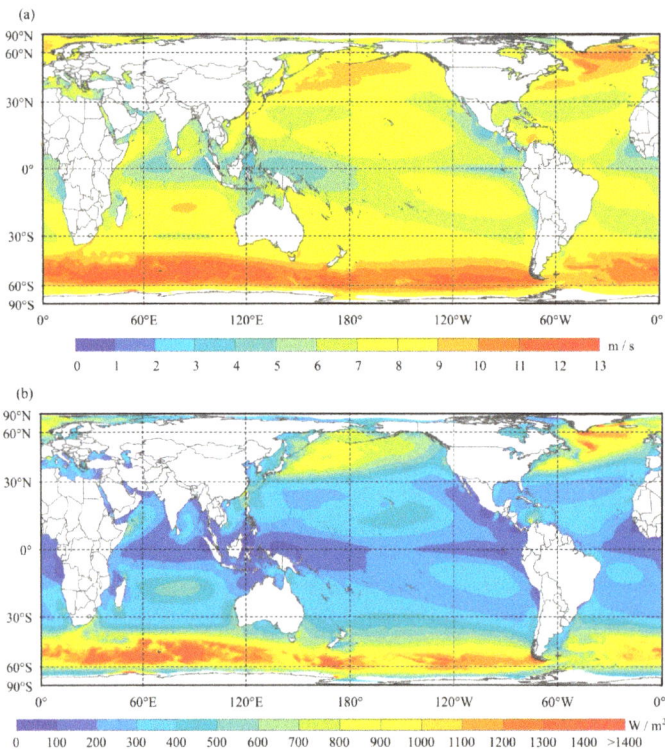

Figure 4. The distribution of global MWS and WPD at 10 m above sea level derived from QuikSCAT+ WindSAT + ASCAT during the period 1999–2015. (**a**) MWS, and (**b**) WPD.

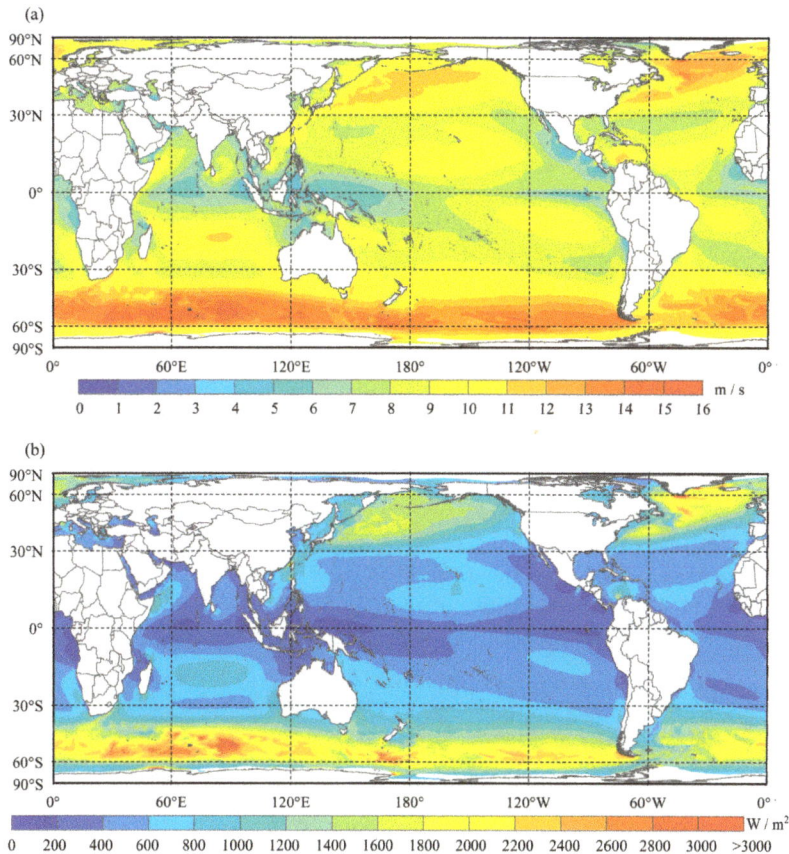

Figure 5. The distribution of global MWS and WPD at 100 m above sea level derived from QuikSCAT + WindSAT + ASCAT during the period 1999–2015. (**a**) MWS, and (**b**) WPD.

5. Discussion

Despite the results provided by this study, we do acknowledge that the uncertainties in ocean wind resources estimation can also be generated from other factors, such as the accuracy of wind vector retrieval algorithms, sensor configurations and the impact of diurnal variability on satellite wind observations. Removal of hard targets at sea (including ships, wind turbines, oil platforms etc.) is also significant for increasing the accuracy of wind retrieval from satellite images [17].

In this study, the number of satellite observations is still limited based on QuikSCAT, WindSAT and ASCAT. In future operational scenarios, there could be more satellite datasets with much improved resolution and thus a higher precision of wind vector retrievals. Based on our results, we hereby recommend the use of more satellite observations (including those from OSCAT, RapidSCAT etc.) which may be evaluated by further analysis.

The distribution of global ocean wind resources at heights of 10 m and 100 m (wind turbine hub height) above sea level has been mapped in this study. However, wind farm wake effects were ignored. A quantification and prediction of the wake effect losses is challenging because of the complex aerodynamic nature of the interdependencies of turbines [40] and therefore deserves greater attention in future studies. Atmospheric stability also plays a significant role in the accuracy of wind speed extrapolation. Therefore, in future research or operational scenarios, the use of information on atmospheric stability is highly recommended to ensure more accurate wind resources assessment at regional and global scales.

6. Conclusions

To develop an understanding of the spatial variability of global ocean wind resources using multiple satellite data, the spatial distribution of mean wind speeds (MWS) and wind power densities (WPD) at 10 m and 100 m above sea level have been mapped in this study using QuikSCAT + WindSAT + ASCAT data.

In this study, MWS and WPD calculated from 39 NDBC buoys during 1999–2015 are first compared with those derived from QuikSCAT, WindSAT and ASCAT, and their different combinations at heights of 10 m and 100 m (wind turbine hub height) above sea level. The results show that for single satellite data, MWS and WPD derived from ASCAT have the lowest RMSE. QuikSCAT and ASCAT overestimated the MWS and WPD derived from buoys, while WindSAT shows a tendency to underestimate MWS and WPD given their respective biases and slopes. This phenomenon may be due to the different sensor configurations and wind retrieval algorithms. Meanwhile, QuikSCAT + WindSAT + ASCAT-derived WPD at heights of 10 m and 100 m show the lowest RMSE and highest correlations, and hence a better accuracy of MWS/WPD may be derived from multiple satellite data than from single satellite data.

Furthermore, we quantified the impacts of data sampling density (number of satellite observations) on the accuracy of satellite-derived MWS and WPD at 10 m and 100 m above sea level. The results show an increase in the accuracy of MWS/WPD estimation with satellite observations, at 10 m and 100 m above sea level.

Acknowledgments: This work was supported by the China Special Fund for Industrial and Scientific Research in the Public Interest (Meteorology), under Grant GYHY201306050. The authors would like to thank the Remote Sensing Systems for providing QuikSCAT, WindSAT and ASCAT data, and the National Data Buoy Center for the provision of NDBC buoy data.

Author Contributions: Jingfeng Huang conceived the original idea of the study, and designed, organized and supervised the entire investigation; Qiaoying Guo collected, processed and analyzed the data, and wrote the article; Xiuzhen Wang and Jian Gao assisted in data collection and preprocessing; Xiazhen Xu, Kangyu Zhang, Zhengquan Li, Weijiao Huang and Weiwei Liu assisted in data processing and analysis; Lamin R. Mansaray assisted in manuscript preparation and revision.

Conflicts of Interest: The authors declare no conflict of interest.

References

1. Patz, J.A.; Campbell-Lendrum, D.; Holloway, T.; Foley, J.A. Impact of regional climate change on human health. *Nature* **2005**, *438*, 310–317. [CrossRef] [PubMed]
2. Snyder, C.W. Evolution of global temperature over the past two million years. *Nature* **2016**, *538*, 226–228. [CrossRef] [PubMed]
3. Stern, P.C.; Sovacool, B.K.; Dietz, T. Towards a science of climate and energy choices. *Nat. Clim. Chang.* **2016**, *6*, 547–555. [CrossRef]
4. Buonocore, J.J.; Luckow, P.; Norris, G.; Spengler, J.D.; Biewald, B.; Fisher, J.; Levy, J.I. Health and climate benefits of different energy-efficiency and renewable energy choices. *Nat. Clim. Chang.* **2015**, *6*, 100–105. [CrossRef]
5. Barthelmie, R.J.; Pryor, S.C. Potential contribution of wind energy to climate change mitigation. *Nat. Clim. Chang.* **2014**, *4*, 684–688. [CrossRef]
6. Lu, X.; McElroy, M.B.; Peng, W.; Liu, S.; Nielsen, C.P.; Wang, H. Challenges faced by China compared with the US in developing wind power. *Nat. Energy* **2016**, *1*. [CrossRef]
7. Zheng, C.W.; Li, C.Y.; Pan, J.; Liu, M.Y.; Xia, L.L. An overview of global ocean wind energy resource evaluations. *Renew. Sustain. Energy Rev.* **2016**, *53*, 1240–1251. [CrossRef]
8. Kucukali, S.; Dinçkal, Ç. Wind energy resource assessment of Izmit in the West Black Sea Coastal Region of Turkey. *Renew. Sustain. Energy Rev.* **2014**, *30*, 790–795. [CrossRef]

9. Mattar, C.; Borvarán, D. Offshore wind power simulation by using WRF in the central coast of Chile. *Renew. Energy* **2016**, *94*, 22–31. [CrossRef]
10. Chadee, X.T.; Clarke, R.M. Large-scale wind energy potential of the Caribbean region using near-surface reanalysis data. *Renew. Sustain. Energy Rev.* **2014**, *30*, 45–58. [CrossRef]
11. Zheng, C.W.; Pan, J. Assessment of the global ocean wind energy resource. *Renew. Sustain. Energy Rev.* **2014**, *33*, 382–391. [CrossRef]
12. Zheng, C.W.; Pan, J.; Li, C.Y. Global oceanic wind speed trends. *Ocean Coast. Manag.* **2016**, *129*, 15–24. [CrossRef]
13. Hasager, C.B.; Nielsen, M.; Astrup, P.; Barthelmie, R.; Dellwik, E.; Jensen, N.O.; Jørgensen, B.H.; Pryor, S.C.; Rathmann, O.; Furevik, B.R. Offshore wind resource estimation from satellite SAR wind field maps. *Wind Energy* **2005**, *8*, 403–419. [CrossRef]
14. Christiansen, M.B.; Koch, W.; Horstmann, J.; Bayhasager, C.; Nielsen, M. Wind resource assessment from C-band SAR. *Remote Sens. Environ.* **2006**, *105*, 68–81. [CrossRef]
15. Hasager, C.B.; Badger, M.; Peña, A.; Larsén, X.G.; Bingöl, F. SAR-Based Wind Resource Statistics in the Baltic Sea. *Remote Sens.* **2011**, *3*, 117–144. [CrossRef]
16. Chang, R.; Zhu, R.; Badger, M.; Hasager, C.; Zhou, R.; Ye, D.; Zhang, X. Applicability of Synthetic Aperture Radar Wind Retrievals on Offshore Wind Resources Assessment in Hangzhou Bay, China. *Energies* **2014**, *7*, 3339–3354. [CrossRef]
17. Hasager, C.B.; Mouche, A.; Badger, M.; Bingöl, F.; Karagali, I.; Driesenaar, T.; Stoffelen, A.; Peña, A.; Longépé, N. Offshore wind climatology based on synergetic use of Envisat ASAR, ASCAT and QuikSCAT. *Remote Sens. Environ.* **2015**, *156*, 247–263. [CrossRef]
18. Chang, R.; Zhu, R.; Badger, M.; Hasager, C.; Xing, X.; Jiang, Y. Offshore Wind Resources Assessment from Multiple Satellite Data and WRF Modeling over South China Sea. *Remote Sens.* **2015**, *7*, 467–487. [CrossRef]
19. Doubrawa, P.; Barthelmie, R.J.; Pryor, S.C.; Hasager, C.B.; Badger, M.; Karagali, I. Satellite winds as a tool for offshore wind resource assessment: The Great Lakes Wind Atlas. *Remote Sens. Environ.* **2015**, *168*, 349–359. [CrossRef]
20. Beaucage, P.; Lafrance, G.; Lafrance, J.; Choisnard, J.; Bernier, M. Synthetic aperture radar satellite data for offshore wind assessment: A strategic sampling approach. *J. Wind Eng. Ind. Aerodyn.* **2011**, *99*, 27–36. [CrossRef]
21. Liu, W.T.; Tang, W.; Xie, X. Wind power distribution over the ocean. *Geophys. Res. Lett.* **2008**, *35*. [CrossRef]
22. Pimenta, F.; Kempton, W.; Garvine, R. Combining meteorological stations and satellite data to evaluate the offshore wind power resource of Southeastern Brazil. *Renew. Energy* **2008**, *33*, 2375–2387. [CrossRef]
23. Capps, S.B.; Zender, C.S. Global ocean wind power sensitivity to surface layer stability. *Geophys. Res. Lett.* **2009**, *36*. [CrossRef]
24. Capps, S.B.; Zender, C.S. Estimated global ocean wind power potential from QuikSCAT observations, accounting for turbine characteristics and siting. *J. Geophys. Res.* **2010**, *115*. [CrossRef]
25. Karamanis, D.; Tsabaris, C.; Stamoulis, K.; Georgopoulos, D. Wind energy resources in the Ionian Sea. *Renew. Energy* **2011**, *36*, 815–822. [CrossRef]
26. Jiang, D.; Zhuang, D.; Huang, Y.; Wang, J.; Fu, J. Evaluating the spatio-temporal variation of China's offshore wind resources based on remotely sensed wind field data. *Renew. Sustain. Energy Rev.* **2013**, *24*, 142–148. [CrossRef]
27. Bentamy, A.; Croize-Fillon, D. Spatial and temporal characteristics of wind and wind power off the coasts of Brittany. *Renew. Energy* **2014**, *66*, 670–679. [CrossRef]
28. Carvalho, D.; Rocha, A.; Gómez-Gesteira, M.; Silva Santos, C. Offshore winds and wind energy production estimates derived from ASCAT, OSCAT, numerical weather prediction models and buoys—A comparative study for the Iberian Peninsula Atlantic coast. *Renew. Energy* **2017**, *102*, 433–444. [CrossRef]
29. Gadad, S.; Deka, P.C. Offshore wind power resource assessment using Oceansat-2 scatterometer data at a regional scale. *Appl. Energy* **2016**, *176*, 157–170. [CrossRef]
30. Remote Sensing Systems. Available online: http://www.remss.com/missions (accessed on 12 April 2016).
31. National Data Buoy Center. Available online: http://www.ndbc.noaa.gov (accessed on 1 September 2016).
32. Badger, M.; Peña, A.; Hahmann, A.N.; Mouche, A.A.; Hasager, C.B. Extrapolating satellite winds to turbine operating heights. *J. Appl. Meteorol. Climatol.* **2016**, *55*, 975–991. [CrossRef]

33. Takeyama, Y.; Ohsawa, T.; Kozai, K.; Hasager, C.B.; Badger, M. Comparison of geophysical model functions for SAR wind speed retrieval in Japanese coastal waters. *Remote Sens.* **2013**, *5*, 1956–1973. [CrossRef]
34. Mohammadi, K.; Shamshirband, S.; Yee, P.L.; Petkovic, D.; Zamani, M.; Ch, S. Predicting the wind power density based upon extreme learning machine. *Energy* **2015**, *86*, 232–239. [CrossRef]
35. Mohammadi, K.; Alavi, O.; Mostafaeipour, A.; Goudarzi, N.; Jalilvand, M. Assessing different parameters estimation methods of Weibull distribution to compute wind power density. *Energy Convers. Manag.* **2016**, *108*, 322–335. [CrossRef]
36. Barthelmie, R.J.; Pryor, S.C. Can satellite sampling of offshore wind speeds realistically represent wind speed distributions? *J. Appl. Meteorol.* **2003**, *42*, 83–94. [CrossRef]
37. Badger, M.; Badger, J.; Nielsen, M.; Hasager, C.B.; Pena, A. Wind class sampling of satellite SAR imagery for offshore wind resource mapping. *J. Appl. Meteorol. Climatol.* **2010**, *490*, 2474–2491. [CrossRef]
38. Atlas, R.; Hoffman, R.N.; Bloom, S.C.; Jusem, J.C.; Ardizzone, J. A multiyear global surface wind velocity dataset using SSM/I wind observations. *Bull. Am. Meteorol. Soc.* **1996**, *77*, 869–882. [CrossRef]
39. Hasager, C.B.; Astrup, P.; Zhu, R.; Chang, R.; Badger, M.; Hahmann, A.N. Quarter-century offshore winds from SSM/I and WRF in the North Sea and South China Sea. *Remote Sens.* **2016**, *8*, 769. [CrossRef]
40. Ritter, M.; Pieralli, S.; Odening, M. Neighborhood effects in wind farm performance: A regression approach. *Energies* **2017**, *10*, 365. [CrossRef]

MDPI

St. Alban-Anlage 66

4052 Basel

Switzerland

Tel. +41 61 683 77 34

Fax +41 61 302 89 18

www.mdpi.com

Remote Sensing Editorial Office

E-mail: remotesensing@mdpi.com

www.mdpi.com/journal/remotesensing

www.ingramcontent.com/pod-product-compliance
Lightning Source LLC
Chambersburg PA
CBHW051719210326
41597CB00032B/5540